THE 12th 'KINGSTON MEETING':
COMPUTATIONAL ASTROPHYSICS

A SERIES OF BOOKS ON RECENT DEVELOPMENTS IN ASTRONOMY AND ASTROPHYSICS

Managing Editor, D. Harold McNamara
Production Manager, Elizabeth S. Holloman

A.S.P. CONFERENCE SERIES PUBLICATIONS COMMITTEE

Sallie Baliunas, Chair
Carol Ambruster
Catharine Garmany
Mark S. Giampapa
Kenneth Janes

© Copyright 1997 Astronomical Society of the Pacific
390 Ashton Avenue, San Francisco, California 94112

All rights reserved

Printed by BookCrafters, Inc.

First published 1997

Library of Congress Catalog Card Number: 97-73640
ISBN 1-886733-43-0

Please contact proper address for information on:

PUBLISHING:
Managing Editor
PO Box 24463
Brigham Young University
Provo, UT 84602-4463
801-378-2298

pasp@astro.byu.edu
Fax: 801-378-2265

ORDERING BOOKS:
Astronomical Society of the Pacific
CONFERENCE SERIES
390 Ashton Avenue
San Francisco, CA 94112 - 1722   USA
415-337-1100, Ext. 112

catalog@aspsky.org
Fax: 415-337-5205

A SERIES OF BOOKS ON RECENT DEVELOPMENTS IN ASTRONOMY AND ASTROPHYSICS

Vol. 1-Progress and Opportunities in Southern Hemisphere Optical Astronomy: The CTIO 25th Anniversary Symposium
ed. V. M. Blanco and M. M. Phillips ISBN 0-937707-18-X

Vol. 2-Proceedings of a Workshop on Optical Surveys for Quasars
ed. P. S. Osmer, A. C. Porter, R. F. Green, and C. B. Foltz ISBN 0-937707-19-8

Vol. 3-Fiber Optics in Astronomy
ed. S. C. Barden ISBN 0-937707-20-1

Vol. 4-The Extragalactic Distance Scale: Proceedings of the ASP 100th Anniversary Symposium
ed. S. van den Bergh and C. J. Pritchet ISBN 0-937707-21-X

Vol. 5-The Minnesota Lectures on Clusters of Galaxies and Large-Scale Structure
ed. J. M. Dickey ISBN 0-937707-22-8

Vol. 6-Synthesis Imaging in Radio Astronomy: A Collection of Lectures from the Third NRAO Synthesis Imaging Summer School
ed. R. A. Perley, F. R. Schwab, and A. H. Bridle ISBN 0-937707-23-6

Vol. 7-Properties of Hot Luminous Stars: Boulder-Munich Workshop
ed. C. D. Garmany ISBN 0-937707-24-4

Vol. 8-CCDs in Astronomy
ed. G. H. Jacoby ISBN 0-937707-25-2

Vol. 9-Cool Stars, Stellar Systems, and the Sun. Sixth Cambridge Workshop
ed. G. Wallerstein ISBN 0-937707-27-9

Vol. 10-The Evolution of the Universe of Galaxies. The Edwin Hubble Centennial Symposium
ed. R. G. Kron ISBN 0-937707-28-7

Vol. 11-Confrontation Between Stellar Pulsation and Evolution
ed. C. Cacciari and G. Clementini ISBN 0-937707-30-9

Vol. 12-The Evolution of the Interstellar Medium
ed. L. Blitz ISBN 0-937707-31-7

Vol. 13-The Formation and Evolution of Star Clusters
ed. K. Janes ISBN 0-937707-32-5

Vol. 14-Astrophysics with Infrared Arrays
ed. R. Elston ISBN 0-937707-33-3

Vol. 15-Large-Scale Structures and Peculiar Motions in the Universe
ed. D. W. Latham and L. A. N. da Costa ISBN 0-937707-34-1

Vol. 16-Atoms, Ions and Molecules: New Results in Spectral Line Astrophysics
ed. A. D. Haschick and P. T. P. Ho ISBN 0-937707-35-X

Vol. 17-Light Pollution, Radio Interference, and Space Debris
ed. D. L. Crawford ISBN 0-937707-36-8

Vol. 18-The Interpretation of Modern Synthesis Observations of Spiral Galaxies
ed. N. Duric and P. C. Crane ISBN 0-937707-37-6

Vol. 19-Radio Interferometry: Theory, Techniques, and Application, IAU Colloquium 131
ed. T. J. Cornwell and R. A. Perley ISBN 0-937707-38-4

Vol. 20-Frontiers of Stellar Evolution, celebrating the 50th Anniversary of McDonald Observatory
ed. D. L. Lambert ISBN 0-937707-39-2

Vol. 21-The Space Distribution of Quasars
ed. D. Crampton   ISBN 0-937707-40-6

Vol. 22-Nonisotropic and Variable Outflows from Stars
ed. L. Drissen, C. Leitherer, and A. Nota   ISBN 0-937707-41-4

Vol. 23-Astronomical CCD Observing and Reduction Techniques
ed. S. B. Howell   ISBN 0-937707-42-4

Vol. 24-Cosmology and Large-Scale Structure in the Universe
ed. R. R. de Carvalho   ISBN 0-937707-43-0

Vol. 25-Astronomical Data Analysis Software and Systems I
ed. D. M. Worrall, C. Biemesderfer, and J. Barnes   ISBN 0-937707-44-9

Vol. 26-Cool Stars, Stellar Systems, and the Sun, Seventh Cambridge Workshop
ed. M. S. Giampapa and J. A. Bookbinder   ISBN 0-937707-45-7

Vol. 27-The Solar Cycle
ed. K. L. Harvey   ISBN 0-937707-46-5

Vol. 28-Automated Telescopes for Photometry and Imaging
ed. S. J. Adelman, R. J. Dukes, Jr., and C. J. Adelman   ISBN 0-937707-47-3

Vol. 29-Workshop on Cataclysmic Variable Stars
ed. N. Vogt   ISBN 0-937707-48-1

Vol. 30-Variable Stars and Galaxies, in honor of M. S. Feast on his retirement
ed. B. Warner   ISBN 0-937707-49-X

Vol. 31-Relationships Between Active Galactic Nuclei and Starburst Galaxies
ed. A. V. Filippenko   ISBN 0-937707-50-3

Vol. 32-Complementary Approaches to Double and Multiple Star Research, IAU Collouquium 135
ed. H. A. McAlister and W. I. Hartkopf   ISBN 0-937707-51-1

Vol. 33-Research Amateur Astronomy
ed. S. J. Edberg   ISBN 0-937707-52-X

Vol. 34-Robotic Telescopes in the 1990s
ed. A. V. Filippenko   ISBN 0-937707-53-8

Vol. 35-Massive Stars: Their Lives in the Interstellar Medium
ed. J. P. Cassinelli and E. B. Churchwell   ISBN 0-937707-54-6

Vol. 36-Planets and Pulsars
ed. J. A. Phillips, S. E. Thorsett, and S. R. Kulkarni   ISBN 0-937707-55-4

Vol. 37-Fiber Optics in Astronomy II
ed. P. M. Gray   ISBN 0-937707-56-2

Vol. 38-New Frontiers in Binary Star Research
ed. K. C. Leung and I. S. Nha   ISBN 0-937707-57-0

Vol. 39-The Minnesota Lectures on the Structure and Dynamics of the Milky Way
ed. Roberta M. Humphreys   ISBN 0-937707-58-9

Vol. 40-Inside the Stars, IAU Colloquium 137
ed. Werner W. Weiss and Annie Baglin   ISBN 0-937707-59-7

Vol. 41-Astronomical Infrared Spectroscopy: Future Observational Directions
ed. Sun Kwok   ISBN 0-937707-60-0

Vol. 42-GONG 1992: Seismic Investigation of the Sun and Stars
ed. Timothy M. Brown   ISBN 0-937707-61-9

Vol. 43-Sky Surveys: Protostars to Protogalaxies
ed. B. T. Soifer
ISBN 0-937707-62-7

Vol. 44-Peculiar Versus Normal Phenomena in A-Type and Related Stars
ed. M. M. Dworetsky, F. Castelli, and R. Faraggiana
ISBN 0-937707-63-5

Vol. 45-Luminous High-Latitude Stars
ed. D. D. Sasselov
ISBN 0-937707-64-3

Vol. 46-The Magnetic and Velocity Fields of Solar Active Regions, IAU Colloquium 141
ed. H. Zirin, G. Ai, and H. Wang
ISBN 0-937707-65-1

Vol. 47-Third Decinnial US-USSR Conference on SETI
ed. G. Seth Shostak
ISBN 0-937707-66-X

Vol. 48-The Globular Cluster-Galaxy Connection
ed. Graeme H. Smith and Jean P. Brodie
ISBN 0-937707-67-8

Vol. 49-Galaxy Evolution: The Milky Way Perspective
ed. Steven R. Majewski
ISBN 0-937707-68-6

Vol. 50-Structure and Dynamics of Globular Clusters
ed. S. G. Djorgovski and G. Meylan
ISBN 0-937707-69-4

Vol. 51-Observational Cosmology
ed. G. Chincarini, A. Iovino, T. Maccacaro, and D. Maccagni
ISBN 0-937707-70-8

Vol. 52-Astronomical Data Analysis Software and Systems II
ed. R. J. Hanisch, J. V. Brissenden, and Jeannette Barnes
ISBN 0-937707-71-6

Vol. 53-Blue Stragglers
ed. Rex A. Saffer
ISBN 0-937707-72-4

Vol. 54-The First Stromlo Symposium: The Physics of Active Galaxies
ed. Geoffrey V. Bicknell, Michael A. Dopita, and Peter J. Quinn
ISBN 0-937707-73-2

Vol. 55-Optical Astronomy from the Earth and Moon
ed. Diane M. Pyper and Ronald J. Angione
ISBN 0-937707-74-0

Vol. 56-Interacting Binary Stars
ed. Allen W. Shafter
ISBN 0-937707-75-9

Vol. 57-Stellar and Circumstellar Astrophysics
ed. George Wallerstein and Alberto Noriega-Crespo
ISBN 0-937707-76-7

Vol. 58-The First Symposium on the Infrared Cirrus and Diffuse Interstellar Clouds
ed. Roc M. Cutri and William B. Latter
ISBN 0-937707-77-5

Vol. 59-Astronomy with Millimeter and Submillimeter Wave Interferometry
ed. M. Ishiguro and Wm. J. Welch
ISBN 0-937707-78-3

Vol. 60-The MK Process at 50 Years: A Powerful Tool for Astrophysical Insight
ed. C. J. Corbally, R. O. Gray, and R. F. Garrison
ISBN 0-937707-79-1

Vol. 61-Astronomical Data Analysis Software and Systems III
ed. Dennis R. Crabtree, R. J. Hanisch, and Jeannette Barnes
ISBN 0-937707-80-5

Vol. 62-The Nature and Evolutionary Status of Herbig Ae / Be Stars
ed. P. S. Thé, M. R. Pérez, and E. P. J. van den Heuvel
ISBN 0-937707-81-3

Vol. 63-Seventy-Five Years of Hirayama Asteroid Families: The role of Collisions in the Solar System History
ed. R. Binzel, Y. Kozai, and T. Hirayama
ISBN 0-937707-82-1

Vol. 64-Cool Stars, Stellar Systems, and the Sun, Eighth Cambridge Workshop
ed. Jean-Pierre Caillault
ISBN 0-937707-83-X

Vol. 65-Clouds, Cores, and Low Mass Stars
ed. Dan P. Clemens and Richard Barvainis   ISBN 0-937707-84-8

Vol. 66- Physics of the Gaseous and Stellar Disks of the Galaxy
ed. Ivan R. King   ISBN 0-937707-85-6

Vol. 67-Unveiling Large-Scale Structures Behind the Milky Way
ed. C. Balkowski and R. C. Kraan-Korteweg   ISBN 0-937707-86-4

Vol. 68-Solar Active Region Evolution: Comparing Models with Observations
ed. K. S. Balasubramaniam and George W. Simon   ISBN 0-937707-87-2

Vol. 69-Reverberation Mapping of the Broad-Line Region in Active Galactic Nuclei
ed. P. M. Gondhalekar, K. Horne, and B. M. Peterson   ISBN 0-937707-88-0

Vol. 70-Groups of Galaxies
ed. Otto G. Richter and Kirk Borne   ISBN 0-937707-89-9

Vol. 71-Tridimensional Optical Spectroscopic Methods in Astrophysics
ed. G. Comte and M. Marcelin   ISBN 0-937707-90-2

Vol. 72-Millisecond Pulsars—A Decade of Surprise, ed. A. A. Fruchter
M. Tavani, and D. C. Backer   ISBN 0-937707-91-0

Vol. 73-Airborne Astronomy Symposium on the Galactic Ecosystem: From Gas to Stars to Dust
ed. M. R. Haas, J. A. Davidson, and E. F. Erickson   ISBN 0-937707-92-9

Vol. 74-Progress in the Search for Extraterrestrial Life,
ed. G. Seth Shostak   ISBN 0-937707-93-7

Vol. 75-Multi-Feed Systems for Radio Telescopes
ed. D. T. Emerson and J. M. Payne   ISBN 0-937707-94-5

Vol. 76-GONG '94: Helio- and Astero-Seismology from the Earth and Space
ed. Roger K. Ulrich, Edward J. Rhodes, Jr., and Werner Däppen   ISBN 0-937707-95-3

Vol. 77-Astronomical Data Analysis Software and Systems IV
ed. R. A. Shaw, H. E. Payne, and J. J. E. Hayes   ISBN 0-937707-96-1

Vol. 78-Astrophysical Applications of Powerful New Databases
ed. S. J. Adelman and W. L. Wiese   ISBN 0-937707-97-X

Vol. 79-Robotic Telescopes: Current Capabilities, Present Developments, and Future Prospects for Automated Astronomy
ed. Gregory W. Henry and Joel A. Eaton   ISBN 0-937707-98-8

Vol. 80-The Physics of the Interstellar Medium and Intergalactic Medium
ed. A. Ferrara, C. F. McKee, C. Heiles, and P. R. Shapiro   ISBN 0-937707-99-6

Vol. 81-Laboratory and Astronomical High Resolution Spectra
ed. A. J. Sauval, R. Blomme, and N. Grevesse   ISBN 1-886733-01-5

Vol. 82-Very Long Baseline Interferometry and the VLBA,
ed. J. A. Zensus, P. K. Diamond, and P. J. Napier   ISBN 1-886733-02-3

Vol. 83-Astrophysical Applications of Stellar Pulsation, IAU Colloquium 155
ed. R. S. Stobie and P. A. Whitelock   ISBN 1-886733-03-1

Vol. 84-The Future Utilisation of Schmidt Telescopes, IAU Colloquium 148
ed. Jessica Chapman, Russell Cannon, Sandra Harrison, and Bambang Hidayat   ISBN 1-886733-05-8

Vol. 85-Cape Workshop on Magnetic Cataclysmic Variables
ed. D. A. H. Buckley and B. Warner   ISBN 1-886733-06-6

Vol. 86-Fresh Views of Elliptical Galaxies
ed. Alberto Buzzoni, Alvio Renzini, and Alfonso Serrano   ISBN 1-886733-07-4

Vol. 87-New Observing Modes for the Next Century
ed. Todd Boroson, John Davies, and Ian Robson   ISBN 1-886733-08-2

Vol. 88-Clusters, Lensing, and the Future of the Universe
ed. Virginia Trimble and Andreas Reisenegger   ISBN 1-886733-09-0

Vol. 89-Astronomy Education: Current Developments, Future Coordination
ed. John R. Percy   ISBN 1-886733-10-4

Vol. 90-The Origins, Evolution, and Destinies of Binary Stars in Clusters
ed. E. F. Milone and J. -C. Mermilliod   ISBN 1-886733-11-2

Vol. 91-Barred Galaxies, IAU Colloquium 157
ed. R. Buta, D. A. Crocker, and B. G. Elmegreen   ISBN 1-886733-12-0

Vol. 92-Formation of the Galactic Halo--Inside and Out
ed. H. L. Morrison and A. Sarajedini   ISBN 1-886733-13-9

Vol. 93-Radio Emission from the Stars and the Sun
ed. A. R. Taylor and J. M. Paredes   ISBN 1-886733-14-7

Vol. 94-Mapping, Measuring, and Modelling the Universe
ed. Peter Coles, Vicent Martinez, and Maria-Jesus Pons-Borderia ISBN 1-886733-15-5

Vol. 95-Solar Drivers of Interplanetary and Terrestrial Disturbances
ed. K.S. Balasubramaniam, S. L. Keil, and R. N. Smartt   ISBN 1-886733-16-3

Vol. 96-Hydrogen-Deficient Stars
ed. C. S. Jeffery and U. Heber   ISBN 1-886733-17-1

Vol. 97-Polarimetry of the Interstellar Medium
ed. W. G. Roberge and D. C. B. Whittet   ISBN 1-886733-18-X

Vol. 98-From Stars to Galaxies: The Impact of Stellar Physics on Galaxy Evolution
ed. Claus Leitherer, Uta Fritze-von Alvensleben, and John Huchra ISBN 1-886733-19-8

Vol. 99-Cosmic Abundances
ed. Stephen S. Holt and George Sonneborn   ISBN 1-886733-20-1

Vol. 100-Energy Transport in Radio Galaxies and Quasars
ed. P. E. Hardee, A. H. Bridle, and J. A. Zensus   ISBN 1-886733-21-X

Vol. 101-Astronomical Data Analysis Software and Systems V
ed. George H. Jacoby and Jeannette Barnes   ISSN   1080-7926

Vol. 102-The Galactic Center, 4th ESO/CTIO Workshop
ed. Roland Gredel   ISBN 1-886733-22-8

Vol. 103-The Physics of Liners in View of Recent Observations
ed. M. Eracleous, A. Koratkar, C. Leitherer, and L. Ho   ISBN 1-886733-23-6

Vol. 104-Physics, Chemistry, and Dynamics of Interplanetary Dust, IAU Colloquium 150
ed. Bo Å. S. Gustafson and Martha S. Hanner   ISBN 1-886733-24-4

Vol. 105-Pulsars: Problems and Progress, IAU Colloquium 160
ed. M. Bailes, S. Johnston, and M.A. Walker   ISBN 1-886733-25-2

Vol. 106-Minnesota Lectures on Extragalactic Neutral Hydrogen
ed. Evan D. Skillman   ISBN 1-886733-26-0

Vol. 107-Completing the Inventory of the Solar System
ed. Terrence W. Rettig and Joseph Hahn   ISBN 1-886733-27-9

Vol. 108-Model Atmospheres and Spectrum Synthesis
ed. S. J. Adelman, F. Kupka, and W. W. Weiss   ISBN 1-886733-28-7

Vol. 109-Cool Stars, Stellar Systems, and the Sun, Ninth Cambridge Workshop
ed. Roberto Pallavicini and Andrea K. Dupree   ISBN 1-886733-29-5

Vol. 110-Blazar Continuum Variability
ed. H. R. Miller, J. R. Webb, and J. C. Noble   ISBN 1-886733-30-9

Vol. 111-Magnetic Reconnection in the Solar Atmosphere
ed. R. D. Bentley and J. T. Mariska   ISBN 1-886733-31-7

Vol. 112-Galactic Chemodynamics 4: The History of the Milky Way System and Its Satellite System
ed. A. Burkert, D. H. Hartmann, and S. A. Majewski   ISBN 1-886733-32-5

Vol. 113-Emission Lines in Active Galaxies: New Methods and Techniques, IAU Colloquium 159
ed. B. M. Peterson, F.-Z. Cheng, and A. S. Wilson   ISBN 1-886733-33-3

Vol. 114-Young Galaxies and QSO Absorption-Line Systems
ed. Sueli M. Viegas, Ruth Gruenwald, and Reinaldo R. de Carvalho   ISBN 1-886733-34-1

Vol. 115-Galactic and Cluster Cooling Flows
ed. Noam Soker   ISBN 1-886733-35-X

Vol. 116-The Second Stromlo Symposium: The Nature of Elliptical Galaxies
ed. M. Arnaboldi, G. S. Da Costa, and P. Saha   ISBN 1-886733-36-8

Vol. 117-Dark and Visible Matter in Galaxies
ed. Massimo Persic and Paolo Salucci   ISBN 1-866733-37-6

Vol. 118-First Advances in Solar Physics Euroconference: Advances in the Physics of Sunspots
ed. B. Schmieder, J. C. del Toro Iniesta, and M. Vázquez   ISBN 1-886733-38-4

Vol. 119-Planets Beyond the Solar System and the Next Generation of Space Missions
ed. David R. Soderblom   ISBN 1-886733-39-2

Vol. 120-Luminous Blue Variables: Massive Stars in Transition
ed. Antonella Nota and Henny J. G. L. M. Lamers   ISBN 1-886733-40-6

Vol. 121-Accretion Phenomena and Related Outflows
ed. D. T. Wickramasinghe, G. V. Bicknell and L. Ferrario   ISBN 1-886733-41-4

Vol. 122-From Stardust to Planetesimals
ed. Yvonne J. Pendleton and A. G. G. M. Tielens   ISBN 1-886733-42-2

Vol. 123-The 12th 'Kingston Meeting': Computational Astrophysics
ed. David A. Clarke and Michael J. West   ISBN 1-886733-43-0

Inquiries concerning these volumes should be directed to the:
Astronomical Society of the Pacific
CONFERENCE SERIES
390 Ashton Avenue
San Francisco, CA 94112-1722 USA
415-337-1100, Ext. 112
catalog@aspsky.org
Fax: 415-337-5205

ASTRONOMICAL SOCIETY OF THE PACIFIC
CONFERENCE SERIES

Volume 123

COMPUTATIONAL ASTROPHYSICS

Proceedings of the 12th 'Kingston meeting'
on Theoretical Astrophysics

Held in Halifax, Nova Scotia, Canada
October 17-19, 1996

Edited by
David A. Clarke and Michael J. West

Dedicated to the father of supercomputing
SEYMOUR CRAY
(1925–1996)

Contents

Titles appearing in **bold face** correspond to invited review talks.

Preface	xix
Meeting Participants	xxiii

OVERVIEW — 1

Computational Astrophysics: The "New Astronomy" for the 21st Century — 3
MICHAEL L. NORMAN

Exceedingly low redshift $(0 \leq z \leq 3 \times 10^{-6})$

I SOLAR SYSTEM DYNAMICS — 15

The Longterm Dynamical Evolution of Orbital Configurations — 17
MARTIN J. DUNCAN

Time Symmetrization Meta-Algorithms — 26
PIET HUT, YOKO FUNATO, EIICHIRO KOKUBO, JUNICHIRO MAKINO, & STEVE MCMILLAN

Variable Time Step Integrators for Long-Term Orbital Integrations — 32
MAN HOI LEE, MARTIN J. DUNCAN, & HAROLD F. LEVISON

II STARS — 39

Abundance Anomalies and Anisotropic Turbulent Transport in Stars — 41
G. MICHAUD, A. VINCENT, & M. MENEGUZZI

Helioseismology by Genetic Forward Modeling — 49
P. CHARBONNEAU & S. TOMCZYK

The Radiation-Magnetohydrodynamics of Accretion-Powered
   Neutron Stars 55
   RICHARD I. KLEIN

On the Existence of Intermediate Shocks 66
   S. A. E. G. FALLE & S. S. KOMISSAROV

Numerical Simulations Can Lead to New Insights 72
   ROBERT F. STEIN, MATS CARLSSON, & ÅKE NORDLUND

Angular Momentum Transport in Simulations of Accretion Disks 78
   JAMES RHYS MURRAY

On Radiative Transfer in SPH 84
   ALEXEI RAZOUMOV

Semi-Empirical Wind Models for λ Velorum (K4 Ib–II) 87
   PHILIP D. BENNETT & GRAHAM M. HARPER

Interaction of the Pulsar-Driven Wind with Supernova Ejecta 94
   BYUNG-IL JUN

The Evolution of Blue Stragglers: Merging SPH and Stellar Evolution
   Codes 100
   ALISON SILLS

Detonation Waves in Type Ia Supernovae 103
   D. J. R. WIGGINS & S. A. E. G. FALLE

Monte-Carlo Simulations of X-Ray Binary Spectra 109
   D. A. LEAHY

Effects of Scattering on Pulse Shapes from Accreting Neutron Stars 112
   D. A. LEAHY

III YSO/ACCRETION/OUTFLOW/ISM 115

Collapse and Outflow: Towards an Integrated Theory of Star
   Formation 117
   RALPH E. PUDRITZ, DEAN E. MCLAUGHLIN, & RACHID OUYED

Time-Dependent Protostellar Accretion and Ejection 129
   R. N. HENRIKSEN

Gas-Dynamical Simulations of the Circumstellar and
   Interstellar Media 136
   JAMES M. STONE

The Role of Weak Magnetic Fields in Kelvin-Helmholtz Unstable
    Boundary Layers   146
    T. W. Jones, Joseph B. Gaalaas, Dongsu Ryu, &
    Adam Frank

The Jeans Condition: A New Constraint on Spatial Resolution in
    Simulations of Self-Gravitational Hydrodynamics   152
    Richard I. Klein, Kelly Truelove, & Christopher F.
    McKee

Warm Interstellar Molecular Hydrogen   159
    P. G. Martin

GRAPESPH with Fully Periodic Boundaries: Fragmentation of
    Molecular Clouds   169
    Ralf Klessen

Origin of Episodic Outflows in YSO and AGN   172
    Rachid Ouyed

Low to moderate redshift $(3 \times 10^{-6} \leq z \leq 2)$

IV GALACTIC DYNAMICS   175

The GRAPE-4, a Teraflops Stellar Dynamics Computer   177
    Piet Hut

On Simulations of Galactic Disks of Stars   189
    Evgeny Griv

A Fokker-Planck Model of Rotating Stellar Clusters   196
    John Girash

Stability Properties of Spherical Models with Central Black Holes 202
    Andres Meza & Nelson Zamorano

Simulation of the Formation of Compact Groups of Galaxies   207
    Nicholas J. White & Alistair H. Nelson

Selected Lagrangian Weights—An Alternative to Smoothed Particle
    Hydrodynamics   211
    Alistair H. Nelson, Peter R. Williams, John P. Sleath
    & Nicholas P. Moore

Galaxy Dynamics by N-Body Simulation   215
    J. A. Sellwood

Galaxy Splashes: The Effects of Collisions Between Gas-Rich Galaxy
  Disks   225
    CURTIS STRUCK

Galaxy Formation in Dissipationless N-Body Models   231
    EELCO VAN KAMPEN

Global Error Measures for Large N-Body Simulations   237
    WAYNE B. HAYES & KENNETH R. JACKSON

V AGN/ACCRETION/OUTFLOW   241

Jet Stability: Numerical Simulations Confront Analytical
  Theory   243
    PHILIP E. HARDEE

A Restarting Jet Revisited: MHD Computations in 3-D   255
    DAVID A. CLARKE

Effect of Magnetic Fields on Mass Entrainment in Extragalactic Jets   262
    ALEXANDER ROSEN, PHILIP E. HARDEE, DAVID A. CLARKE,
    & AUDRESS JOHNSON

Jet Turbulence and High-Resolution 3-D Simulations   268
    C. LOKEN

Modern Schemes for Solving Hyperbolic Conservation Laws
  of Interest in Computational Astrophysics on Parallel
  Machines   274
    DINSHAW S. BALSARA

Relativistic Jet Simulations and VLBI Maps   284
    PHILIP HUGHES, AMY MIODUSZEWSKI, & COMER DUNCAN

Simulations of Jet Production in Magnetized Accretion Disk Coronae   290
    DAVID L. MEIER, SAMANTHA EDGINGTON, PATRICK GODON,
    DAVID G. PAYNE, & KEVIN R. LIND

Numerical Simulations of Rotating Accretion Flows near a Black Hole   296
    DONGSU RYU & SANDIP K. CHAKRABARTI

High to infinite redshift $(2 \leq z \leq \infty)$

VI COSMOLOGY AND NUMERICAL RELATIVITY   303

The Binary Black Hole Grand Challenge Project   305
    M. W. CHOPTUIK

Head-on Collisions of Two Black Holes ... 314
D. W. Hobill, P. Anninos, H. E. Seidel, L. L. Smarr, & W.-M. Suen

The Lyman Alpha Forest Within the Cosmic Web ... 323
J. Richard Bond & James W. Wadsley

SPH P^{3}MG Simulations of the Lyman Alpha Forest ... 332
J. W. Wadsley & J. Richard Bond

Simulating Cosmic Structure at High Resolution: Towards a Billion Particles? ... 340
H. M. P. Couchman

Nature of the Low Column Density Lyman Alpha Forest ... 351
Michael L. Norman, Peter Anninos, Yu Zhang, & Avery Meiksin

Reheating of the Universe and Population III ... 357
Jeremiah P. Ostriker & Nickolay Y. Gnedin

Simulating X-Ray Clusters with Adaptive Mesh Refinement ... 363
Greg L. Bryan & Michael L. Norman

The Effect of Substructure on the Final State of Matter in an X-Ray Cluster ... 369
Eric Tittley & H. M. P. Couchman

Author Index ... 373

Preface

As in all fields of science, computational methods have grown in prominence in astrophysics, particularly over the past fifteen years. As discussed by Michael Norman in his introductory talk to the meeting on page 3, computational astrophysics is now roughly at the same stage that theoretical astronomy was 100 years ago, and it will undoubtedly continue to grow in importance into the next century. And while there have been innumerable meetings devoted (either explicitly or implicitly) to observational and theoretical techniques in astronomy and astrophysics, there have been relatively few devoted to the subject of computational astrophysics.

Still, there have been a few, and these include "Numerical Astrophysics", a meeting organised in 1984 by Joan Centrella and her organising committees in Philadelphia[1], and "Numerical Simulations in Astrophysics: Modelling the Dynamics of the Universe", a meeting organised in 1993 by J. Franco and his organising committees in Mexico City[2]. A more topical meeting was "Astrophysical Radiation Hydrodynamics", organised in 1985 by Karl-Heinz Winkler and Michael Norman[3], and one of the themes for the 1994 summer meeting of the American Astronomical Society in Minnesota was "The Grand Challenges of Computational Astrophysics".

The organisers chose computational astrophysics as the theme for the "12th 'Kingston meeting' on Theoretical Astrophysics" to reflect the explosive growth of computer simulations in astronomy. This is the first meeting of its kind to be held in Canada, and while Canadian participation was high, the meeting was by any standards international. Of the 52 papers appearing in this volume, 34 are from speakers at institutions outside the country. These proceedings, then, give a good reflection of the interests of computational astrophysicists from around the world at the close of the twentieth century. This meeting brought together computational and theoretical astrophysicists from solar system dynamics to cosmology and relativity, and at nearly every redshift in between. And while

---

[1] *Numerical Astrophysics: A Festschrift in Honor of James R. Wilson*, 1985, eds. J. M. Centrella, J. M. LeBlanc, & R. L. Bowers (Boston: Jones and Bartlett)

[2] *Proceedings of Numerical Simulations in Astrophysics: Modelling the Dynamics of the Universe*, 1994, eds. J. Franco, L. Aguilar, S. Lizano, & E. Daltabuit (Cambridge: Cambridge University Press)

[3] *Astrophysical Radiation Hydrodynamics (Proceedings of the NATO Advanced Research Workshop)*, 1986, eds. K.-H. A. Winkler & M. L. Norman (Munich: M. P. I. für Physik und Astrophysik)

the subjects were indeed varied, the interest was universal. The overwhelming consensus from the participants was an appreciation of the chance to learn from experts in all areas of astrophysics, and an astonishment at how much overlap there is in computational methodology among apparently disparate fields. Above all, we hope these proceedings manage to convey these aspects of the meeting to the reader.

There were many high points to the meeting both scientifically, and socially. The science, contained in this volume, speaks for itself. Socially, the meeting had many memorable moments, including the formal surroundings of the banquet dinner at the *Waegwoltic Club* overlooking the "Northwest Arm" in Halifax' west end, the pub crawl which, for most, began and ended in a good old Maritime sing-along at the *Lower Deck*, the closing party, and the tours of Peggy's Cove and Cape Split. Nova Scotia showed its visitors one of its nicest autumns in recent years, and it was a meeting these organisers will not soon forget.

For those outside the Canadian astronomical community, the *'Kingston meeting' on Theoretical Astrophysics* is now held more or less biannually, and focuses on areas of current interest in astrophysics. While the meetings are hosted by institutions across Canada, they began as the brain-child of Dick Henriksen at Queen's University in the early 1980's. The first three meetings were indeed held in Kingston, Ontario, from which the series derives its name. First and foremost, therefore, we would like to thank Dick for his perseverance in making these meetings happen in the early days, and for his foresight in realising the importance of such a venue to the Canadian astrophysical community.

We are deeply indebted to our Scientific Organising Committee: Dick Henriksen (Queen's), Michael Norman (Illinois), Scott Tremaine (Toronto, soon to be Princeton), and, at the early stages of organisation, Don Vandenberg (Victoria). These four, along with DC, were responsible for gathering the invited speakers and organising and chairing the individual sessions. It was also Scott, then as director of CITA, who first asked if we would be interested in hosting the 1996 'Kingston meeting'. We thank also Dick Bond and George Mitchell for chairing two of the sessions at the meeting.

We would like to thank the other members of the Local Organising Committee, namely Malcolm Butler (Saint Mary's), who worked tirelessly throughout the organisation of the meeting, George Mitchell (Saint Mary's) and Alan Coley (Dalhousie). We thank David Guenther for creating the meeting poster. Thanks are also due to the students and staff at Saint Mary's and Dalhousie who helped with the many tasks involved in making this meeting a success. These include Kevin Douglas, Marcia Kissner, Beverly Miskolczi, Shawn Mitchell, Steven Short, Rene Tanaja, and Elfrie Waters all of Saint Mary's University, and Andrew Billyard of Dalhousie University. In particular, we thank Dave Lane for his many contributions to the logistics of the meeting.

No meeting can run without sponsors, and we were very fortunate in raising sufficient capital to keep the registration fees down and to provide travel support for many of our participants. We therefore express our profound gratitude to our sponsors, which include CITA (who have been supporting 'Kingston meetings' since the late 1980's), the Fields Institute for Research in Mathematical Sciences, and Saint Mary's University, particularly Dr. Colin Dodds (V.P. Academic and Research), Dr. David Richardson (Dean of Science), and the Department of Astronomy and Physics. We also thank David Richardson for giving the opening address where he reminded us about the similarities between lichens and astronomy (whoda thunkit?), and in general for his tireless support of astronomy and astrophysics at Saint Mary's.

Finally, we thank the invited speakers, whose talks and written submission serve as an excellent review to experts and novices alike, and all the participants who made this meeting such a success.

David A. Clarke and Michael J. West
Halifax, Nova Scotia, May 1997

Cover photo: Pictured is the ENIAC computer, the first large-scale general-purpose electronic computer, whose fiftieth anniversary in 1996 coincided with the 12th 'Kingston meeting' on Theoretical Astrophysics. The ENIAC was built from 17,468 vacuum tubes, and was capable of performing 5,000 additions and 300 multiplications per second.

Photo credit: John Mauchly Papers, Special Collections, University of Pennsylvania Library, reproduced with permission, for which we express our sincere appreciation.

Camera-ready copy was provided to the ASP by the editors using Leslie Lamport's LaTeX, based on Donald E. Knuth's TeX. Other macros were provided by the ASP. The editors express their gratitude to these and other developers for placing these invaluable packages in the public domain.

Past and Future 'Kingston meetings'

	Year	University	Topic	Organisers
1	1981	Queen's		R. Henriksen
2	1982	Queen's		R. Henriksen
3	1983	Queen's		R. Henriksen
4	1984	Laval	Galactic Dynamics	S. Pineault
5	1985	Manitoba		M. Clutton-Brock
6	1986	Victoria	Large Scale Structure in the Universe	D. Hartwick
7	1987	Saint Mary's		G. Mitchell
8	1988	Calgary	Cosmological Constraints from Stellar Evolution	S. Kwok
9	1989	Victoria	Formation of the Galaxy	S. van den Bergh
10	1991	Queen's	Star and Planet Formation	M. Duncan & R. Henriksen
11	1993	Montréal	Accretion Discs	P. Bastien
12	1996	Saint Mary's	Computational Astrophysics	D. Clarke
13	1998	McMaster	TBA	R. Pudritz

Meeting Participants

Dinshaw S. Balsara, NCSA, University of Illinois at Urbana-Champaign, Urbana, IL 61820, U.S.A. u10956@ncsa.uiuc.edu

Philip D. Bennett, CASA, University of Colorado, Campus Box 389, Boulder, CO 80309, U.S.A. pbennett@casa.colorado.edu

Andrew P. Billyard, Department of Physics, Dalhousie University, Halifax, NS B3H 3J5, Canada. jaf@cs.dal.ca

J. Richard Bond, CITA, University of Toronto, 60 St. George Street, Toronto, ON M5S 1A7, Canada. bond@cita.utoronto.ca

Greg Bryan, Department of Physics, MIT, MIT 6-207, Cambridge, MA 02139, U.S.A. gbryan@arcturus.mit.edu

Malcolm N. Butler, Department of Astronomy and Physics, Saint Mary's University, Halifax, NS B3H 3C3, Canada. mbutler@ap.stmarys.ca

Paul Charbonneau, National Center for Atmospheric Research, High Altitude Observatory, 3450 Mitchell Lane, Boulder, CO 80301, U.S.A. paulchar@hao.ucar.edu

Matt Choptuik, Department of Astronomy, University of Texas, RLM 15.308, Austin, TX 78712-1083, U.S.A. matt@infeld.ph.utexas.edu

David A. Clarke, Department of Astronomy and Physics, Saint Mary's University, Halifax, NS B3H 3C3, Canada. dclarke@ap.stmarys.ca

Maurice Clement, Department of Astronomy, University of Toronto, 60 St. George Street, Toronto, ON M5S 3H8, Canada. mclement@astro.utoronto.ca

Alan Coley, Department of Mathematics, Statistics, and Computing Science, Dalhousie University, Halfiax, NS B3H 3J5, Canada. aac@cs.dal.ca

Hugh Couchman, Department of Physics and Astronomy, University of Western Ontario, London, ON N6A 3K7, Canada. couchman@coho.astro.uwo.ca

Peter Damiano, Department of Physics, University of Alberta, Edmonton, AB T6G 2J1, Canada. damiano@space.ualberta.ca

Kevin Douglas, Department of Astronomy and Physics, Saint Mary's University, Halifax, NS B3H 3C3, Canada. kdouglas@ap.stmarys.ca

Martin Duncan, Physics Department, Queen's University at Kingston, Kingston, ON K7L 3N6, Canada. duncan@astro.queensu.ca

Katherine Durrell, Department of Physics and Astronomy, McMaster University, 1280 Main Street W., Hamilton, ON L8S 4M1, Canada. kingk@jabba.physics.mcmaster.ca

L. Jonathan Dursi, Department of Computing Science, University of Waterloo, Waterloo, ON N2L 3G1, Canada. ljdursi@yoho.uwaterloo.ca

Sam A. E. G. Falle, Department of Applied Mathematics, University of Leeds, Woodhouse Lane, Leeds, Yorkshire, LS2 9JT, U.K. sam@amsta.leeds.ac.uk

Todd Fuller, Department of Astronomy and Physics, University of Western Ontario, London, ON N6A 3K7, Canada. tfuller@phobos.astro.uwo.ca

John Girash, Center for Astrophysics, Harvard University, 60 Garden Street, MS 10, Cambridge, MA 02143, U.S.A. girash@skyron.harvard.edu

Nick Gnedin, Astronomy Department, University of California at Berkeley, Berkeley, CA 94720, U.S.A. gnedin@arcturus.mit.edu

Evgeny Griv, Physics Department, Ben-Gurion University of the Negev, P.O. Box 653, Beersheva 84105, Israel. griv@chen.bgu.ac.il

Philip E. Hardee, Department of Physics and Astronomy, University of Alabama, Gallalee Hall, Tuscaloosa, AL 35487, U.S.A. hardee@venus.astr.ua.edu

Wayne Hayes, Department of Computer Science, University of Toronto, 10 King's College Road, Toronto, ON M5S 3G4, Canada. wayne@cs.utoronto.ca

Richard N. Henriksen, Department of Physics, Queen's University at Kingston, Kingston, ON K7L 3N6, Canada. henriksn@astro.queensu.ca

David Hobill, Department of Physics and Astronomy, University of Calgary, Calgary, AB T2N 1N4, Canada. hobill@acs.ucalgary.ca

Philip Hughes, Department of Astronomy, University of Michigan, Dennison Building, Ann Arbor, MI 48109, U.S.A. hughes@astro.lsa.umich.edu

Piet Hut, Institute for Advanced Study, Princeton University, Princeton, NJ 08540, U.S.A. piet@sns.ias.edu

Werner Israel, Department of Physics, University of Alberta, Edmonton, AB T6G 2J1, Canada. israel@euclid.phys.ualberta.ca

Thomas W. Jones, Department of Astronomy, University of Minnesota, 116 Church Street S.E., Minneapolis, MN 55455, U.S.A. twj@msi.umn.edu

Byung-Il Jun, Department of Astronomy, University of Minnesota, 116 Church Street S.E., Minneapolis, MN 55455, U.S.A. jun@msi.umn.edu

Richard Klein, Department of Astronomy, University of California at Berkeley, Berkeley, CA 94720-3411, U.S.A. klein@radhydro.berkeley.edu

Ralf Klessen, Max-Planck-Institut für Astronomie, Königstuhl 17, Heidelberg, 69117, Germany. klessen@mpia-hd.mpg.de

Denis Leahy, Department of Physics and Astronomy, University of Calgary, 2500 University Drive N.W., Calgary, AB T2N 1N4, Canada. leahy@iras.ucalgary.ca

Man Hoi Lee, Department of Physics, Queen's University at Kingston, Kingston, ON K7L 3N6, Canada. mhlee@astro.queensu.ca

Chris Loken, Department of Astronomy, Dept. 4500, New Mexico State University, Las Cruces, NM 88003, U.S.A. cloken@nmsu.edu

Peter G. Martin, CITA, University of Toronto, 60 St. George Street, Toronto, ON M5S 3H8, Canada. pgmartin@cita.utoronto.ca

David L. Meier, Jet Propulsion Laboratory, 238-332, 4800 Oak Grove Drive, Pasadena, CA 91109, U.S.A. dlm@cena.jpl.nasa.gov

Andres Meza, Departamento de Fisica, Universidad de Chile, Avda. Blanco Encalada 2008, Santiago, Casilla 487-3, Chile. ameza@cec.uchile.cl

Georges Michaud, Département de physique, Université de Montréal, C.P. 6128 Centre-Ville, Montréal, PQ H3C 3J7, Canada. michaudg@cerca.umontreal.ca

Beverly Miskolczi, Department of Astronomy and Physics, Saint Mary's University, Halifax, NS B3H 3C3, Canada. bmiskolczi@ap.stmarys.ca

Romas Mitalas, Department of Physics and Astronomy, The University of Western Ontario, London, ON N6A 3K7, Canada. mitalas@uwovax.uwo.ca

George F. Mitchell, Department of Astronomy and Physics, Saint Mary's University, Halifax, NS B3H 3C3, Canada. gmitchell@ap.stmarys.ca

James Murray, CITA, University of Toronto, 60 St. George Street, Toronto, ON, M5S 3H8, Canada. jmurray@cita.utoronto.ca

Alistair Nelson, Department of Physics and Astronomy, University of Cardiff, Cardiff, Wales CF2 3YB, U.K. nelsona@cf.ac.uk

Michael Norman, Department of Astronomy, University of Illinois at Urbana-Champaign, 1002 W. Green Street, Urbana, IL 61801, U.S.A. norman@ncsa.uiuc.edu

Rachid Ouyed, Department of Physics and Astronomy, McMaster University, Hamilton, ON L8S 4M1, Canada. ouyed@physics.mcmaster.ca

Marc Pinsonneault, Department of Astronomy, Ohio State University, 174 W. 18th Avenue, Columbus, OH 43210, U.S.A. pinsono@payne.mps.ohio-state.edu

Ralph Pudritz, Department of Physics and Astronomy, McMaster University, Hamilton, ON L8S 4M1, Canada. pudritz@jabba.physics.mcmaster.ca

Kevin P. Rauch, CITA, University of Toronto, 60 St. George Street, Toronto, ON M5S 3H8, Canada. raunch@cita.utoronto.ca

Alexei Razoumov, Department of Physics and Astronomy, University of British Columbia, 129-2219 Main Mall, Vancouver, BC V6T 1Z4, Canada. razoumov@astro.ubc.ca

Örnolfur E. Rögnvaldsson, Theoretical Astrophysics Center, Juliane Maries Vej 30, DK-2100 Köbenhavn Ä, Denmark. ossi@tac.dk

Alex Rosen, Department of Physics and Astronomy, University of Alabama, Tuscaloosa, AL 35487, U.S.A. rosen@eclipse.astr.ua.edu

Dongsu Ryu, Department of Astronomy, University of Washington, FM-20, Seattle, WA 98195, U.S.A. ryu@sirius.chungnam.ac.kr

Jerry Sellwood, Department of Physics and Astronomy, Rutgers University, P.O. Box 849, Piscataway NJ 08855-0849, U.S.A. sellwood@physics.rutgers.edu

Steven Short, Department of Astronomy and Physics, Saint Mary's University, Halifax, NS B3H 3C3, Canada. sshort@ap.stmarys.ca

Alison Sills, Department of Astronomy, Yale University, P.O. Box 208101, New Haven, CT 06520-8101, U.S.A. asills@astro.yale.edu

Bob Stein, Department of Physics and Astronomy, Michigan State University, East Lansing, MI 48824, U.S.A. stein@pa.msu.edu

James M. Stone, Department of Astronomy, University of Maryland, College Park, MD 20742-2421, U.S.A. jstone@ophir.astro.umd.edu

Curtis Struck, Department of Physics and Astronomy, Iowa State University, 12 Physics Building, Ames, IA 50011, U.S.A. curt@iastate.edu

Rene Tanaja, Department of Astronomy and Physics, Saint Mary's University, Halifax, NS B3H 3C3, Canada. rtanaja@ap.stmarys.ca

Eric Tittley, Department of Physics and Astronomy, University of Western Ontario, London, ON N6A 3K7, Canada. etittley@phobos.astro.uwo.ca

Scott Tremaine, CITA, University of Toronto, 60 St. George Street, Toronto, ON M5S 1A7, Canada. tremaine@cita.utoronto.ca

Eelco van Kampen, Theoretical Astrophysics Center, Juliane Maries Vej 30, DK-2100 Copenhagen, Denmark. evk@tacsg1.tac.dk

James Wadsley, Department of Astronomy, University of Toronto, 60 St. George Street, Toronto, ON M5S 1A1, Canada. wadsley@cita.utoronto.ca

Michael J. West, Department of Astronomy and Physics, Saint Mary's University, Halifax, NS B3H 3C3, Canada. mwest@ap.stmarys.ca

David Wiggins, Department of Applied Mathematics, University of Leeds, Woodhouse Lane, Leeds, Yorkshire, LS2 9JT, U.K. dwigg@amsta.leeds.ac.uk

Meeting Participants

Meeting Participants

Pictured, from left to right (and from top to bottom in the event two people have the same horizontal position): Rene Tanaja (Saint Mary's), Maurice Clement (Toronto), Rachid Ouyed (McMaster), Beverly Miskolczi (Saint Mary's), Hugh Couchman (Western Ontario), David Clarke (Saint Mary's), Marc Pinsonneault (Ohio State), Paul Charbonneau (NCAR), Ralph Pudritz (McMaster), Philip Hardee (Alabama), Evgeny Griv (Ben-Gurion), Greg Bryan (MIT), Philip Hughes (Michigan), Kevin Douglas (Saint Mary's), Byung-Il Jun (Minnesota), Werner Israel (Alberta), James Wadsley (CITA), Nickolay Gnedin (Berkeley), Michael Norman (Illinois), Curtis Struck (Iowa State), Alison Sills (Yale), ?, Dinshaw Balsara (Illinois), Sam Falle (Leeds), Romas Mitalas (Western Ontario), Richard Klein (Berkeley), Dick Henriksen (Queen's), Bob Stein (Michigan State), Tom Jones (Minnesota), David Wiggins (Leeds), Steve Short (Saint Mary's), John Girash (CfA), Andres Meza (Chile), Eric Tittley (Western Ontario), Alistair Nelson (Cardiff), Todd Fuller (Western Ontario), Man Hoi Lee (Queen's), Dick Bond (CITA), Dongsu Ryu (Chungnam), Alexei Razoumov (British Columbia), Martin Duncan (Queen's), Dave Meier (CalTech), David Hobill (Calgary), Michael West (Saint Mary's), Alex Rosen (Alabama), Andrew Billyard (Dalhousie), Chris Loken (New Mexico State), Alicia Sintes (Dalhousie), Peter Martin (CITA), Jim Stone (Maryland), Katherine Durrell (McMaster), Wayne Hayes (Toronto), Jerry Sellwood (Rutgers), Ralf Klessen (Max Planck Institute), George Mitchell (Saint Mary's), Kevin Raunch (CITA), Eelco van Kampen (TAC), Phil Bennett (Colorado).

Missing, or not identified in the photograph: Malcolm Butler (Saint Mary's), Matt Choptuik (Texas), Alan Coley (Dalhousie, photographer), Peter Damiano (Alberta), Jonathan Dursi (Waterloo), Piet Hut (Princeton), Denis Leahy (Calgary, entirely blocked by DAC, with apologies!), Georges Michaud (Montréal), James Murray (CITA), Örnolfur Rögnvaldsson (TAC), Scott Tremaine (CITA).

OVERVIEW

Computational Astrophysics: The "New Astronomy" for the 21st Century

Michael L. Norman

Laboratory for Computational Astrophysics, Astronomy Department, and NCSA, University of Illinois at Urbana-Champaign, Urbana, IL 61801, U.S.A.

Abstract. I discuss the role computer simulation has played in astronomical research, reviewing briefly the origins of the field only to place into perspective the enormous strides which have been achieved in recent decades. I will highlight areas where computational astrophysics has already made a scientific impact, and attempt to discover the conditions which lead to real progress. Finally, I will prognosticate on what the future may hold in store for the second "New Astronomy" revolution already well underway.

1. Historical Perspective

The "New Astronomy" ... seems assured of a most brilliant future.
G. E. Hale

Modern astrophysics as we know it began as a quiet revolution in the mid-19th century with the development of the science of spectroscopy by Kirchhoff, Fraunhofer and others. When combined with the revolutionary discoveries about the nature of matter provided by atomic and molecular physics in the first half of the twentieth century, astronomers suddenly had a new, powerful analytic tool to diagnose the cosmos—a tool in many ways more important to astrophysics than the telescope itself. For without spectroscopic measurements, astronomers would not have been able to detect the expanding universe, map the structure of our Milky Way galaxy, discover the quasars, or confirm the hot Big Bang origin of our universe. Indeed, spectroscopy underlies most of our twentieth century advances in astronomy.

Recognizing the potential of spectroscopy to transform astronomy into a quantitative *physical science*, solar physics pioneer and astrophysics founding father George Ellery Hale devoted his life's energies to realizing its potential. He called astrophysics the "New Astronomy" for the 20th century (Wright, 1966). In promoting this new way of doing astronomy, Hale made no small plans. Among his career accomplishments were the founding of the Yerkes, Mount Wilson, and Mount Palomar Observatories, the California Institute of Technology, the US National Academy of Sciences (NAS), and the Astrophysical Journal (Wright, 1966; Osterbrock, 1995).

In like fashion, digital computers have been quietly transforming astronomy and astrophysics since their invention in the mid-20th century. There is no doubt that computers are ubiquitous and indispensable tools for observer and theorist alike. In theoretical astrophysics, computer simulation allows us to probe additional dimensions of structure, dynamics and evolution for any astrophysical system including the universe as a whole (Norman, 1996). The breadth of topics presented at this meeting confirm this. In recognition of the growing importance of computers in astronomy, and especially theoretical astrophysics, the NAS Decade Survey of Astronomy and Astrophysics (the "Bahcall Report", Bahcall, 1991) devoted an entire chapter to it.

In my talk, I will focus on the role computer simulation has played in astronomical research, reviewing briefly the origins of the field only to place into perspective the enormous technical strides which have been achieved in recent decades. I will highlight areas where computational astrophysics has already made a scientific impact, and attempt to discover the conditions which lead to real progress. Finally, I will prognosticate on what the future may hold in store for the second "New Astronomy" revolution already well underway.

2. Early Pioneers

The immediate imitation in the laboratory, under experimental conditions subject to easy trial, of solar and stellar phenomena, not only tends to clear up obscure points, but prepares the way for developing along logical lines the train of reasoning started by the astronomical works. G. E. Hale

While Hale was, of course, advocating what we now call laboratory astrophysics, he could just as well have been describing computational astrophysics. For computational astrophysics has become a theoretical laboratory for experimenting with astrophysical systems in much the same way laboratory astrophysicists experiment with astrophysical plasmas. While it is true the latter deals with reality and the former deals only with simulated realities, there are many similarities in goals and methodology. These include a concern about precision of measurement, the exploration of relevant parameter regimes, and the importance of qualifying the scope of validity of the results.

The view of the computer as a numerical laboratory took some time to emerge. Stellar evolutionists Martin Schwarzschild in the U.S. and Rudolph Kippenhahn in Germany were among the earliest pioneers to use digital computers to solve astrophysical problems. Both were at institutions where electronic computers were being designed and built shortly after WWII (Princeton and Gottingen). Both men made seminal contributions to stellar evolution theory in the 1950's and 1960's. The principal computational task in stellar evolution calculations is to solve the equations of stellar structure—four coupled ordinary differential equations—subject to certain boundary conditions and mass and composition constraints. In keeping with the parlance of the day, computers were viewed as numerical integrators—tools to evaluate a quadrature or compute a ballistic trajectory—rather than a tool for experimentation. The latter, more grandiose view of computers was held by the visionary John von Neumann. However, it is unclear whether his writings on the subject (Goldstine & von Neumann, 1963) reached the attention of the first computational astrophysicists.

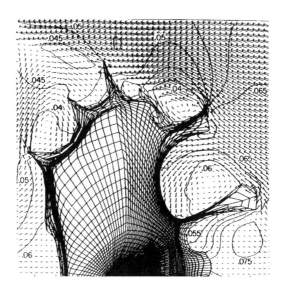

Figure 1. State-of-the-art numerical hydrodynamic simulation ca. 1975 of an interstellar cloud compressed by a passing interstellar shock wave. The two dimensional calculation utilized an innovative coupled Eulerian-Lagrangian grid of 28,000 cells and taxed the resources of a CDC 7600. From Woodward (1976), reproduced by permission.

The notion of computational astrophysics as experimental astronomy came about naturally when astrophysicists began to simulate dynamical systems in astronomy, principally stellar or hydrodynamical systems. These simulations are generally motivated by the question "What happens if?" more so than "What is the solution to these equations?". Remarkably, the earliest N-body experiment pre-dated digital computers by half a decade. Erik Holmberg simulated the tidal interaction of two galaxies with an analog computer consisting of an array of movable light bulbs and photocells each representing a point mass (Holmberg, 1941). Numerical integration was accomplished by placing the analog point masses on a mat inscribed with a Cartesian grid, and moving them about by hand according to the local gravitational acceleration determined by measuring the flux and direction of light incident on the stars' photocells (flux, like gravity, falls off as $1/r^2$). Each galaxy was represented by 37 point masses, arranged in circular rings.

Numerical stellar dynamics entered the modern era with the pioneering calculations of Aarseth, who in 1963 carried out the first N-body simulation of the dynamical evolution of a cluster of galaxies (Aarseth, 1963). Another pioneering effort was that of Juri and Alar Toomre, whose calculations of galactic encounters convincingly established a tidal origin for intergalactic tails and "antennae" in peculiar galaxies (Toomre & Toomre, 1972). While the Toomres' simulations assumed the test particle approximation, Aarseth's calculations were fully self-consistent, the force on each particle determined by direct summation over particle pairs. The success of these calculations launched a world-wide industry in gravitational N-body simulations as the tool of choice to study stellar, galactic

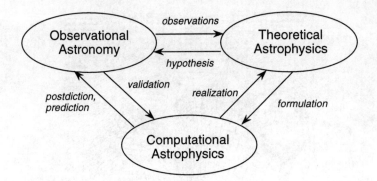

Figure 2. Synergies between observational, theoretical and computational astrophysics.

and more recently, cosmic dynamics. A good review of the early development in this field can be found in Aarseth & Lecar (1975).

Pioneers in astrophysical hydrodynamics employed computers in the late 1960's to simulate the birth and death of stars, and processes occurring in the interstellar medium. They include Richard Larson, who first self-consistently simulated star formation in 1-D spherical symmetry (Larson, 1969); W. David Arnett and independently James Wilson, who constructed the first detailed simulations of core collapse supernova explosions (Arnett, 1967; Wilson, 1971), rotating relativistic stars (Wilson, 1972), and hydromagnetically-driven winds (Wilson & LeBlanc, 1970); and Paul Woodward, who carried out the first multi-dimensional simulation of the implosion of an interstellar cloud by a shock wave (Woodward, 1976; *cf.*, Figure 1). Grids of several hundred points in 1-D, or several thousand in 2-D characterized these early simulations, in contrast to the millions used in today's large scale simulations.

3. Role of Computational Astrophysics

It is customary to distinguish sharply between observational and experimental sciences, including astronomy in the former. ... that distinction between these two methods of research is not so fundamental as it might appear. G. E. Hale

Computational astrophysics interacts synergistically with observation and theory, and borrows elements from each, as illustrated in Figure 2. The interplay between observation and theory is standard and will not be belabored here. The interplay between computational astrophysics and the other two methodologies is perhaps less clear, and therefore I will discuss it briefly. Theory interacts with simulation in a number of essential ways. First, theory provides the mathematical formulation which is used in the construction of a numerical model, as well as defines the parameter space of solutions to be searched. Second, desirable analytic properties (*e.g.*, conservation laws) of the solution can be incorporated into improved numerical algorithms. Third, analytic solutions provide excellent test problems for code validation. In fact, it is often the case that failure to reproduce an analytic result stimulates the critical thinking required to invent more accurate algorithms. Finally, when analyzing the results of a numerical simulation, especially one involving many complex physical processes such as our examples, one attempts to construct simplifying analytic models which nonetheless capture the essential physics.

For their part, simulations provide realizations of theoretical models which are in general too complex to be solved analytically. These realizations are in essence the laboratory data upon which the correctness of theoretical models is tested. This is done in two ways. First, it is often the case in astrophysics that one is uncertain whether all of the relevant physics has been included. A failure of the simulation to reproduce the observed phenomena may indicate missing physics, bad numerics, bad observational data, or any combination thereof. Second, simulations build physical intuition by providing the modeler direct experience with the complex phenomena embodied in the governing equations. Improved physical intuition generally precedes the formulation of improved theoretical models.

Finally, the simulation must confront observations. Observations are the final validation of a model's correctness. In the early phases of model building, one strives merely to "postdict" the available observational data. As the quality of the data improves, models are either revised or rejected. A correct model will not only remain consistent as new observations accumulate, but will also predict and in fact suggest new observations to come.

4. Progress in Computational Astrophysics

Little progress can be made without powerful means.—G. E. Hale

Progress in computational astrophysics, as in other branches of computational science, has been paced by the development of and access to: (1) high-performance computer architectures, and (2) accurate and efficient algorithms. As shown in Figure 3, the peak speed of the fastest available supercomputer at any given time has advanced steadily since the ENIAC, the first fully electronic computer, was built 50 years ago. To a good approximation, speed has increased exponentially with time since WWII, with an e-folding time of about two years. This trend is expected to continue well into the next decade through continuing advances in VLSI circuitry and massively parallel architectures (Brenner, 1986).

Equally important has been the development and dissemination of robust, accurate and efficient algorithms for hydrodynamical and N-body simulations. Particularly influential have been, in hydrodynamics, the von Neumann and Richtmyer 1-D Lagrangian algorithm (Richtmyer & Morton, 1967); higher order-accurate Godunov algorithms for multidimensional gas dynamics (*e.g.*, Colella & Woodward, 1984); Smoothed Particle Hydrodynamics (Monaghan, 1992), the ZEUS codes for radiation magnetohydrodynamics (Stone & Norman, 1992b), and the Hawley, Smarr and Wilson algorithm for general relativistic hydrodynamics (Hawley *et al.*, 1984). On the N-body side, the following have received wide application: Aarseth's direct summation methods (Aarseth, 1971); the PM and P^3M algorithms of Hockney & Eastwood (1988), and the tree code of Barnes & Hut (1986) and Hernquist (1987). These algorithms, used separately

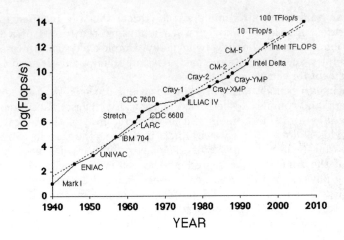

Figure 3. Growth in peak supercomputer performance *vs.* time (adapted from Kaufmann & Smarr, 1993). An exponential fit $2^{3t/5}$ is shown as the dotted line.

or in combination (*e.g.*, TREESPH, Hernquist & Katz, 1989), underly a large fraction of current computational astrophysics research.

Gains in algorithmic efficiency have roughly kept pace with hardware improvements. Consequently astrophysicists who have kept abreast of the latest algorithms and supercomputers have been able roughly to double the complexity of their simulations every year in recent decades. Until the mid 1980's, however, access to supercomputers was limited to a small cadre of researchers at defense laboratories or at a few well-endowed academic institutions. The establishment of the NSF Supercomputing Centers in 1985 opened up access to state-of-the-art supercomputers to the entire U.S. academic community. This development, as well as the subsequent creation of state and regional supercomputing centers, open supercomputing facilities at DOE and NASA labs, the emergence of powerful and affordable workstations, and the growth of the Internet have all played a role in increasing the number of computational astrophysicists roughly one hundred fold. As a result, computational astrophysics research has enjoyed a decade of unprecedented growth and progress.

The impact has been a broadening and a deepening of computational astrophysics research. As a result of improved hardware and algorithms, numerical simulations in the established areas of N-body and astrophysical fluid dynamics have matured in at least three significant ways. This is illustrated schematically in Figure 4, and by way of examples in Figures 5–7. Generally, there has been a progression from lower to higher dimensional simulations; from lower to higher resolution simulations; and from simple physical models to complex models embodying many physical processes. In any given field, progress tends to proceed along one axis at a time until a qualitatively new threshold of complexity and physical realism is reached. Research is carried out within the new paradigm, often community wide, until it is replaced by another advance, typically along an orthogonal axis.

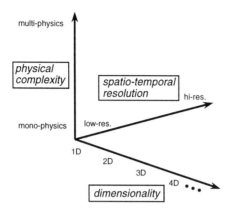

Figure 4. Progress in numerical modeling occurs along (at least) three axes in a conceptual phase space: dimensionality, spatio-temporal resolution, and physical complexity.

5. Case Study

Scores of problems suggest themselves for solution ... G. E. Hale

To illustrate, consider simulations of a shock wave interacting with an interstellar cloud. This set piece problem has received considerable study over the years because it is a fixture in the McKee and Ostriker theory of the hot interstellar medium (McKee & Ostriker, 1977). A central question is how long the cloud survives after being hit by the shock wave. The earliest simulation carried out by Woodward (1976) correctly predicted the existence of Rayleigh-Taylor and Kelvin-Helmholtz fluid instabilities which grow on the cloud boundary and ultimately destroy the cloud. However, Lagrangian mesh tangling in the cloud interior terminated the calculation before cloud destruction was complete. The fully Eulerian calculations (MacLow et al., 1994) shown in Figure 5 do not suffer from this defect. However, concerns about the effects of numerical resolution and assumed axisymmetry on the disruptive instabilities now come to the fore.

The steady improvements in supercomputers and algorithms in the 1980's permitted for the first time serious convergence studies to be made on multi-dimensional simulations. Convergence studies and their related validation test suites (see e.g., Stone et al., 1992) are now a standard part of good numerical methodology, and are an indication of a maturing field. The convergence study in Figure 5 illustrates that although the bulk deformation of the cloud can be captured with 50-100 cells per cloud radius, much higher resolution is required to capture smaller scale modes of instability which contribute to cloud disruption (Klein et al., 1994).

The cloud is ultimately shredded by Kelvin-Helmholtz instabilities of the vortex sheet created by the interaction of the shock wave with the cloud's boundary. At the high Reynold's numbers of astrophysical fluids, the fully non-linear development of the K-H instability leads to turbulence, which is inherently 3-D. The first 3-D simulation of the shock-cloud interaction (Stone & Norman, 1992),

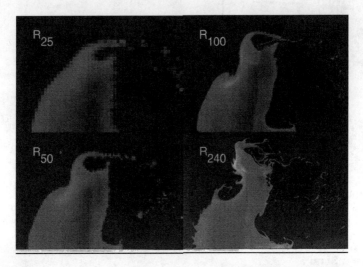

Figure 5. Convergence study of 2-D shock-cloud interaction simulations. The subscript in the figure label R_n refers to the number of cells used to resolve the initial cloud radius (from MacLow *et al.*, 1994)

Figure 6. Two snapshots from a 3-D simulation of a shock-cloud interaction. Shown is the line-of-sight integrated fluid vorticity magnitude (from Stone & Norman, 1992a).

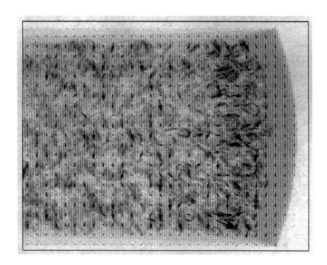

Figure 7. A numerical observation of a 3-D magnetic supernova remnant. Polarized radio surface brightness is shown in grey-scale, and polarization B-vectors are shown as vectors (from Jun & Norman, 1996)

shown in Figure 6, verifies this expectation, and provides the first realistic picture for the late phases of cloud disruption. Three-dimensional simulations such as these have only recently become practical with large memory parallel supercomputers. Although comprehensive 3-D convergence studies are barely feasible today because of computer limitations, isolated examples exist (Kang et al., 1994; Jun et al., 1995).

The simulations thus far have assumed ideal, adiabatic gas dynamics. Real interstellar clouds are magnetized, cool radiatively when shocked, etc. In order to engage observations in a meaningful way, simulations must also mature along the third axis of physical complexity. At the very least, additional physics describing the emissivity of the material must be added to the model so that observables can be computed, a step I call "numerical observations." Figure 7 shows an example of a numerical observation of a radio supernova remnant simulated by Byung-Il Jun and myself (Jun & Norman, 1996). A 3-D numerical MHD simulation of a young supernova remnant was carried out to compute self-consistently the complex structure of the magnetic field in the turbulent shell. The numerical observation is made by integrating the Stokes parameters along rays passing through the remnant, assuming radio synchrotron emission. The simulation reproduces the observed radial magnetic polarization in young supernova remnants, although the simulation is not predictive with regard to the radio luminosity since the relativistic electron population is not computed self-consistently. We have also not been able to converge on the amount of magnetic field amplification in the turbulent shell. Further model maturation is required to remove these defects.

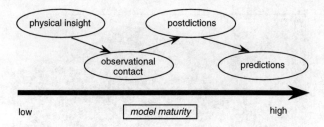

Figure 8. The primary contribution of a numerical model changes as it matures.

6. Scientific Impact

The scientific contribution of a numerical model changes as it matures. This is illustrated in Figure 8. In a few rare cases, such as the galaxy encounter simulations of the Toomres, an early, crude simulation will provide a key insight into the nature of an astrophysical phenomenon. A recent example is the still somewhat controversial role of neutrino-driven convection in core collapse supernova explosions (Norman, 1996). Such simulations change the way we think about things, and constitute important scientific contributions in their own right. Often, these simulations can be reduced to a cartoon or a mathematical toy model *post facto*, and become enshrined in elementary astronomy textbooks.

More often than not, early models miss essential physics inherent in a phenomenon, or are of insufficient resolution to simulate it accurately. Such models require substantial maturation before observations are engaged in any meaningful way. However, once all the physics is included and computers are adequate to the task, rapid progress generally follows. Stellar evolution theory is the premier example of this. The equations of stellar structure were known in the 1930's, the missing piece of physics—nuclear energy generation—was supplied in the 1940's, serviceable opacities and numerical algorithms were in place by the end of the 1950's, and adequate computer power was available in the 1960's. Scientific progress swiftly followed. An important factor for progress was the abundant, high-quality observational stellar data exhibiting clear statistical trends. Stellar evolution calculations first rationalized these data into a coherent theory, and then proceeded to make testable predictions. Frontier areas remain in stellar evolution theory. However, these invariably involve dynamical phenomena (*e.g.*, star formation), three-dimensional physics (*e.g.*, convection, symbiotic binary stars), or other complications, which up the computational ante.

Reflecting on other areas of computational astrophysics which have made a definite scientific impact, I find the following conditions must be met: (1) complete physical model; (2) good numerical algorithms; (3) adequate computing power; (4) unambiguous observational data; (5) insensitivity to initial conditions where they are unknown, *or*, sensitivity to initial conditions where they are known (or at least knowable). When these conditions are met, simulations have predictive power and are able to rationalize observations into a proper theory. Areas where this is occurring include dynamic stages of stellar evolution, stellar and galactic dynamics, galaxy interactions, and cosmological structure formation. Rapidly maturing areas include accretion disks, jets and outflows from young stars, morphology of planetary nebulae, Type Ia and Type II supernova mechanisms, accretion onto compact objects, galaxy formation and evolution, and the structure of the Lyman α forest. Areas which still lack one or more of these criteria include, in my opinion, star formation, structure of the interstellar medium, active galactic nuclei, astrophysical dynamos, turbulent convection in stars, astrophysical particle acceleration, and radio galaxies.

7. Future Outlook

Questions of all degrees of complexity remain to be answered, and every day sees their number increased. G. E. Hale

Supercomputing hardware performance is expected to continue improving at its historical rate, with computing speed doubling roughly every 20 months, for at least the next decade. Supercomputers with peak speeds in excess of a teraflop (10^{12} Flop/s) will be operational at the Department of Energy weapons labs by 1998. Ten years after that we can expect to have 100 teraflops. My own center, the NCSA, will have a teraflop computer available to academic researchers by the turn of the century. This guarantees that computational astrophysics will continue to develop and mature. Teraflop supercomputers will permit models of realistic complexity to be simulated in 3-D at high resolution (*e.g.*, $1,024^3$.) This kind of computer power will benefit, in particular, research in cosmological structure formation, stellar convection, star formation, supernova phenomena, and the relativistic dynamics of coalescing binary neutron stars and black holes.

However, recent history has shown that more significant than supercomputers to the spread of computational astrophysics research are affordable workstations and PCs. These computers follow the same speedup trend, but lag supercomputer performance by about 10 years. For example, by the year 2000, affordable workstations with 1 gigaflop/s processors and 1 gigabyte of RAM will be on the market. This means that the average university astronomer will have the equivalent power of a Cray-2 supercomputer on their desktop. What will this machine be used for, and where will the software come from?

Since 1993, the Laboratory for Computational Astrophysics (LCA) which I direct has developed and distributed astrophysics simulation software to the international research community (visit our web site: lca.ncsa.uiuc.edu.) Two programs for astrophysical fluid dynamics called ZEUS-2D and ZEUS-3D, developed by Jim Stone, David Clarke and myself, have attracted a large following. Curiously, most users prefer to install the software on their local workstations rather than use a supercomputer. This limits the size and sophistication of the calculations they can do. However, with gigaflop workstations just around the corner, that will change. High-resolution 2-D simulations and medium-resolution 3-D simulations will be routinely doable on desktop machines. Moreover, with the emergence of the Web as a metaphor for information access and the coming of higher network bandwidths, we can envision an era of high-end web computing where the users' PC becomes an interface to a simulation running on a remote supercomputer. Work has begun on a web-based LCA computational workbench to make this idea a reality.

Acknowledgments. I gratefully thank my early mentors Jim Wilson, Paul Woodward, Larry Smarr and Karl-Heinz Winkler for kindling my interest in computational astrophysics.

References

Aarseth, S. 1963, MNRAS, 126, 223.
Aarseth, S. 1971, Astrophys. & Space Sci., 14, 118.
Aarseth, S., & Lecar, M. 1975, ARA&A, 13, 1.
Arnett, W. D. 1967, Canadian J. Phys., 45, 1621.
Bahcall, J. 1991, *The Decade of Discovery in Astronomy and Astrophysics*, (Washington: National Academy Press), 91.
Barnes, J., & Hut, P. 1986, Nature, 324, 446.
Brenner, A. 1996, Physics Today, 49, No. 10, 24.
Colella, P., & Woodward, P. R. 1984, J. Comput. Phys., 54, 174.
Goldstine, H., & von Neumann, J. 1963, *John von Neumann Collected Works, Volume V*, (New York: Pergamon), 1.
Hawley, J. F., Smarr, L. L. & Wilson, J. R. 1984. ApJ, 277, 296.
Hernquist, L. 1987, ApJ, 64, 715.
Hernquist, L., & Katz, N. 1989, ApJ, 70, 419.
Hockney, R. W., & Eastwood, J. W. 1988, *Computer Simulation Using Particles*, (Bristol: Institute of Physics), 120.
Holmberg, E. 1941, ApJ, 94, 385.
Jun, B.-I., & Norman, M. L. 1996, ApJ, 472, 245.
Jun, B.-I., Norman, M. L., & Stone, J. M. 1995, ApJ, 453, 332.
Kang, H., Ostriker, J. P., Cen, R., Ryu, D., Hernquist, L., Evrard, A. E., Bryan, G. L., & Norman, M. L. 1994. ApJ, 430, 80.
Kaufmann III, W. J., & Smarr, L. L. 1993, *Supercomputing and the Transformation of Science*, (New York: W. H. Freeman), 33.
Klein, R. I., McKee, C. F., & Colella, P. 1994, ApJ, 420, 213.
Larson, R. B. 1969, MNRAS, 145, 271.
MacLow, M.-M., McKee, C. F., Klein, R. I., Stone, J. M., & Norman, M. L. 1994, ApJ, 433, 757.
McKee, C. F., & Ostriker, J. P. 1977, ApJ, 218, 148.
Monaghan, J. J. 1992, ARA&A, 30, 543.
Norman, M. L. 1996. Physics Today, 49, No. 10, 42.
Osterbrock, D. E. 1995, ApJ, 438, 1.
Richtmyer, J. D., & Morton, K. W. 1967, *Difference Methods for Initial Value Problems*, (New York: Interscience).
Stone, J. M., Hawley, J. F., Evans, C. R., & Norman, M. L. 1992, ApJ, 388, 415.
Stone, J. M., & Norman, M. L. 1992a, ApJ, 390, L17.
Stone, J. M., & Norman, M. L. 1992b, ApJS, 80, 753.
Toomre, A., & Toomre, J. 1972, ApJ, 178, 623.
Wilson, J. R. 1971, ApJ, 163, 209.
Wilson, J. R. 1972, ApJ, 176, 195.
Wilson, J. R., & LeBlanc, J. M. 1970, ApJ, 161, 541.
Woodward, P. R. 1976, ApJ, 207, 484.
Wright, H. 1966, *Explorer of the Universe: A Biography of George Ellery Hale*, (New York: Dutton).

Part I

SOLAR SYSTEM DYNAMICS

The Longterm Dynamical Evolution of Orbital Configurations

Martin J. Duncan

Physics Department, Queen's University at Kingston, Kingston, ON K7L 3N6, Canada

Abstract. Numerical experiments over the past decade have made it increasingly apparent that most orbits in the solar system exhibit the chaotic behaviour found in many non-linear Hamiltonian systems. In the case of the solar system, however, the time scale for this underlying chaos to produce "macroscopic" manifestations (*e.g.*, orbit crossings, close encounters, ejections, *etc.*) can exceed millions or billions of orbital periods. As a result, the long-term orbital evolutions of the planets, moons, comets and asteroids are still poorly understood although their collective behaviour is central to understanding the origin and present state of our solar system. Chaotic motion also presumably has implications for the architecture of planetary systems around other stars. I review recent advances in the numerical techniques used to perform long-term simulations of such systems and present the results of a few (hopefully representative) examples. I also indicate some of the numerical and conceptual challenges facing us in the future.

1. Numerical Methods

Most orbits in the solar system can be studied by modelling the system as one governed by a non-linear Hamiltonian. In recent years, symplectic integrators (from the Greek word "symplegma" or "tangled") have become the dominant method for long-term solar system simulations since they generally preserve much of the geometric structure of phase space (see, *e.g.*, Yoshida, 1993; Sanz-Serna & Calvo, 1994). Their strength also lies in the fact that for reasonable step size, the integrated trajectories are solutions of a Hamiltonian which is usually very close to the true Hamiltonian. In their most powerful form for solar system simulations without close encounters (Wisdom & Holman, 1991), they are elegant generalizations of well-known "leapfrog" integrators which, in this case, alternate between moving particles along Keplerian ellipses and kicking their velocities via the mutual perturbations among the orbiting bodies (hereafter called satellites). For weakly-interacting satellites, these symplectic methods are typically an order of magnitude faster than conventional methods. Treatments of close approaches between satellites can also be done symplectically, as discussed in Levison & Duncan (1994) and Lee *et al.* (these proceedings). Space does not permit a detailed exposition on symplectic methods—interested readers may consult the references cited above as well as a very readable review by Tremaine (1995).

2. Stability of Moons and Planets

Many of the results from studies of long-term stability in the solar system can be understood by utilizing the tools of non-linear Hamiltonian dynamics. Summaries of these issues are to be found in the proceedings edited by Ferraz-Mello (1992) and the reviews by Duncan & Quinn (1993) and Tremaine (1995). In what follows, I will emphasize even more recent results using examples which are hopefully representative but undoubtedly biased. We begin with the simplest nontrivial case—that of two low-mass satellites orbiting a central body. Many decades of previous mathematical work on this problem was summarized by Gladman (1993), who recast topological constraints based on total energy and angular momentum conservation into a simple rule of thumb: if two nearly coplanar satellites on nearly circular orbits have semimajor axes separated by more than 3.5 R_H, they cannot suffer a mutual close encounter. The quantity R_H is the mutual Hill radius defined by

$$R_H = 0.5\,(a_1 + a_2) \left[\frac{m_1 + m_2}{3M}\right]^{1/3},$$

where M is the mass of the central body, m_1, m_2 are the masses of the two satellites and a_1, a_2 are their semimajor axes. Stewart (1997) has devised a mapping for the two-planet case that explains much of the behaviour seen in the integrations of Gladman. Two groups have performed stability studies of systems of three or more satellites and independently found approximate scaling laws for the time scale t_c for two orbits to cross:

- Chambers et al. (1996) studied systems of three or more equal-mass satellites with equal spacing (relative to their mutual Hill radii) and showed that for a given satellite/primary mass ratio, the logarithm of the crossing time was roughly proportional to the spacing. Unlike the two-planet case, low-mass bodies spaced by as much as 10 mutual Hill radii were found to be unstable on million-orbit time scales in these larger N systems.

- Duncan & Lissauer (1997a) studied the stability of eight of the small inner Uranian satellites discovered by the Voyager spacecraft, using orbital elements from Owen & Synnott (1987) and masses calculated from Voyager size estimates assuming densities equal to that of the Uranian moon Miranda. Since the satellite masses are rather uncertain, for each of the runs all of the satellite masses were multiplied by the same numerical factor m_f and the system was integrated until orbit-crossing occurred. The results reveal quite a tight power-law relationship between m_f and the crossing time t_c of the form $t_c = \beta m_f^\alpha$, where the values of the constants α and β depend on the system; values of α ranging from -4.1 to -3 were found for systems based on the eight inner satellites.

The mass scaling seems to be related to the trends found by Chambers et al. (1996), as can be seen by considering sets of runs of three equally-spaced,

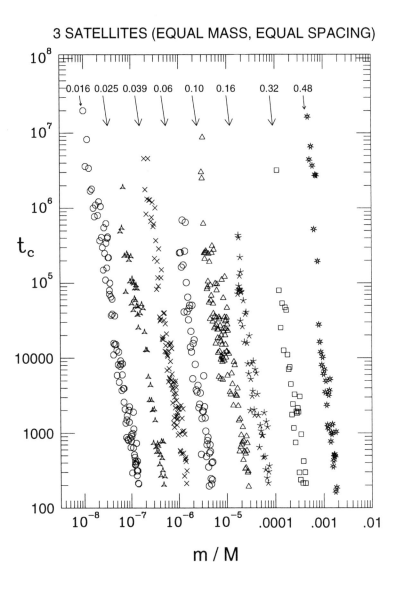

Figure 1. Time to orbit-crossing, t_c, for systems of three equally-spaced satellites of equal mass m initially on near-circular, low-inclination orbits about a central mass M. Time is measured in units of the orbital period of the innermost satellite. Each set of runs is labelled by the spacing of the satellites, in units of the semimajor axis of the innermost one. The results for a given set can be approximated as $t_c = \beta \, (m/M)^\alpha$, where the values of the constants α and β depend on the system; values of α ranging from -8 to -4 are found here.

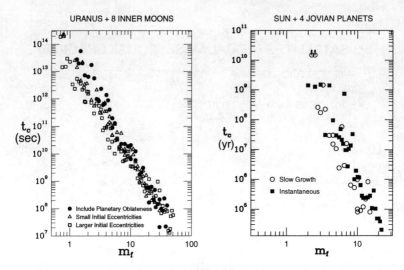

Figure 2. **Left: Inner Uranian Satellites.** The orbit-crossing time, t_c, is shown as a function of the mass enhancement factor, m_f, for three sets of runs using initial conditions based on the 8 inner Uranian satellites. The sets differ as labelled (see Duncan & Lissauer, 1997a, for details). All three sets appear to be well-modelled by power laws, albeit with slightly differing slopes. The crossing times for large satellite masses do not appear to depend systematically on whether Uranus' quadrupole moment is taken into account or on small increases in initial satellite eccentricities. However, for smaller satellite masses, larger initial eccentricities destabilize the system, whereas including Uranus' higher order gravitational moments enhances the stability.

Figure 3. **Right: Jovian Planets.** The orbit-crossing time, t_c, is shown as a function of the mass enhancement factor, m_f, for runs based on the Sun and the four Jovian planets. In one set of runs, all of the planetary masses were instantaneously multiplied by a common factor m_f at the beginning of each run, while in the other set the planetary masses were linearly increased over a time scale of 10^4 yr rather than instantly changed. Within the fairly large scatter, there appear to be no significant differences between the two sets.

equal-mass satellites. Rather than fixing the planetary masses and varying the spacing (as they did), we can fix the spacings and vary the satellite/primary mass ratio to generate each set of runs. As shown in Figure 1, the resulting curves show a power-law dependence of crossing time on mass ratio, with exponent α ranging from -8 to -4. For comparison, a similar plot for some of the inner Uranian satellite sets is shown in Figure 2, where the exponent is found to range from -4.1 to -3.4 for the three sets shown.

A third example of mass scaling comes from studies of the system of the Sun and the four giant planets (see Figure 3, with further discussion in Duncan & Lissauer, 1997b). Here the results also indicate a rough correlation for $m_f > 3$, in this case with exponent α of roughly -6. The orbit crossings observed in this case always involved Uranus and Neptune. Thus it appears that if the four outer planets had all ended up in essentially their current orbits but each with three times their current mass, Uranus and Neptune probably would have had a close encounter within a billion years after the formation of the system.

What is typically seen in all of the longer-lived (but ultimately unstable) runs described above is that the semimajor axes of the satellites remain very nearly constant while the eccentricities vary chaotically, often with a relatively sudden jump in eccentricities leading to orbit crossing (which terminates the run). This behaviour is demonstrated for the case of the Uranian satellites in Figure 4, which displays the temporal evolution of the three satellites most often involved in the orbit-crossing behaviour. Such behaviour leads one to speculate that these systems are diffusing chaotically, perhaps because of the overlap of weak high-order resonances, and often until a sufficiently strong resonance is encountered that pumps the eccentricities to orbit crossing. The mass-scaling would then be a reflection of correlations between the diffusion time scales and the strengths of the perturbations producing the chaotic sea. This means, of course, that as the masses are altered, the locations and widths of the resonances subtly shift, so that ultimately the structure and interconnectivity of the chaotic zones will change. Indeed, changes in the exponent of the m_f–t_c relation are seen in a given set of runs for some of the simulations described above. Thus, none of these scaling results should be overinterpreted and certainly should not be extrapolated too far from the regions in which they are found to apply. Nonetheless, mass-scaling clearly can play a useful role as a diagnostic of the stability of many orbital configurations.

3. The Asteroid and Kuiper Belts

Since the discovery of its first member in 1992, the trans-Neptunian Kuiper belt has been transformed from a theoretical construct invoked to explain the origin of short-period comets to a *bona fide* member of the solar system (*e.g.*, Jewitt & Luu, 1995; Weissman & Levison, 1997). Like the asteroid belt, the structure of the Kuiper belt has almost certainly been sculpted by long-term dynamical instabilities associated with overlapping mean-motion and secular resonances (*e.g.*, Duncan *et al.*, 1995; Morbidelli *et al.*, 1995). However, in both the outer parts of the asteroid belt and the inner parts of the Kuiper belt, there are regions which appear to be dynamically stable on billion-year time scales (given the current architecture of the solar system) but are in fact sparsely populated (see Duncan, 1994, for a review). In particular, objects of sufficiently low eccentricities and inclinations appear to be stable in certain bands. The most plausible explanation for the absence of bodies in these regions is that they were stirred (and in some cases swept up) during the late stages of planet formation. Mechanisms for depleting the outer asteroid belt include excitation by the sweeping of secular resonances (Heppenheimer, 1980), stirring by planetary embryos long since ejected (Wetherill, 1991), the effects of nebular gas drag (Ida & Lin, 1996), and the sweeping of mean-motion resonances caused by planetary migration (Liou & Malhotra, 1996). For the inner Kuiper belt, similar mechanisms may be operating, especially the trapping into mean-motion resonances via planetary migrations (Malhotra, 1995), and the sweeping of secular resonances (Levison *et al.*, 1997).

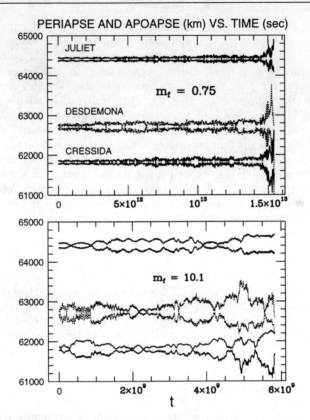

Figure 4. **Evolution of the three most active inner Uranian moons. Top:** Radial distances (in km) of the periapses and apoapses of Cressida, Desdemona and Juliet are shown as a function of time in seconds for the run with $m_f = 0.75$. Note that the semimajor axes remain very nearly constant while the eccentricities vary chaotically, with a relatively sudden jump in eccentricities leading to orbit-crossing which terminates the run. **Bottom:** Same as above but for the run with $m_f = 10.1$. Note that the time scale is reduced by over three orders of magnitude.

Readers interested in our rapidly evolving understanding of the Kuiper belt may wish to read the review of Weissman & Levison (1997). However, an additional aspect of Kuiper belt dynamics has very recently emerged from long-term simulations that may have important observational consequences. This section concludes with a discussion of these long-term integrations.

Duncan *et al.* (1995) followed the evolution of thousands of massless test particles in low- to moderate-eccentricity orbits in the Kuiper belt for times up to the age of the solar system. It was shown that while objects in the Kuiper belt can remain in low- to moderate-eccentricity orbits for time scales comparable to the age of the solar system, some slowly leak into Neptune-encountering orbits. Levison & Duncan (1997) subsequently presented numerical orbital integrations of these "leaking" particles as they evolved from Neptune-encountering orbits. This work showed that the orbital element distribution of objects evolving inward

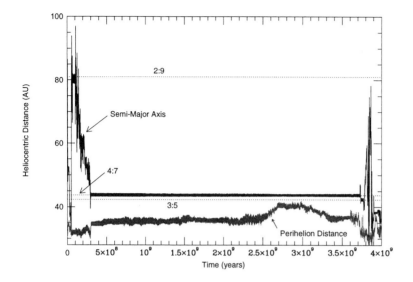

Figure 5. The temporal behaviour of a comet in the scattered disk. The black curve shows the behaviour of the comet's semi-major axis. The gray curve shows the perihelion distance. The three dotted curves show the location of the 2:9, 4:7, and 3:5 mean motion resonances with Neptune.

far enough to become visible comets is dynamically indistinguishable from what are known observationally as Jupiter-family comets. However, a surprising result was that, although the median dynamical lifetime of the comets is 4.5×10^7 yr measured from the time of the first encounter with Neptune, about 5% of the particles survived the length of the integration (10^9 yr).

Recently, Duncan & Levison (1997) have extended the earlier integrations of those objects which began with a Neptune encounter and found that $\sim 1\%$ of such comets can be stored in an extended scattered disk for the age of the solar system. (By "scattered disk", we mean a disk of bodies which have had at least one close encounter with Neptune and which we find to be made up of bodies on eccentric orbits with perihelia typically just beyond Neptune.) Many of the long-lived particles in our simulations spent extended periods temporarily trapped in mean-motion resonances with Neptune—a protection mechanism since they then do not have close encounters with Neptune during these periods. To illustrate some of the processes that are important in protecting these particles, Figure 5 shows the semi-major axis and perihelion distance as a function of time for a particle that was trapped for a long time in a mean-motion resonance with Neptune and had its perihelion distance increased in the process. This object initially underwent a random walk in semi-major axis because of encounters with Neptune. At about 7×10^7 yr it was temporarily trapped in Neptune's 2:9 mean motion resonance for about 5×10^7 yr. It then performed a random walk in semi-major axis until about 3×10^8 yr, when it was trapped in the 4:7 mean motion resonance, where it remained until 3.7×10^9 yr. Notice the increase in the perihelion distance near the time of capture. While trapped in this resonance, the particle's eccentricity became as small as 0.04. After leaving the 4:7 resonance, it was trapped temporarily in Neptune's 3:5 mean motion resonance and then went through a random walk in semi-major axis for the remainder of the simulation. [It should be noted that the process of temporary capture into mean-motion resonances was also observed in the 10^8 yr integrations of Holman & Wisdom (1993).]

The phenomenon of "sticky" behaviour near resonances is apparently common in chaotic scattering in systems which contain KAM surfaces (*i.e.*. nonchaotic regions) and leads to power-law rather than exponential time decay of survivors from a population of bodies starting near a KAM surface (for a nontechnical review see Yalcinkaya & Lai, 1995). In the current context it leads to the possibility that the source of some or all of the Jupiter-family comets could be an extended disk of comets containing bodies which encountered Neptune during the late stages of formation of the outer solar system. [Such a scattered disk was also discussed by Fernandez & Ip (1983) and Torbett (1989).] The recent discovery of two trans-Neptunian objects which are dynamically very similar to the theoretical objects described here makes this possibility all the more plausible (*cf.*, Duncan & Levison, 1997, for references and further discussion). However, it is still too early to determine the extent of the contribution of the scattered disk to the Jupiter-family population.

4. Challenges for the Future

It appears that there is very little room between the giant planets (Gladman & Duncan, 1990) or between the terrestrials (Laskar, 1994; 1997) to squeeze in other planets that could survive the age of the solar system. While the architectures of planetary systems around other stars are likely to be diverse, it seems likely that the populations of most planetary systems have been winnowed down by billions of years of chaotic "natural selection". Simulations of the stability of putative planetary systems are currently underway which may help shed some light on these issues. However, the parameter space and time scales are sufficiently enormous that at best only general guidelines are likely to emerge. A more satisfying alternative to this "top-down" method would be one from the "bottom-up", *i.e.*, large-scale simulations of planet formation involving large numbers of planetesimals. A complete simulation starting from dust grains in a protostellar disk is, of course, almost unimaginably complex. However, tentative attempts are being made to study various stages in the planet-forming process, with the earlier stages being treated via hydrodynamic and/or "particle-in-a box" approaches. The late stages of planet formation are amenable to direct N-body simulations using, *e.g.*, multiple time scale symplectic techniques (Lee *et al.*, these proceedings). Simulations currently in the offing should shed considerable light on issues such as planetesimal scattering and accretion, planetary migration, resonance sweeping, *etc.*, of relevance to the asteroid and Kuiper belts as well as planetary spacings. In addition, hybrid simulations patching the various phases together may emerge in the next few years. In any event, it is clear in this field (as in most of computational astrophysics) that there are sufficient numerical and conceptual challenges to keep us all busy for many years to come.

Acknowledgments. I am grateful for the continuing financial support of the Natural Science and Engineering Research Council of Canada. Many of the simulations described here were performed using the software package SWIFT (written by Hal Levison and me) and were done in collaboration with Hal Levison or Jack Lissauer, to whom I am also grateful for comments on the manuscript. These simulations were supported in part by NASA Origins of Solar Systems program and some were performed on the SSONIC computer system, which was purchased using funds from the NSF and Southwest Research Institute.

References

Chambers, J. E., Wetherill, G. W., & Boss, A. P. 1996, Icarus, 119, 261.
Duncan, M. J. 1994, in *Circumstellar Dust Disks and Planet Formation*, eds. R. Ferlet, & A. Vidal-Madjar, (Paris: Editions Frontières), 245.
Duncan, M. J., & Levison, H. F. 1997, submitted to Science.
Duncan, M. J., Levison, H. F., & Budd, S. M. 1995, AJ, 110, 3073.
Duncan, M. J., & Lissauer, J. J. 1997a, Icarus, 125, 1.
Duncan, M. J., & Lissauer, J. J. 1997b, preprint.
Duncan, M. J., & Quinn, T. 1993, ARA&A, 31, 265.
Fernandez, J. A., & Ip, W.-H. 1983, Icarus, 54, 377.
Ferraz-Mello, S. (editor) 1992, *Chaos, Resonance and Collective Dynamical Phenomena in the Solar System* (Dordrecht: Kluwer).
Gladman, B. J. 1993, Icarus, 106, 247.
Gladman, B. J., & Duncan, M. J. 1990, AJ, 100, 1669.
Heppenheimer, T. A. 1980, Icarus, 41, 76.
Holman, M., & Wisdom, J. 1993, AJ, 105, 1987.
Ida, S., & Lin, D. N. C. 1996, AJ, 112, 1239.
Jewitt, D. C., & Luu, J. X. 1995, AJ, 109, 1987.
Laskar, J. 1994, A&A, 287, L9.
Laskar, J. 1997, A&A, 317, L75.
Levison, H. F., & Duncan, M. J. 1994, Icarus, 108, 18.
Levison, H. F., & Duncan, M. J. 1997, Icarus, in press.
Levison, H. F., Weissman, P. R., & Duncan, M. J. 1997, in *Asteroids, Comets and Meteors*
Liou, J. C., & Malhotra, R. 1996, BAAS, 28, 10.46.
Malhotra, R. 1995, AJ, 110, 420.
Morbidelli, A., Thomas, F., & Moons, M. 1995, Icarus, 118, 322.
Owen Jr., W. M., & Synnott, S. P. 1987, AJ, 93, 1268.
Sanz-Serna, J. M., & Calvo, M. P. 1994, *Numerical Hamiltonian Problems* (London: Chapman & Hall).
Stewart, G. 1997, preprint.
Torbett, M. 1989, AJ, 98, 1477.
Tremaine, S. 1995, in *Frontiers of Astrophysics, Proceedings of the Rosseland Centenary Symposium*, eds. P. B. Lilje, & P. Maltby, (Oslo: Institute of Theoretical Astrophysics), 85.
Weissman, P. R., & Levison, H. F. 1997, in *Pluto*, eds. D. J. Tholen, & S. A. Stern. (Tucson: University of Arizona Press).
Wetherill, G. W. 1991, Science, 253, 535.
Wisdom, J., & Holman, M. 1991, AJ, 104, 2022.
Yalcinkaya, T., & Lai, Y.-C. 1995, Comp. in Phys., 9, 511.
Yoshida, H. 1993, Cel. Mech., 56, 27.

Time Symmetrization Meta-Algorithms

Piet Hut

Institute for Advanced Study, Princeton, NJ 08540, U.S.A.

Yoko Funato, Eiichiro Kokubo

Department of Earth Science and Astronomy, University of Tokyo, 3-8-1 Komaba, Meguro-ku, Tokyo 153, Japan

Junichiro Makino

Department of General Systems Study, University of Tokyo, 3-8-1 Komaba, Meguro-ku, Tokyo 153, Japan

Steve McMillan

Department of Physics and Atmospheric Science, Drexel University, Philadelphia, PA 19104, U.S.A.

Abstract. We present two types of meta-algorithm that can greatly improve the accuracy of existing algorithms for integrating the equations of motion of dynamical systems. The first meta-algorithm takes an integrator that is time-symmetric only for constant time steps, and ensures time symmetry even in the case of varying time steps. The second meta-algorithm can be applied to any self-starting integration scheme to create time symmetry, both for constant and for variable time steps, even if the original scheme was not time-symmetric. Our meta-algorithms are most effective for Hamiltonian systems or systems with periodic solutions. If the system is not Hamiltonian (for example, if some dissipative force exists), our methods are still useful so long as the dissipation is small.

1. Introduction

Many problems in computational physics are governed by underlying equations that are intrinsically time-symmetric. In particular, for any simulation in Hamiltonian dynamics, we can run the movie of our computation equally well forwards as backwards, and in both cases obtain a physically allowable solution. Clearly, it is desirable to use an integration algorithm that reflects this time symmetry as a built-in property. In that case, intuitively speaking, particles can no longer fly "off the tracks", so to speak, when moving through a curve. An example of particular interest to this conference is astrophysical particle simulations, in which particles may correspond directly to physical units, such as molecules or stars, or may form tracers used to approximate the solution of a set of underlying equations with continuous variables.

The basic idea is this: if an algorithm would make a particle spiral out systematically in a given situation, it would have to do so equally in the forward and backward direction. Time reversal, however, would force an inward motion, and the conclusion is that the net spiral out (or spiral in) has to be zero, at least when averaged over a full period in a periodic system. The error in energy thus has to be periodic as well, and cannot build up from orbit to orbit. Of course, errors are still made for finite integration step size, but they show up as errors in phase. In many applications, phase errors are preferable over errors in quantities such as the total energy, that should be conserved.

Most algorithms that are in common use for orbit integration in simulations are not time-symmetric. Even those that are, typically lose their symmetry property as soon as one allows variable time steps. In the remainder of this paper, we offer two types of meta-algorithm for constructing a larger class of time-symmetric algorithms. In § 2 and § 3, we show how to restore time symmetry for algorithms that were symmetric, but lose that property when one uses variable time steps. In § 4 and § 5, we show how to create time symmetry even for those algorithms that were never time symmetric to begin with, not even in the constant time step case.

2. Building a Better Leapfrog

A celebrated example of an integration scheme with built-in time symmetry is the leapfrog scheme, also known as the Verlet method. It is widely used in many applications of particle simulations, such as in molecular dynamics, plasma physics, fluid dynamics and stellar dynamics (Hockney & Eastwood, 1988; Barnes & Hut, 1986). The time symmetry is manifest in the interleaved representation, which gave rise to the "leapfrog" name:

$$r_1 = r_0 + v_{\frac{1}{2}}\delta t, \qquad (1a)$$

$$v_{\frac{3}{2}} = v_{\frac{1}{2}} + a_1 \delta t, \qquad (1b)$$

where r can stand for the position vector of a single particle or the combined vector $\mathbf{r}_1, \mathbf{r}_2, \ldots, \mathbf{r}_N$ representing a system of N particles. The quantity $v = dr/dt$ is the velocity and $a(t) = a(r(t)) = dv/dt$ the acceleration. The subscripts after the various quantities indicate the time at which they apply, in units of the time step, i.e. $v_{\frac{1}{2}} = v(t + \frac{1}{2}\delta t)$.

It is convenient to map the standard interleaved description into a form in which all variables are defined at the same instant in time:

$$r_1 = r_0 + v_0 \delta t + \tfrac{1}{2} a_0 (\delta t)^2, \qquad (2a)$$

$$v_1 = v_0 + \tfrac{1}{2}(a_0 + a_1)\delta t. \qquad (2b)$$

Starting from $\{r_0, v_0, a_0\}$, one first computes r_1, then $a_1(r_1)$, by evaluating the appropriate expression dictated by the system under consideration, and finally v_1. While equations (2) look as though they have lost their explicit time symmetry, they are still equivalent to the original equations (1), as can be verified by direct substitution (Barnes & Hut, 1989). However, if the time step δt is allowed to vary, through a functional dependence $\delta t = h(r, v)$ for example, time symmetry is lost.

Time symmetry can be restored if we force the time step to be a symmetric function of the begin point and end point of each time step, as was first shown by Hut et al. (1995). For example, we can use

$$\delta t = \tfrac{1}{2}[h(r_0, v_0) + h(r_1, v_1)], \tag{3}$$

With this choice of time step, the combined set of equations (2) and (3) has become implicit: in order to determine $\{r_1, v_1\}$ from $\{r_0, v_0\}$, we need to know the time step size δt, which in turn is dependent on $\{r_1, v_1\}$. We can solve this problem by starting with $\delta t = h(r_0, v_0)$ as a first approximation. This will give us approximate values for $\{r_1, v_1\}$, from which we can determine a more accurate value for δt using equation (3). If necessary, this iteration process can be repeated several times, but in practice one or two iterations are generally sufficient to reach time symmetry to machine accuracy.

3. Restoring Time Symmetry

The notion of time symmetry restoration for variable time steps, given above for the particular case of the leapfrog scheme, carries over to any class of integration schemes that is explicitly time-symmetric for constant time steps, as shown by Hut et al. (1995). Another example, given in the same paper, concerns the following natural fourth-order generalization of the leapfrog scheme:

$$r_1 = r_0 + \tfrac{1}{2}(v_1 + v_0)\delta t - \tfrac{1}{12}(a_1 - a_0)(\delta t)^2, \tag{4a}$$

$$v_1 = v_0 + \tfrac{1}{2}(a_1 + a_0)\delta t - \tfrac{1}{12}(j_1 - j_0)(\delta t)^2, \tag{4b}$$

which is a truncated form of the Hermite scheme (Makino, 1991a). Here the jerk $j = da/dt$ is calculated directly by differentiation of the expression for the force (thereby introducing a dependency on velocity as well as position in the case of Newtonian gravitational forces). This set of equations is manifestly time symmetric for constant time steps. Unlike the original second-order leapfrog given in equations (2), the scheme in equations (4) is implicit, even for constant time steps. Applying the same symmetrization procedure given in equation (3) leads, after iteration, to a fully time-symmetric fourth-order generalization of the leapfrog. In practice, iteration can give a quick convergence, leading to substantially improved accuracy for a fixed amount of computer time (Hut et al., 1995).

4. Creating Time Symmetry

Even if no time symmetry is present in a given algorithm, it is possible to construct an implicit version that is manifestly time symmetric, for the general class of self-starting (i.e., one-step) integration schemes. The meta-algorithm that performs this feat is described in detail by Funato et al. (in preparation), as a generalization of the prescription offered by Hut et al. (1995). Here we give a brief outline of the main idea behind the treatment.

Let us start with the ordinary differential equation

$$\frac{dy}{dx} = f(x,y), \tag{5}$$

and suppose we are given a self-starting integration scheme, expressed as

$$y_{i+1} = y_i + F(x_i, y_i; h_i), \tag{6}$$

where x_i and y_i are the values of the variables x and y after the i-th step, and h_i is the step size at x_i.

Our meta-algorithm can be expressed as follows:

$$y_{i+1} = y_i + \tilde{F}(x_i, y_i; \tilde{h}_i) = y_i + \frac{1}{2}\left[F(x_i, y_i; \tilde{h}_i) - F(x_{i+1}, y_{i+1}; -\tilde{h}_i)\right], \tag{7}$$

where $\tilde{h}_i$ can be constructed in the form of a function $\tilde{h}_i = f(h_i, h_{i+1})$ that is symmetric in its arguments: $f(x,y) = f(y,x)$. For example, we could take simply

$$\tilde{h}_i = \tfrac{1}{2}[h(x_i, y_i) + h(x_{i+1}, y_{i+1})], \tag{8}$$

or a root mean square, or any other symmetric combination. Equation (7) gives an implicit formula for y_{i+1}. As before, we can solve this equation by iteration, starting with the original non-symmetric scheme as the initial trial function.

While this recipe is surprisingly simple, we can do even better. Instead of taking a given estimate for the step size, we can use the difference

$$\Delta F_i \equiv \tilde{F}(x_i, y_i, \tilde{h}_i) - F(x_i, y_i, \tilde{h}_i), \tag{9}$$

to estimate the local truncation error. This information can be used to implement a form of adaptive step size control. See Funato et al. (in preparation) for further details.

5. Building a Better Runge-Kutta

As an example application of our more general meta-algorithm, we have constructed a time-symmetric version of the popular fourth-order Runge-Kutta integration scheme. Figure 1 shows the behavior of the errors in the relative energy and angular momentum for a binary orbit with initial eccentricity $e = 0.9$ (the values plotted are determined at apocenter). The dashed and solid curves show the time evolution of the errors for the standard Runge-Kutta scheme and for the symmetrized Runge-Kutta scheme, respectively. In both cases, variable step sizes have been used. The number of time steps is comparable in both cases (around 600 per orbit).

Figure 1 shows that no discernible secular error is produced, even after 1,000 orbital periods, for the run integrated by the time-symmetric fourth-order Runge Kutta Method. In contrast, the error increases linearly for the run integrated by the standard fourth-order Runge Kutta Method. Further quantitative details will be provided in the cost/performance analysis by Funato et al. (in preparation).

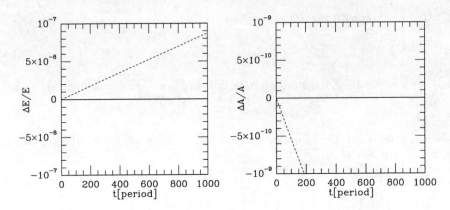

Figure 1. Effects of our meta-algorithm applied to a fourth-order Runge-Kutta scheme. Plotted is the growth of the relative error in energy (left) and angular momentum (right) for a Kepler orbit with eccentricity 0.9. Dashed and full lines correspond to the standard Runge-Kutta scheme and the time-symmetrized Runge-Kutta version, respectively.

6. Discussion

We have reviewed two types of meta-algorithm, based upon a time-symmetrization procedure. The first meta-algorithm preserves time symmetry that would otherwise be lost when integration step sizes are allowed to change during integration. The second meta-algorithm creates time symmetry, even for those algorithms where no symmetry was present in the equal-step-size case.

We mention here briefly a few recent applications of these ideas. McMillan & Hut (1996) have constructed a fully-automated package for performing gravitational three-body scattering experiments, where the central orbit integrator is built along the principles outlined by Hut et al. (1995). They found that the symmetrization meta-algorithm gave a significant speed-up to the fourth-order Hermite scheme used. Most importantly, they found that the fraction of rejected experiments was diminished greatly, compared to the standard integration scheme. The problem here is that some resonant scattering experiments can stay in an intermediate state for a very long time, before finally decaying into a final state. No matter how accurate an initial time step criterion has been chosen, there will always be a small fraction of such lingering states that will ultimately lead to an unacceptable build-up of errors (both integration errors and round-off errors). Time symmetrization, while not circumventing this problem, can greatly alleviate the situation.

Another application has been discussed by Funato et al. (1996a; 1996b). They have applied time symmetrization to the Kustaanheimo-Stiefel regularization method, a sophisticated way to "unfold" the singularity of the three-dimensional Kepler problem by mapping each point in three-dimensional space to a unit circle in an auxiliary four-dimensional space. The combination of these two powerful techniques has resulted in the most accurate way yet designed to integrate orbits near collision singularities.

For some applications, specific adaptations of our general meta-algorithm can make good use of the known constraints inherent in the underlying problem. One example we have recently explored is large-scale simulations of planetary formation. The problem is that close encounters and physical collisions of planetesimals make it absolutely necessary to use individually variable time steps (Aarseth, 1985). All standard choices for highly accurate integration schemes, such as the symplectic schemes, lose their desirable properties once we allow individual particles to change their integration time step length at will. In contrast, our meta-algorithm shows a way out, as demonstrated by Kokubo & Makino (in preparation). They made use of the fact that most planetesimals have nearly circular orbits ($e \ll 0.01$), which means that their time step is practically constant when block time steps (McMillan, 1986; Makino, 1991b) are used, even if we allow the use of variable time steps. Even though the time step size of a particle shrinks significantly during a close encounter leading to a break-down of time-symmetry, this occurs only for a very small fraction of time for a typical particle. Since most of the integration error actually comes from the integration of nearly unperturbed orbits around the sun, reserving strict time symmetry for unperturbed orbits turns out to be a good compromise, leading to high overall accuracy. Details will be provided by Kokubo & Makino (in preparation).

Acknowledgments. This work was supported in part by the National Science Foundation under grants ASC-9612029 and AST-9308005, and by a Grant-in-Aid for Specially Promoted Research (04102002) of the Ministry of Education, Science, Sports and Culture, Japan.

References

Aarseth, S. J. 1985, in *Multiple Time Scales*, eds. J. U. Brackhill & B. I. Cohen (New York: Academic), 377.
Barnes, J. E., & Hut, P. 1986, Nature, 324, 446.
Barnes, J. E., & Hut, P. 1989, ApJS, 70, 389 [equation (A4)].
Funato, Y., Hut, P., McMillan, S., & Makino J., 1996a, ApJ, 112, 1697.
Funato, Y., Makino, J., Hut, P. & McMillan, S. 1996b, in *Dynamical Evolution of Star Clusters*, I.A.U. Symp. 174, eds. P. Hut and J. Makino, (Dordrecht: Kluwer), 367.
Hockney, R. W., & Eastwood, J. W. 1988, *Computer Simulation Using Particles*, (New York: Adam Hilger)
Hut, P., Makino, J., & McMillan, S. 1995, ApJ, 443, L93.
Makino, J. 1991a, ApJ, 369, 200.
Makino, J. 1991b, PASJ, 43, 859.
McMillan, S. L. W. 1986, in *The Use of Supercomputers in Stellar Dynamics*, eds. S. L. W. McMillan & P. Hut, (Berlin: Springer), 156.
McMillan, S. L. W., & Hut, P. 1996, ApJ, 467, 348.

12th 'Kingston meeting': Computational Astrophysics
ASP Conference Series Vol. 123, 1997
David A. Clarke & Michael J. West (eds.)

Variable Time Step Integrators for Long-Term Orbital Integrations

Man Hoi Lee & Martin J. Duncan

Department of Physics, Queen's University at Kingston, Kingston, ON K7L 3N6, Canada

Harold F. Levison

Space Science Department, Southwest Research Institute, Boulder, CO 80302, U.S.A.

Abstract. Symplectic integration algorithms have become popular in recent years in long-term orbital integrations because these algorithms enforce certain conservation laws that are intrinsic to Hamiltonian systems. For problems with large variations in time scale, it is desirable to use a variable time step. However, naïvely varying the time step destroys the desirable properties of symplectic integrators. We discuss briefly the idea that choosing the time step in a time symmetric manner can improve the performance of variable time step integrators. Then we present a symplectic integrator which is based on decomposing the force into components and applying the component forces with different time steps. This multiple time scale symplectic integrator has all the desirable properties of the constant time step symplectic integrators.

1. Symplectic Integrators

Long-term numerical integrations play an important role in our understanding of the dynamical evolution of many astrophysical systems. (See Duncan, these proceedings, for a review of solar-system integrations.) An essential tool for long-term integrations is a fast and accurate integration algorithm. Symplectic integration algorithms (SIA) have become popular in recent years because the Newtonian gravitational N-body problem is a Hamiltonian problem and SIA enforce certain conservation laws that are intrinsic to Hamiltonian systems (see Sanz-Serna & Calvo, 1994, for a general introduction to SIA).

For an autonomous Hamiltonian system, the equations of motion are

$$d\boldsymbol{w}/dt = \{\boldsymbol{w}, H\}, \qquad (1)$$

where $H(\boldsymbol{w})$ is the explicitly time-independent Hamiltonian, $\boldsymbol{w} = (\boldsymbol{q}, \boldsymbol{p})$ are the $2d$ canonical phase-space coordinates, $\{\ ,\ \}$ is the Poisson bracket, and $d\ (= 3N)$ is the number of degrees of freedom. The formal solution of equation (1) is

$$\boldsymbol{w}(t) = \exp(t\{\ , H\})\boldsymbol{w}(0). \qquad (2)$$

If the Hamiltonian H has the form $H_A + H_B$, where H_A and H_B are separately integrable, we can devise a SIA of constant time step τ by approximating $\exp(\tau\{\,,H\})$ as a composition of terms like $\exp(\tau\{\,,H_A\})$ and $\exp(\tau\{\,,H_B\})$. For example, a second-order SIA is

$$\exp(\frac{\tau}{2}\{\,,H_A\})\exp(\tau\{\,,H_B\})\exp(\frac{\tau}{2}\{\,,H_A\}). \qquad (3)$$

For the gravitational N-body problem, we can write $H = T(\mathbf{p}) + V(\mathbf{q})$, where $T(\mathbf{p})$ and $V(\mathbf{q})$ are the kinetic and potential energies respectively. Then the second-order SIA in equation (3) becomes,

$$\mathbf{p}_{n+\frac{1}{2}} = \mathbf{p}_n + \frac{\tau}{2}\mathbf{F}(\mathbf{q}_n),\ \mathbf{q}_{n+1} = \mathbf{q}_n + \tau\mathbf{v}(\mathbf{p}_{n+\frac{1}{2}}),\ \mathbf{p}_{n+1} = \mathbf{p}_{n+\frac{1}{2}} + \frac{\tau}{2}\mathbf{F}(\mathbf{q}_{n+1}), \quad (4)$$

where $\mathbf{F} = -\partial V/\partial \mathbf{q}$ and $\mathbf{v} = \partial T/\partial \mathbf{p}$; this is the familiar leapfrog integrator. For solar-system type integrations, a central body (the Sun) is much more massive than the other bodies in the system, and it is better to write $H = H_{\text{Kep}} + H_{\text{int}}$, where H_{Kep} is the part of the Hamiltonian that describes the Keplerian motion of the bodies around the central body and H_{int} is the part that describes the perturbation of the bodies on one another. Symplectic integrators using this decomposition of the Hamiltonian were introduced by Wisdom & Holman (1991), and they are commonly called mixed variable symplectic (MVS) integrators.

The constant time step SIA have the following desirable properties:

(1) As their names imply, SIA are symplectic, *i.e.*, they preserve $d\mathbf{p} \wedge d\mathbf{q}$.

(2) For sufficiently small τ, SIA solve almost exactly a nearby "surrogate" autonomous Hamiltonian problem with $\tilde{H} = H + H_{\text{err}}$. For example, the second-order SIA in equation (3) has

$$H_{\text{err}} = \frac{\tau^2}{12}\{\{H_A, H_B\}, H_B + \frac{1}{2}H_A\} + O(\tau^4). \qquad (5)$$

Consequently, we expect that the energy error is bounded and the position (or phase) error grows linearly.

(3) Many SIA [*e.g.*, equation (3)] are time reversible. Note, however, that there are algorithms that are symplectic but not reversible.

Figure 1 shows the energy error $\Delta E/E$ of an integration of the $e = 0.5$ Kepler orbit using the constant time step leapfrog integrator in equation (4). [In this and all subsequent figures, the orbits are initially at the apocenter $r_a = a(1 + e)$, where a and e are the semi-major axis and eccentricity respectively.] Figure 1 illustrates that there is no secular drift in $\Delta E/E$.

2. Simple Variable Time Step

For problems with large variations in time scale (caused by close encounters or high eccentricities), it is desirable to use a variable time step. A common practice is to set the time step using the phase-space coordinates $\mathbf{w}_n$ at the beginning of the time step: $\tau = h(\mathbf{w}_n)$. If the SIA discussed in § 1 are implemented with this simple variable time step scheme, they are still symplectic if we assume that the

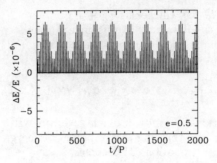

Figure 1. Energy error $\Delta E/E$ of an integration of the Kepler problem using the constant time step leapfrog integrator [equation (4)]. The eccentricity $e = 0.5$ and time step $\tau = P/4000$ (P = orbital period).

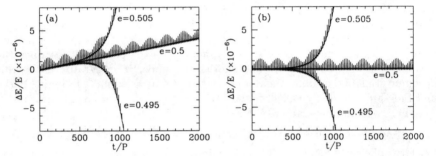

Figure 2. a) Same as Figure 1, but using the leapfrog integrator [equation (4)] with the simple variable time step $\tau = \tau_0(r/r_a)^{3/2}$, where $\tau_0 = P/2000$. The errors for two neighboring initial conditions are also shown (see text). b) Same as a), but using a symmetrized time step.

sequence of time steps determined for a particular initial condition are also used to integrate neighboring initial conditions (see Skeel & Gear, 1992, for another point of view). However, tests have shown that this and similar simple variable time step schemes [and also $\tau = h(t)$] destroy the desirable properties of the integrators (e.g., Gladman et al., 1991; Calvo & Sanz-Serna, 1993). Figure 2a shows the energy error of an integration with $e = 0.5$ and $h(r) = \tau_0(r/r_a)^{3/2}$. Although the error is initially smaller than that in Figure 1 (the integrations shown in Figures 1 and 2 use nearly the same number of time steps per orbit), it shows a linear drift. In Figure 2a, we also show the errors for two neighboring initial conditions ($e = 0.5 \pm 0.005$). They were integrated using the time steps determined for the $e = 0.5$ orbit. Note that these errors grow even faster.

The degradation in performance results because properties (2) and (3) listed in § 1 are no longer true. The algorithm is not time reversible because the time step depends only on the coordinates at the beginning of the time step. Since H_{err} depends explicitly on the time step τ [see, e.g., equation (5)], a variable time step changes the surrogate Hamiltonian problem that the integrator is solving from step to step. Thus the solution from $t = 0$ to t_n after n steps is not in general the solution of a nearby autonomous Hamiltonian problem.

3. Time Symmetrization

Hut et al. (1995; also Funato et al., 1996; Hut et al., these proceedings) pointed out that the performance of a variable time step integrator can be improved if time reversibility is restored by choosing the time step in a time-symmetric manner: $\tau = \tau(\boldsymbol{w}_n, \boldsymbol{w}_{n+1})$ with $\tau(\boldsymbol{w}_n, \boldsymbol{w}_{n+1}) = \tau(\boldsymbol{w}_{n+1}, \boldsymbol{w}_n)$. For example, $\tau = [h(\boldsymbol{w}_n) + h(\boldsymbol{w}_{n+1})]/2$. We have tested this idea in detail by i) integrating a series of problems (pendulum, Kepler orbits, restricted three-body problem) using a symmetrized variable time step leapfrog integrator, and ii) integrating the restricted three-body problem using a symmetrized MVS integrator. We have found that time symmetrization usually reduces the drift in energy (or the Jacobi constant) to negligible levels. Figure 2b shows an integration of the $e = 0.5$ Kepler orbit using the symmetrized leapfrog integrator (cf., Figure 2a).

In Figure 2b, we also show the errors for two neighboring initial conditions that were integrated using the time steps determined for the $e = 0.5$ orbit. There is no improvement in the error growth for the neighboring initial conditions. Integrating an orbit using the time steps determined for another orbit may seem to be somewhat artificial, but similar situations do occur in realistic integrations. For example, if we integrate an N-body system using a shared time step scheme, the time step used by all of the particles is determined by the few having the strongest close encounter. In practice, since the symmetrized time step criterion depends on $\boldsymbol{w}_{n+1}$, a symmetrized integrator also has the disadvantage that it requires iteration and can be significantly slower than the original integrator (unless the original integrator is also implicit).

4. Multiple Time Scale Symplectic Integrators

In this section, we describe a "variable time step" SIA that has all the desirable properties of the constant time step SIA. The algorithm is based on ideas proposed by Skeel & Biesiadecki (1994; see also MacEvoy & Scovel, 1994). It is also similar to the individual time step scheme of Saha & Tremaine (1994).

For simplicity, let us consider in particular the Kepler problem with $T(\boldsymbol{p}) = |\boldsymbol{p}|^2/2$ and $V(\boldsymbol{r}) = -1/r$. We choose a set of cutoff radii $r_1 > r_2 > \cdots$ and decompose the potential V into V_i, or equivalently the force $\boldsymbol{F}$ into $\boldsymbol{F}_i = -\partial V_i/\partial \boldsymbol{r}$, such that i) $\boldsymbol{F} = \sum_{i=0}^{\infty} \boldsymbol{F}_i$, ii) $\boldsymbol{F}_i$ (except $\boldsymbol{F}_0$) is zero at $r > r_i$, and iii) $\boldsymbol{F}_i$ is "softer" than $\boldsymbol{F}_{i+1}$. The force $\boldsymbol{F}_i$ is to be applied with a time step τ_i. If we assume that $\tau_i/\tau_{i+1} = M_{i+1}$ is an integer, we can apply the second-order SIA in equation (3) recursively to obtain the following second-order algorithm:

$$\exp(\tau_0\{\ , H\}) \approx \exp(\frac{\tau_0}{2} K_0) \exp[\tau_0(D + K_1 + K_2 + \cdots)] \exp(\frac{\tau_0}{2} K_0)$$

$$\approx \exp(\frac{\tau_0}{2} K_0) \left[\exp(\frac{\tau_1}{2} K_1) \exp[\tau_1(D + K_2 + K_3 + \cdots)] \right.$$

$$\left. \times \exp(\frac{\tau_1}{2} K_1) \right]^{M_1} \exp(\frac{\tau_0}{2} K_0) \qquad (6)$$

$$\vdots$$

where $D = \{\ , T\}$ and $K_i = \{\ , V_i\}$. Hereafter, we adopt $M_i = 2$.

The multiple time scale algorithm [equation (6)] has an overall time step τ_0, but it is effectively a variable time step scheme because the recursion terminates at level i if $K_{i+1} + K_{i+2} + \cdots = 0$ during a substep of length τ_i. For example, if the particle is in the region $r_1 > r > r_2$ during an overall step, equation (6) reduces to

$$\exp(\frac{\tau_0}{2}K_0 + \frac{\tau_0}{4}K_1) \exp(\frac{\tau_0}{2}D) \exp(\frac{\tau_0}{2}K_1) \exp(\frac{\tau_0}{2}D) \exp(\frac{\tau_0}{4}K_1 + \frac{\tau_0}{2}K_0). \quad (7)$$

Since equation (6) is based on the recursive application of equation (3), it has all the desirable properties of a constant time step SIA. It is obviously symplectic and time reversible. We can also derive the surrogate autonomous Hamiltonian solved by this integrator. For example, if we decompose V into two levels V_0 and V_1 only, the error Hamiltonian is [cf., equation (5)]

$$H_{\text{err}} = \frac{\tau_0^2}{12}\{\{V_0, T\}, T + \frac{1}{2}V_0\} + \frac{\tau_1^2}{12}\{\{V_1, T\}, T + \frac{1}{2}V_1\} + \frac{\tau_0^2}{12}\{\{V_0, T\}, V_1\}$$
$$+ O(\tau_0^4). \quad (8)$$

After some experiments, we found two force decompositions that work well. If we write $\boldsymbol{F}_{c,0} = \tilde{\boldsymbol{F}}_{c,0}$ and $\boldsymbol{F}_{c,i} = \tilde{\boldsymbol{F}}_{c,i} - \tilde{\boldsymbol{F}}_{c,i-1}$ for $i \neq 0$, one of the decompositions is

$$\tilde{\boldsymbol{F}}_{c,i-1} = \begin{cases} -\boldsymbol{r}/r^3 & \text{if } r \geq r_i, \\ -\left[9\left(r/r_i\right)^2 - 5\left(r/r_i\right)^6\right]\boldsymbol{r}/4r_i^3 & \text{if } r < r_i. \end{cases} \quad (9)$$

An alternative decomposition uses

$$\tilde{\boldsymbol{F}}_{p,i-1} = \begin{cases} -\boldsymbol{r}/r^3 & \text{if } r \geq r_i, \\ -f\left(\frac{r_i-r}{r_i-r_{i+1}}\right)\boldsymbol{r}/r^3 & \text{if } r_{i+1} \leq r < r_i, \\ 0 & \text{if } r < r_{i+1}, \end{cases} \quad (10)$$

where $f(x) = 2x^3 - 3x^2 + 1$. Unlike the forces suggested by Skeel & Biesiadecki (1994), these forces have continuous first derivatives and decrease rapidly (or

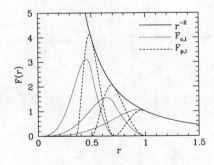

Figure 3. Decomposition of the force $F(r) = 1/r^2$ into $F_{c,i}$ or $F_{p,i}$. The $i = 0, 1, 2$ components, with $r_1 = 1$ and $r_i/r_{i+1} = \sqrt{2}$, are shown.

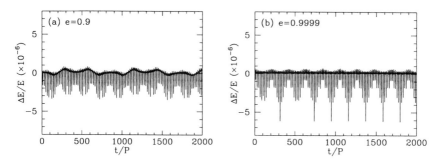

Figure 4. Energy error $\Delta E/E$ of two integrations of the Kepler problem using the multiple time scale symplectic integrator with $F_{c,i}$. The overall time step $\tau_0 = P/2000$, $M_i = 2$, $r_1 = \sqrt{2}\,a$, and $r_i/r_{i+1} = \sqrt{2}$.

exactly) to zero at $r \ll r_i$ (see Figure 3). We shall not provide the details here, but we can understand why these properties are important from an analysis of H_{err}. In Figure 4 we show the energy error $\Delta E/E$ of two integrations using the multiple time scale symplectic integrator with $F_{c,i}$. As expected, there is no secular drift in $\Delta E/E$. Note also that, with the chosen integration parameters ($\tau_i \propto r_i^2$), the maximum error is almost independent of the pericentric distance (which changes by 10^3 in the two cases shown). This again agrees with the expectation from an analysis of H_{err}.

One of the goals of this study is to develop a variable time step integrator for solar system integrations. We have developed a second-order multiple time scale MVS integrator based on the algorithm described in this section (see Levison & Duncan, 1994, for another approach). We are currently testing this integrator in detail. Initial results indicate that the integrator is fast and accurate and has all the desirable properties of the constant time step symplectic integrators.

Acknowledgments. This work was supported in part by NASA, NSERC, SwRI, and a CITA National Fellowship (to MHL).

References

Calvo, M. P., & Sanz-Serna, J. M. 1993, SIAM J. Sci. Comput., 14, 936.
Funato, Y., Hut, P., McMillan, S., & Makino, J. 1996, AJ, 112, 1697.
Gladman, B., Duncan, M., & Candy, J. 1991, Celest. Mech. Dynam. Astr., 52, 221.
Hut, P., Makino, J., & McMillan, S. 1995, ApJ, 443, L93.
Levison, H. F., & Duncan, M. J. 1994, Icarus, 108, 18.
MacEvoy, W., & Scovel, J. C. 1994, preprint.
Saha, P., & Tremaine, S. 1994, AJ, 108, 1962.
Sanz-Serna, J. M., & Calvo, M. P. 1994, *Numerical Hamiltonian Problems*, (London: Chapman & Hall).
Skeel, R. D., & Biesiadecki, J. J. 1994, Ann. Numer. Math., 1, 191.
Skeel, R. D., & Gear, C. W. 1992, Physica D, 60, 311.
Wisdom, J., & Holman, M. 1991, AJ, 102, 1528.

Part II
STARS

Abundance Anomalies and Anisotropic Turbulent Transport in Stars

G. Michaud[1] & A. Vincent[1]

Département de physique, Université de Montréal, Montréal, PQ H3C 3J7, Canada

M. Meneguzzi[2]

Service d'Astrophysique, Centre d'Études de Saclay, 91191 Gif sur Yvette, France

Abstract. Abundance anomalies (*e.g.*, Li in the Sun, metals in Am or Ap stars, hydrogen in white dwarfs) strictly constrain turbulence below superficial stellar convection zones. They limit the apparent Reynolds number to values less than 100, which normally correspond to laminar flows. A direct numerical simulation of particle transport has been carried out for anisotropic flows. The anisotropy of turbulence strongly reduces the efficiency of the scalar transport, and hence the vertical diffusion coefficient, D_{Tz}, for a given turbulent vertical velocity field. It is found that for an anisotropy, A, of horizontal to vertical velocities, $D_{Tz} \sim \nu \mathrm{Re}/A^2$. This may explain why the constraints obtained from abundance anomalies in stars lead to apparent Reynolds numbers smaller than 100.

1. Abundance Anomalies and Particle Transport in Stars

Abundance anomalies are present at the surface of most classes of stars. On the lower main sequence, underabundances of Li are generally present, while on the upper main sequence (*e.g.*, in HgMn stars) overabundances by factors of up to 10^6 have been reported for mercury and some other heavy metals along with isotopic anomalies of mercury and thallium (Leckrone *et al.*, 1996; Wahlgren *et al.*, 1995). Most metals are affected (Takada-Hidai, 1991). In the AmFm stars, anomalies are observed but are somewhat smaller (Cayrel *et al.*, 1991). Some 20% of main sequence stars with $7{,}000 < T_{\mathrm{eff}} < 20{,}000$ K show abundance anomalies while nearly all the cooler ones have abundance anomalies of Li. In the Sun, Li is underabundant by a factor of 100 or so, while helioseismology shows that the abundance of He is affected by gravitational settling below the surface convection zone (Babu *et al.*, 1996; Bahcall *et al.*, 1995; Guzik & Cox, 1992; Proffitt, 1994; Richard *et al.*, 1996). None of those abundance anomalies

---

[1] also, CERCA, 5160 bd Décarie, Montréal, PQ, H3X 2H9, Canada

[2] also, ASCI-CNRS (Bat. 506), Université de Paris Sud, 91405 Orsay cedex, France

are explained by *standard* stellar evolution unless gravitational settling and/or turbulent transport are included. Similarly, most giant stars (Sneden, 1991) show abundance anomalies of Li, C, N, O or other heavy metals that require particle transport in order to be understood (Iben, 1991). In white dwarfs, the very large H abundance in most DAs requires He settling (Schatzman, 1958), while the presence of metals in many requires a transport process competing with gravitational settling (Fontaine & Wesemael, 1991). In horizontal branch stars, the underabundance of He, the anomalies of some metals, and the overabundance of ^{3}He point to the importance of atomic diffusion (Heber, 1991).

The exact nature of the particle transport processes involved is still poorly understood. It is now reasonably well established that atomic diffusion plays a role, but by itself it generally leads to larger anomalies than are observed so that other transport processes must be competing with atomic diffusion. It appears most likely that, in some cases, turbulent transport plays an important role (see Schatzman, 1977) but it is a poorly understood process requiring a number of arbitrary parameters for its description (see, for instance, Michaud & Charbonneau, 1991).

In § 2, we briefly describe how upper limits to turbulent transport coefficients can be obtained from abundance anomalies as well as the implications for the Reynolds number. The numerical simulations of particle transport in anisotropic flows will then be described and applied to transport coefficients. The implications for turbulence in radiative zones of stars will finally be discussed.

2. Constraints on Turbulence from Abundance Anomalies

In order to determine the surface evolution of the abundance of chemical species, one solves for each transport equation at the same time as the star evolution (see, *e.g.*, Michaud & Charbonneau, 1991):

$$\rho \frac{\partial c}{\partial t} = \frac{1}{r^2} \frac{\partial}{\partial r} \left[\rho r^2 (D + D_T) \frac{\partial c}{\partial r} - \rho v_g c \right] - S(\rho, T) c, \qquad (1)$$

where D is the atomic diffusion coefficient, D_T is a turbulent diffusion coefficient, S is a sink (or source) term caused by nuclear reactions, and v_g is the advective part of the atomic diffusion velocity including gravitational settling, thermal diffusion and radiation driven diffusion. Mass loss could also be included in v_g.

If turbulence is small, the transport is dominated by gravitational settling and radiation driven diffusion. The radiative acceleration calculations can now be improved thanks to the availability of the large OP and OPAL data bases (see Gonzalez *et al.*, 1995; Seaton, 1997). Radiative acceleration dominates over gravity in upper main sequence stars (Michaud *et al.*, 1976) and hot horizontal branch stars (Michaud *et al.*, 1983). Elsewhere, gravity generally dominates. Calculations show that at least some turbulence is required to explain the Li abundance observations but at the same time it gives strict upper limits to its value. One obtains a relation between the Reynolds number, Re, and D_T since one may write, using a simple mixing length model:

$$D_T = l_0 v_0, \qquad (2)$$

and then for the Reynolds number:

$$\text{Re} = \frac{l_0 v_0}{\nu} = \frac{D_T}{\nu}, \quad (3)$$

where v_0 is a root mean square turbulent velocity and l_0 is a characteristic length of turbulent motions. The constraints that the abundances put on D_T have been determined by Michaud & Fontaine (1984) for white dwarfs, Proffitt & Michaud (1991) for the sun, and Charbonneau & Michaud (1991) for AmFm stars. Using the Li abundance in solar type stars of clusters, such as reviewed by Michaud & Charbonneau (1991), one obtains the value from Li in clusters. These lead to the constraints on the Reynolds number that are shown in Table 1. Such low Reynolds numbers usually characterize laminar flows, not turbulent ones. Since stars typically have rotational velocities of the order of 1–100 km s^{-1}, it would seem surprising that differential rotation should vary by much less than a few meters per second over one pressure scale height (some 10^9 cm) so that using equation (3), one then expects Re $\simeq 10^9$ even for $\nu \simeq 100$. This is so much larger than the upper limits determined in Table 1 that it requires an explanation.

In a series of papers starting with Pinsonneault et al. (1989), the Yale group has shown that to explain simultaneously the rotation of solar type stars and the Li abundance observed in them, one had to assume that

$$D_T \leq \frac{1}{50} \nu \text{Re}. \quad (4)$$

This has remained unexplained but adds to the observations pointing to a need for a better understanding of turbulent transport in stratified fluids and of its interaction with atomic diffusion.

Table 1. Constraints on Re

Class of Objects	Observational constraint	Reynolds number
Sun	Li and Be abundances	20
Li in clusters	Li and Be abundances	15
AmFm stars	Underabundances	$10^2 - 10^3$
White dwarfs	H abundance	3

3. Transport Properties of Anisotropic Flows

Stratified fluids may be found in stars, the Earth's atmosphere and ocean, and various industrial flows. In the Earth's ocean, turbulence is known to be anisotropic, the transport being much more rapid in the horizontal than the vertical direction. The basic properties of such flows, in particular how they interact with atomic diffusion, are not well understood.

It has been shown by Charbonneau & Michaud (1991) and by Zahn (1992) that strong horizontal turbulence reduces the vertical transport caused by meridional circulation. This may be understood by considering the integral of the flux crossing a horizontal surface on the sphere:

$$\oint c\rho \mathbf{U} \cdot d\mathbf{S}. \tag{5}$$

If there is very strong horizontal turbulence, so that c on the surface is homogenized horizontally much more rapidly than it is modified by meridional circulation, then one may take c out of the integral. The integral then gives zero because of the conservation of mass across the surface.

This argument does not apply only to meridional circulation. It may be applied to any strongly anisotropic *advective* flux. This effect is currently not included in turbulence transport models of anisotropic flows. A more complete understanding of the basic transport properties of anisotropic flows is needed but some aspects can be understood only through 3-D numerical simulations: the transport is vertical but the homogenizing motions take place in the two other spatial directions (horizontal) so that the three spatial directions must necessarily be simulated.

In a first approach, Vincent et al. (1996, hereafter VMM) chose to consider a fluid stratified in the z-direction by gravity and study the relation between anisotropy and transport. To that end, they used stochastic turbulent flows in order to vary the anisotropy in a controlled way and determine the effect of flow anisotropy on the transport, independent of other flow characteristics. In a further study, it is planned to determine the extent to which anisotropy is affected by other flow properties, in shear flows for instance.

In a direct numerical simulation, turbulent transport is not modeled by D_T as in equation (1), but instead all turbulent motions are followed in detail. Neglecting the source term, equation (1) is then replaced by:

$$\rho \frac{\partial c}{\partial t} = -\nabla \cdot (\rho c \mathbf{V}) - \rho \mathbf{U} \cdot \nabla c + \nabla \cdot (D\rho \nabla c), \tag{6}$$

where $\mathbf{V}$ is the advective part of the diffusion velocity while $\mathbf{U}$ is an advective turbulent motion and must satisfy the conservation equation:

$$\nabla \cdot (\rho \mathbf{U}) = 0. \tag{7}$$

In the simulations, a trace element distributed in a Gaussian manner about the central horizontal plane of a box was used. The development of the transport under the influence of turbulent motions was then followed. Gravitational settling may be removed from the simulation (Galilean invariance).

The space discretization was done using a Fourier spectral method and calculations were carried out in wave number ($\mathbf{k}$) space. The isotropic turbulence was assumed to have a Kolmogorov spectrum. A Gaussian spectrum was drawn within each $\mathbf{k}$ interval with the energy within each interval fixed by the Kolmogorov spectrum. This determines the width of the Gaussian distribution of modes drawn for each interval. In the largest simulations, a total of $256 \times 256 \times 128$ modes were drawn. Within each $\mathbf{k}$ interval (see Figure 6 of VMM), the number of modes is large enough that the random drawing of modes reproduces very well the Kolmogorov spectrum except for $k < 8$, where random fluctuations appear. Anisotropy was introduced by adding a $E_2 k^{-3}$ spectrum

Isotropic　　　　　　　　Anisotropic

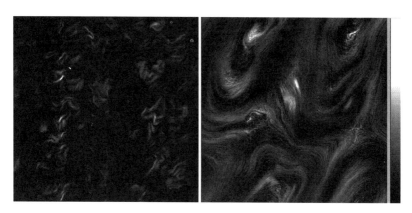

Figure 1. The left panel shows the concentration gradients in the isotropic case while the right panel shows the concentration gradients in the anisotropic simulation. The smallest gradients are darkest. The scale shown on the right is linear. In the anisotropic case, the gradients are larger, mainly horizontal, and in filaments. They lead to additional horizontal atomic diffusion of the concentration fluctuations.

of horizontal motions and drawing modes in a similar way as for the isotropic spectrum. The value of the anisotropy, A, defined by

$$A = \frac{(\langle v_x^2 \rangle + \langle v_y^2 \rangle)^{0.5}}{(2\langle v_z^2 \rangle)^{0.5}}, \tag{8}$$

is fixed by the constant E_2. The characteristic length of the turbulence field is given by:

$$l_0 = \pi \int_0^\infty \frac{E(k)}{k} dk. \tag{9}$$

The simulations are characterized by the value of the turbulent Reynolds number defined by equation (3), A (or equivalently E_2), and the turbulent Peclet number ($\mathrm{Pe_T}$):

$$\mathrm{Pe_T} = \frac{D_\mathrm{T}}{D} = \frac{l_0 v_0}{D}. \tag{10}$$

The resolution of the simulation limits the value which may be used for Re and $\mathrm{Pe_T}$.

It is useful to define a vertical Peclet number (Pe_z):

$$\mathrm{Pe}_z = \frac{l_{0z} v_{0z}}{D}, \tag{11}$$

where v_{0z} is the rms vertical velocity and l_{0z} is the vertical integral scale.

The simulations were done with isotropic flows and then repeated by adding a 2-D spectrum leading to various values of the anisotropy but keeping the vertical motions constant. At the beginning of the simulation, the relative concentration of the trace element, $c(z)$, is constant in each horizontal plane and so depends only on z. The horizontally averaged vertical flux, $\overline{\rho c v_z}$, carried by the vertical motion is then zero when the simulations are started. As the simulation proceeds (see Figure 7 of VMM), the vertical flux increases, at first independent of the value of A (compare the continuous and short dash curves at $t = 4 \times 10^{-6}$). At maximum flux carried vertically, the vertical flux is some 30 times larger in the isotropic than in the anisotropic simulation ($A = 33$) even though the vertical turbulent velocities are the same in both cases. The reduction of the vertical flux in the presence of the anisotropy is caused by the horizontal homogenization that the rapid horizontal motions cause. The large horizontal motions cause the appearance of large horizontal gradients (see Figure 1) that ultimately lead to horizontal atomic diffusion and so preclude the growth of fluctuations of c from its average in any given horizontal plane. This limits the vertical flux. Increasing the atomic diffusion has a similar effect since fluctuations are then limited by their diffusion. This may be seen in Figure 5 of VMM where the vertical turbulent diffusion coefficient,

$$D_{Tz}(z) = -\frac{\overline{\rho c v_z}}{\partial \overline{\rho c}/\partial z}, \tag{12}$$

is shown to depend on the Peclet number (Pe). Since Pe $\propto 1/D$, the vertical turbulent diffusion coefficient is seen to be inversely proportional to the atomic diffusion coefficient. The reduction of fluctuations leads to a smaller vertical flux whether the reduction of fluctuations is caused by the increased anisotropy or by an increased atomic diffusion coefficient. This also implies that different atomic species have different D_{Tz} since they have different atomic diffusion coefficients and this affects D_{Tz}.

The dependence of vertical transport on anisotropy and Pe is clearly seen on Figure 10 of VMM. As soon as $A > 3$, the ratio $D_{Tz}/l_{0z}v_{0z}$ becomes proportional to A^{-1} if Pe is large enough. At large values of A, the ratio becomes independent of Pe. This is clearly consistent with the preceding analysis.

4. Implications for Particle Transport in Stars

The first implication is the dependence of the vertical turbulent diffusion coefficient on atomic species. It is generally assumed that turbulent transport is the same for all species since a simple model such as the one leading to equation (2) leads turbulence transport to depend only on turbulent motions. As seen above, this is not always the case. When $D > D_{Tz}$, atomic diffusion may reduce the effective D_{Tz}. For the effect to be non-negligible, large differences between the D values of elements are needed. This could conceivably occur only between neutral and ionized elements and in stars, in practice, neutral helium must be involved.

A more general effect is the modification of the relation between the Reynolds number and D_{Tz} caused by the anisotropy, A. Given equation (8), one has $v_z \simeq v_0/A$. One also has $l_z \simeq l_0$ (see Figure 2). The results of the simulations

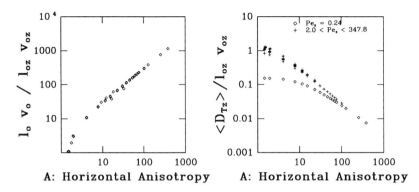

Figure 2. Left panel: the variation with A of the ratio $l_0 v_0 / l_{0z} v_{0z}$. Right panel: the variation of D_{Tz} for a given value of $l_{0z} v_{0z}$. Both combine to give the dependence of D_{Tz} on A for a given Re.

give, as discussed above (see Figure 2):

$$D_{Tz}(z) \simeq \frac{l_{0z} v_{0z}}{A}. \tag{13}$$

One then obtains, using equation (3):

$$D_{Tz}(z) \simeq \nu \frac{\mathrm{Re}}{A^2}, \tag{14}$$

which shows that the simple relation obtained from mixing length arguments between D_T and Re [see equation (3)] is modified by a factor A^2 in anisotropic flows. In the case of the sun for instance, the upper limit given in Table 1 becomes Re $\simeq 2 \times 10^5$ for $A = 100$. Even if D_{Tz} is small, R_e can be large enough for turbulence to exist for plausible values of the anisotropy.

One of our longer term aims is to arrive at a simulation of meridional circulation and the particle transport to which it leads. Three dimensions are necessary for such a simulation to capture the process shown above to be important. Only in three dimensions is it possible to simulate the transport in the z-direction at the same time as the horizontal motions that reduce its efficiency. In spite of the symmetry of the problem, a 2-D simulation of meridional circulation misses some of the most important aspects of particle transport. A 3-D simulation is therefore essential.

Acknowledgments. We thank Jacques Richer for help with many aspects of this paper.

References

Babu, S., Christensen-Dalsgaard, J., Schou, J., Thompson, M. J., & Tomczyk, S. 1996, ApJ, 460, 1064.
Bahcall J. N., Pinsonneault, M. H., & Wasserburg, G. J. 1995, Rev. Mod. Phys., 67, 781.

Cayrel, R., Burkhart, C., & Van't Veer, C. 1991, in *IAU Symposium 145, Evolution of Stars: The Photospheric Abundance Connection*, eds. G. Michaud & A. Tutukov (Dordrecht: Kluwer), 99.
Charbonneau, P., & Michaud, G. 1991, ApJ, 370, 693.
Fontaine, G., & Wesemael, F. 1991, in *IAU Symposium 145, Evolution of Stars: The Photospheric Abundance Connection*, eds. G. Michaud & A. Tutukov (Dordrecht: Kluwer), 421.
Gonzalez, J.-F., LeBlanc, F., Artru, M.-C., & Michaud, G. 1995, A&A, 297, 223.
Guzik, J. A., & Cox, A. N. 1992, ApJ, 386, 729.
Heber, U. 1991, in *IAU Symposium 145, Evolution of Stars: The Photospheric Abundance Connection*, eds. G. Michaud & A. Tutukov (Dordrecht: Kluwer), 363.
Iben, I. 1991, in *IAU Symposium 145, Evolution of Stars: The Photospheric Abundance Connection*, eds. G. Michaud & A. Tutukov (Dordrecht: Kluwer), 257.
Leckrone, D. S., Johansson, S. G., Kalus, G., Wahlgren, G. M., Brage, T., & Proffitt, C. R. 1996, ApJ, 462, 937.
Michaud, G., & Charbonneau, P. 1991, Space. Sci. Rev., 57, 1.
Michaud, G., Charland, Y., Vauclair, S., & Vauclair, G. 1976, ApJ, 210, 447.
Michaud, G., & Fontaine, G. 1984, ApJ, 283, 787.
Michaud, G., Vauclair, G., & Vauclair, S. 1983, ApJ, 267, 256.
Pinsonneault, M. H., Kawaler, S. D., Sofia, S., & Demarque, P. 1989, ApJ, 338, 424.
Proffitt, C. R. 1994, ApJ, 425, 849.
Proffitt, C. R., & Michaud, G. 1991, ApJ, 380, 238.
Richard O., Vauclair S., Charbonnel C., & Dziembowski W. A. 1996 A&A, 312, 1000.
Schatzman, E. 1958, *White Dwarf*, (Amsterdam: North Holland).
Schatzman, E. 1977, A&A, 56, 211.
Seaton, M. J. 1997, MNRAS, submitted.
Sneden, C. 1991, in *IAU Symposium 145, Evolution of Stars: The Photospheric Abundance Connection*, eds. G. Michaud & A. Tutukov (Dordrecht: Kluwer), 235.
Takada-Hidai, M. 1991, in *IAU Symposium 145, Evolution of Stars: The Photospheric Abundance Connection*, eds. G. Michaud & A. Tutukov (Dordrecht: Kluwer), 137.
Vincent, A., Michaud, G., & Meneguzzi, M. 1996, Phys. Fluids, 8, 1312 (VMM).
Wahlgren, G. M., Leckrone, D. S., Johansson, S. G., Rosberg, M., & Brage, T. 1995, ApJ, 444, 438.
Zahn, J.-P. 1992, A&A, 265, 115.

Helioseismology by Genetic Forward Modeling

P. Charbonneau & S. Tomczyk

High Altitude Observatory, NCAR, P.O. Box 3000, Boulder, CO 80307, U.S.A.

Abstract. Genetic Forward Modeling is a genetic algorithm-based technique which can be used to perform helioseismic inversions. After a brief description of the method is given, some of its operational advantages are illustrated in the context of a specific helioseismological problem, namely the determination of the rotation rate in the deep solar core ($r/R_\odot < 0.5$). Use of the technique in conjunction with the LOWL 2-year frequency splitting data set suggests that the solar core rotates rigidly down to $r/R_\odot \sim 0.1$.

1. Genetic Algorithms

Genetic Algorithms (hereafter GA) are a class of heuristic search techniques inspired by the biological process of evolution by means of natural selection (Holland, 1975). GA can be used to construct extremely robust numerical optimization methods that very often outperform most other algorithms on global optimization problems characterized by multimodal and/or ill-behaved search spaces (Goldberg, 1989; Davis, 1991). For a gentle introduction in the astronomical/astrophysical context, see Charbonneau (1995).

Consider a model defined by a set of parameters $\mathbf{a} \equiv \{a_1, a_2, ..., a_N\}$, and a goodness-of-fit measure (or "fitness") $f(\mathbf{a})$ computable for any given realization of the model. The task is to find the set of parameters $\mathbf{a}^*$ corresponding to the model that maximizes $f(\mathbf{a})$. A top-level view of a generic GA-based method for this task is as follows: (1) construct an initial random "population", *i.e.*, a group of parameters sets $\mathbf{a}^m$ where each a_n^m is chosen randomly and evaluate the fitness of each population member; (2) construct a new population by breeding selected individuals (selection probability proportional to fitness) from the old population; (3) evaluate the fitness of each member of the new population; (4) replace the old population by the new population. Repeat steps (2), (3) and (4) until the fittest individual reaches or exceeds the desired goodness-of-fit level.

In the course of the iteration, each population member acts as a "trial solution", whose goodness-of-fit determines the degree to which it will contribute to the subsequent "generations". Superficially, this may look like some peculiar variation on the Monte Carlo theme. The crucial difference lies with step (2): *breeding*; parameter sets defining each population member are first encoded in the form of a linear string. Once two "parents" have been selected for breeding, their defining strings are subjected to biologically-inspired operations of crossover (exchange of string segments between each parent) and mutation (random alteration of a small number of string elements). The resulting "offspring" strings are then decoded into two new parameter sets which subsequently become available as "parents" as the iteration proceeds. It can be shown that crossover and mutation, operating in conjunction with fitness-based selection, greatly enhance the searching capabilities of the algorithm, as compared to simply applying a perturbation operator to a "good" trial solution from the previous iteration (see Goldberg, 1989, chapter 2).

2. Helioseismology

Acoustic noise generated by the turbulent convective motions pervading the outer $\simeq$ 30% of the Sun reverberates throughout the solar interior, much in the same way as earthquake-generated sound waves travel throughout the Earth's interior. In the Sun, however, the modes are trapped because of effective inward reflection near the surface and upward deflection (refraction) in the interior (Ulrich, 1970). Their surface manifestation takes the form of coherent radial velocity variations of low but observable amplitude (Leighton *et al.*, 1962). For the purpose of analysis, the net surface velocity signal is usually decomposed into spherical harmonics. The angular degree l of the harmonics is related to the radial depth of the acoustic cavity associated with a given resonant mode (and its radial overtones). Roughly speaking, low-l modes travel deep into the solar interior while high-l modes travel closer to the surface. Modes having different l values thus effectively sample different regions of the solar interior (*e.g.*, Christensen-Dalsgaard *et al.*, 1985).

The frequency *splittings* for a pair of prograde/retrograde modes of identical radial and angular degree (n, l) contain information relating to rotation. Under the assumption of axisymmetry, the frequency splittings $[a(n, l)]$ are related to the internal rotation profile $[\Omega(r, \theta)]$ through the integral equation

$$a(n,l) = \frac{1}{2\pi} \int \int K_{nl}(r,\theta)\Omega(r,\theta)\, r\, dr\, d\theta, \qquad (1)$$

which expresses the fact that the magnitude of the splittings perceived by an observer at rest is a *global* measure of the rotation-induced frequency shift accumulated by the wave as it travels through the solar interior. It is customary to solve equation (1) with a given solar structural model, so that the integration kernels K_{nl} are known quantities. The data thus consist of a discrete set of frequency splittings coefficients $a^*(n, l)$ with associated errors ε_{nl}, and the task is to invert the RHS of equation (1) to extract $\Omega(r, \theta)$. In what follows we use frequency splittings obtained from the LOWL instrument (Tomczyk *et al.*, 1995a).

3. Genetic Forward Modeling

The GA-based solution procedure is a direct transcription of the general procedure outlined in § 1. It is assumed here that the quantity to be solved, the rotation rate $\Omega(r, \theta)$, is defined by a discrete set of parameters $\{c\}$, in a manner as yet unspecified. This could be done through a direct 2-D spatial discretization on a pre-defined mesh, or through some complicated functional relationship.

The point to note is that the specific form of the relationship between $\Omega(r,\theta)$ and its defining parameters (linear vs. non-linear, etc.) does not affect the structure of the algorithm.

INITIALIZATION: Construct random population $\Omega_m(r,\theta)$, $m = 1, ..., M$
for $k = 1, 2, ...$ **do begin**
    COMPUTE for each individual:

$$a_m(n,l) = \frac{1}{2\pi} \int \int K_{nl}(r,\theta)\Omega_m(r,\theta) r \, dr \, d\theta$$

$$\chi_m^2(a_m) = \sum_{n,l} \left(\frac{a_m(n,l) - a^*(n,l)}{\varepsilon_{nl}} \right)^2$$

    BREED selected population members (selection based on fitness $\equiv 1/\chi^2$) to construct new population
    REPLACE old population by new population
    TEST for solution convergence
enddo

The use of a χ^2 above is merely illustrative; any other statistical estimator can be substituted without altering the overall algorithm. The goodness-of-fit measure need not even be differentiable with respect to the parameters defining $\Omega(r,\theta)$, as no gradient information is required by the algorithm. Note finally that the fitness calculation simply involves the computation of the integral appearing on the RHS of equation (1), as opposed to discretization and inversion of the RHS; the GA-based method is a form of *forward modeling*. All results presented below were obtained using the GA-based general purpose optimization FORTRAN subroutine **pikaia** (Charbonneau, 1995, Appendix; Charbonneau & Knapp, 1995).

4. The Rotation Rate of the Deep Solar Core

Only the lowest l-modes penetrate the deep solar core ($r/R_\odot < 0.5$), and even those modes actually spend little of their time at great depths, in view of the rapid inward increase of the sound speed. As a consequence, the kernels K_{nl} in equation (1) are not well localized spatially at great depths. This, in turn, makes the extraction of rotational information pertaining to the deep solar core a particularly challenging task. Most formal inversion methods suffer from significant loss of accuracy below $r/R_\odot \simeq 0.4$, and have effectively no sensitivity below $r/R_\odot \simeq 0.2$ (*e.g.*, Tomczyk *et al.*, 1995b, Figure 4). In this section we describe some of our latest inversion experiments using genetic forward modeling, designed specifically to extract rotational information below $r/R_\odot = 0.5$.

Formal inversions have demonstrated that the surface latitudinal gradient in the Sun's rotation (characterized by equatorial acceleration) is maintained throughout the convective envelope ($0.7 \leq r/R_\odot \leq 1.0$), and vanishes across a thin shear layer located immediately beneath its base (Brown *et al.*, 1989).

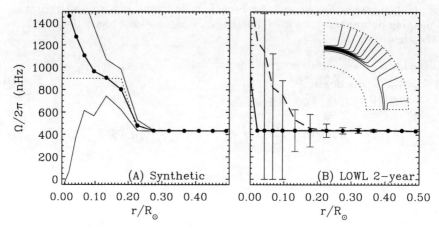

Figure 1. 1-D rotation profiles in the deep solar core obtained through genetic forward modeling. (A) solution for a synthetic data set generated using the rotation profile plotted as a dotted line. (B) results obtained using the LOWL 2-years data set. The dots correspond to the mesh points. The inset on part (B) shows the solar-like profile used in $r/R_\odot > 0.5$ (isocontours $340 \leq \Omega/2\pi \leq 460$ nHz with 10 nHz spacing).

Deeper down, the rotation rate Ω shows little, if any, angular dependence down to the depths where the inversions become inaccurate. Consequently, we *assume* that Ω is only a function of depth below $r/R_\odot = 0.5$, discretize it on a 1-D radial mesh, and solve only for the rotation rate on that mesh assuming piecewise linear variation between mesh points. In the outer half of the Sun we *enforce* a solar-like $\Omega(r,\theta)$ profile. We exclude from the data set all frequency splittings associated with modes having their inner turning point above $r/R_\odot = 0.5$.

In order to ascertain the sensitivity of our method, we have carried out a series of calculations using synthetic splitting data including distinct noise realizations at the level of the LOWL 2-year data set, generated from various artificial rotation curves having different profiles with depth below $r/R_\odot = 0.5$. Figure 1(A) shows an example of one such rotation curve (dotted line), along with a solution obtained by genetic forward modeling (thick solid line).

The solid line is in fact an average of 10 sequences, each performed for 10 distinct noise realizations. Notice how the solution reproduces the abrupt increase in rotation rate at $r/R_\odot = 0.2$, and succeeds in reproducing the magnitude of the increase within about 20%. Results for the various rotation curves used for testing indicates that the method is quite accurate down to $r/R_\odot \simeq 0.2$, and retains useful sensitivity down to $r/R_\odot \simeq 0.1$. The thin lines are 1-σ uncertainties estimated *a posteriori* from the 100 different runs by calculating the rms deviation about the average solution.

Figure 1(B) is a similar calculation, but this time using real splitting data, specifically the LOWL 2-years data set. Again retaining only the modes with lower turning point below $r/R_\odot = 0.5$ leaves 203 modes with angular degrees $1 \leq l \leq 24$ and n in the range 10–24 (depending on the l value). The solution is an average of 20 runs, and the 1-σ error bars are estimated from the behavior of the method on the synthetic test cases under different noise realizations. The solutions of Figure 1 were obtained using the monotonicity constraint $d\Omega/dr \leq 0$. It is far from trivial to incorporate such classes of constraints in formal inversion methods. Here this is carried out in a most straightforward manner, by encoding the *increments* in rotation rate from one mesh point to the next lower point, with positivity imposed on the increment and starting at the outermost mesh point.

Strictly speaking, a χ^2 measure is a valid statistical estimator only if the data errors are normally distributed about the "true" value. In the presence of systematic errors (*i.e.*, correlated trends in the errors), minimizing χ^2 can yield a misleading fit to the data. Consider instead the following modified version of the χ^2 goodness-of-fit estimate:

$$H(a_m) = \sum_{n,l} \begin{cases} 0, & a^* - \varepsilon \leq a_m \leq a^* + \varepsilon; \\ 1, & \text{otherwise.} \end{cases} \qquad (2)$$

The penalty associated with a synthetic splitting (a_m) lying more than $\pm\varepsilon$ away from the data (a^*) is 1, no matter how far off a_m actually is; contrast this to the χ^2 measure, where the penalty is proportional to $(a_m - a^*)^2$. Clearly both estimators will respond very differently to out-liers (truly deviant data points) and systematic error correlations in the data.

The dashed line on Figure 1(B) is a twenty-run average solution obtained by genetic forward modeling under the same parameter settings and monotonicity constraint as before, but using equation (2) instead of χ^2 for a goodness-of-fit measure. The good agreement down to $r/R_\odot \simeq 0.15$ between the two sets of solutions on Figure 1 indicates that the solutions are not artefacts of the specific statistical estimator being minimized by the algorithm.

We note finally that even though our monotonicity constraint effectively biases the algorithm towards finding rotation curves that increase inward, the solution of Figure 1(B) shows no hint of a rapidly rotating core, in marked contradiction with the predictions of a popular class of models for the rotational evolution of the Sun (*e.g.*, Pinsonneault *et al.*, 1989; Chaboyer *et al.*, 1995).

5. Conclusion: Computational Aspects

The helioseismological example discussed above illustrates the flexibility of GA-based modeling techniques, in terms of incorporating constraints and easily switching from one statistical estimator to another (in this latter case, it involved changing two, and only two, lines of code—really). Less obvious from the example is the inherent robustness of the GA-based method with respect to its internal parameter settings (selection strategy, mutation rate, *etc.*). While there obviously exist ranges of parameter settings that will not produce convergence in any reasonable amount of time, the range of settings where the exact choice of internal parameter settings has little effect on the performance of the algorithm is quite wide. Even more important, such "robust" settings usually remain robust across problem space.

On the downside, the method can be rather demanding in CPU time. In the helioseismology example discussed herein, the computation of synthetic frequency splittings given a 2-D rotation curve [*i.e.*, computing the RHS of equation

(1)], is not a computationally trivial task, and it must be carried out $n_p \times n_g$ times, where n_p is the population size and n_g the number of generations over which the solution is left to evolve ($n_p \times n_g = 25{,}000$ for the solutions of Figure 1). The ease with which the technique can be parallelized offsets this difficulty at least partly; typically, most of the CPU times goes into fitness evaluation, which can be performed completely independently for all individuals in the population, so that the n_p fitness evaluations required within each generational iteration can be performed concurrently. At any rate, once the dust settles, a correct solution obtained at significant CPU-time expenditure remains infinitely better than a cheaper solution that is just plain wrong.

One thing that GA-based methods do not do efficiently at all is produce high accuracy solutions. This is because the required "fine tuning" of the population occurs primarily through the agency of the mutation operator, by definition a slow process. If high accuracy is required in the solution parameters, then rather than running the GA-based method over many thousands of generations, it is far more advantageous to turn to a different optimization technique once the global extremum has been located. To this end, an algorithm such as the simplex method is particularly easy to combine with a GA-based method, since both are essentially forward techniques. The resulting hybrid algorithm combines the good exploratory capabilities of GA with the superior convergence behavior of more conventional methods.

References

Brown, T. M., Christensen-Dalsgaard, J., Dziembowski, W. A., Goode, P., Gough, D. O., & Morrow, C. A. 1989, ApJ, 343, 526.
Chaboyer, B., Demarque, P., & Pinsonneault, M. H. 1995, ApJ, 441, 865.
Charbonneau, P. 1995, ApJS, 101, 309.
Charbonneau, P., & Knapp, B. 1995, *A User's Guide to PIKAIA 1.0, NCAR Technical Note 418+IA*.
Christensen-Dalsgaard, J., Gough, D. O., & Toomre, J. 1985, Science, 229, 923.
Davis, L. 1991, *Handbook of Genetic Algorithms*, (New York: Van Nostrand Reinhold).
Goldberg, D. E. 1989, *Genetic Algorithms in Search, Optimization & Machine Learning*, (Reading: Addison-Wesley).
Holland, J. H. 1975, *Adaptation in Natural and Artificial Systems* (Ann Arbor: University of Michigan Press).
Leighton, R. B., Noyes, R. W., & Simon, G. W. 1962, ApJ, 135, 474.
Pinsonneault, M. H., Kawaler, S. D., Sofia, S., & Demarque, P. 1989, ApJ, 338, 424.
Tomczyk, S., Streander, K., Card, G., Elmore, D., Hull, H., & Cacciani, A. 1995a, Sol. Phys., 159, 1.
Tomczyk, S., Schou, J., & Thompson, M. J. 1995b, ApJ, 448, L57.
Ulrich, R. K. 1970, ApJ, 162, 993.

The Radiation-Magnetohydrodynamics of Accretion-Powered Neutron Stars

Richard I. Klein[1]
Department of Astronomy, University of California, Berkeley, CA 94720-3411, U.S.A.

Abstract. Accretion-powered magnetized neutron stars form a unique laboratory for the study of Super-Eddington accretion flows since the magnetosphere focuses the flow onto the small magnetic poles of the neutron star. The resulting local conditions correspond to optically-thick accretion rates up to 1,000 times the local Eddington limit. Under such circumstances, the free fall of the accreting plasma on the surface is primarily halted by the pressure of radiation released in the slowly settling mound at the base of the flow near the surface of the neutron star. The dynamical structure is highly inhomogeneous giving rise to highly turbulent flow that results in the development of photon bubbles in the accretion column. The resulting problem is described by time-dependent, multi-dimensional radiation-magnetohydrodynamics and represents a major challenge in computational physics. We begin by reviewing early attempts at solution using analytical approximations. We then describe both analytic and numerical two-dimensional solutions in the limit of time-independent flow and examine steady-state stability of the structure. We discuss the approach to a full attack on the time-dependent multi-dimensional problem and recent results leading to the discovery of Photon Bubble Oscillations in accretion-powered neutron stars. These results are used to provide the first intepretation of the recent discovery by RXTE of sub-millisecond variability in neutron stars and implications for probing the magnetic field of a neutron star.

1. Introduction

The flow of fully-ionized plasma onto the polar caps of a strongly magnetized, rotating neutron star is well accepted as the basic model for pulsating X-ray sources in binary systems and is understood to be the mechanism whereby accretion-powered pulsars shine (Pringle & Rees, 1972). Much of the effort during the past fifteen years has focused on studies of the detailed physical processes applied to simplified models of the radiating region whose inferred optical depths are inconsistent with the density found in dynamical calculations (Davidson, 1973; Basko & Sunyaev, 1976) for the brighter ($L_X \geq 10^{36}$ erg s^{-1}) sources. For

---

[1] also, Lawrence Livermore National Laboratory, University of California, Livermore, CA 94550, U.S.A.

these sources, the radiation pressure becomes locally super-Eddington, resulting in a strong radiative deceleration of the inflowing plasma (Davidson, 1973; Basko & Sunyaev, 1976; Arons *et al.*, 1987; Klein & Arons, 1991; Arons, 1992; Klein *et al.*, 1996a). Since radiation escapes from the sides of the column, the inhomogenious pressure distribution creates strong inhomogeneities throughout the accretion column. In the super-Eddington regime, the calculations of Klein *et al.* (1996a) show the formation of a radiation-dominated shock, below which the plasma settles with an approximate hydrostatic balance maintained between radiation pressure and gravity. This delicate balance engenders time variability in the flow since a heavy fluid (plasma) is balanced against gravity by the radiation pressure exerted by the lighter fluid (photons). Stability investigations of these models (Hameury *et al.*, 1980; Wang, 1982; Arons, 1992) have suggested the possibility of time variability in the flow present on millisecond time scales because of instability-driven radiation dissipation, even for steady accreting flow ($\dot{M}/|d\dot{M}/dt|_{\text{diff}} \sim 1$ ms). A numerical investigation (Klein & Arons, 1989) suggests the same thing.

The physical conditions that pertain to the accretion column of an accreting X-ray pulsar (AXP) can be viewed as a highly non-classical stellar atmosphere with extreme conditions. For most AXP in the galaxy, the accretion flow is optically thick (not thin) and the accretion is upheld. This implies a strong coupling of radiation to matter in a strong radiation pressure gradient; thus strong dynamical effects are caused by the radiation field. The investigation of phenomena in the super-Eddington regime requires the self-consistent solution of the full two-dimensional time-dependent non-LTE equations of radiation hydrodynamics governing the accretion of matter in a highly magnetized plasma ($B \sim 10^{12}$ G) onto the polar caps. Preliminary results solving the full radiation hydrodynamics have been presented by Klein & Arons (1989; 1991) in which they suggest that the settling mound does indeed develop a new form of "turbulence" in which "photon bubbles" form in the plasma and transport energy to the surface of the accretion column. The linear theory of this phenomenon (Arons, 1992) suggests the formation of small-scale, radiation-filled holes in the medium. More recently, Klein *et al.* (1996a) presented results for initially uniformly filled accretion columns for two models with intrinsic luminosity $L_X = 3 \times 10^{37}$ erg s^{-1} and $L_X = 10^{38}$ erg s^{-1}. Their results indicate that when this instability occurs in concert with continued accretion onto the polar cap, the formation of a statistically steady state shows clear evidence for substantial coalescence of the photon bubbles to become relatively large, rising, optically-thin pockets occupying much of the volume of the settling mound. They are filled with hot ($T \sim 10$ keV) radiation embedded in optically-thick plasma. After the photon bubbles develop non-linear growth, they result in signficant, observable quasi-periodic fluctuations which Klein *et al.* (1996a) called "Photon Bubble Oscillations" (PBO). Before we discuss the time-dependent coupled radiation hydrodynamics of accretion, we will first briefly review previous work on steady state models.

2. Analytic and Steady State Models of Accretion on Magnetized Neutron Stars

Davidson (1973) was the first to explore accretion at the magnetic pole of a neutron star and attempted to describe a simplified, steady-state scattering model which retained some of the notable features of the physics at the base of the accretion column. His key approximations are: (i) the effective optical depth is caused by electron scattering, rather than frequency averaged absorption; (ii) a frequency and polarization-independent, isotropic electron scattering coefficient is adopted, constraining the magnetic field to $\simeq 10^{12}$ G for 10–100 keV X-rays; (iii) the emission of radiation is prompt; that is the infall kinetic energy is converted directly to X-rays as the gas is decelerated by radiation pressure; and (iv) the gas is principally decelerated by radiation pressure. Davidson (1973) arrived at a simple analytic solution with these assumptions. Let ρ and $\mathbf{v}$ be the gas density and velocity at a point in an accretion cylinder of radius R. The momentum density (matter flux) $\mathbf{S} = \rho\mathbf{v}$ is assumed to be uniform within the cylinder. In terms of $\mathbf{S}$ and the electron scattering κ he finds:

$$\nabla^2 Q = -(\kappa/c)\mathbf{S}\cdot\nabla(6\sqrt{Q} - 5Q) = (\kappa S/c)\partial/\partial z(6\sqrt{Q} - 5Q), \qquad (1)$$

where $Q = v^2/v_0^2$ and v_0 is the initial infall velocity of the gas. This equation can be easily solved numerically with appropriate boundary conditions ($Q = 1$ at $z = +\infty$ and $Q = 0$ at $z = 0$). As Davidson (1973) has shown, one obtains a mound-like distribution of gas by plotting contours of constant Q. This simple solution was the first indication that the accretion column has a free-fall region above a deceleration region that allows matter to fall freely almost to the surface of the star near the edge of the column, but is greatly decelerated to $v \ll v_0$ in the central parts of the column. The solution also demonstrates that for reasonable values of the matter flux, the height of the mound is comparable to the radius. As we will see, these important properties of the accretion column have been borne out by detailed radiation hydrodynamic calculations.

The question of the limiting luminosity of an accreting strongly magnetized neutron star was the subject of an important paper by Basko & Sunyaev (1976). In the case of spherically-symmetric accretion, the upper bound for the luminosity can be readily shown to be $L_X \leq L_{\text{EDD}} = 1.26\times 10^{38}(\kappa_T/\kappa)(M/M_\odot)$ erg s^{-1}, where κ_T is the electron scattering opacity. Observations show evidence, however, that several X-ray sources in the Galaxy and in the LMC and SMC have $L_X \geq 10^{38}\, L_{\text{EDD}}$ (Margon & Ostriker, 1973). In fact, the source SMC X-1 has $L_X \sim 10^{39} > L_{\text{EDD}}$. Is there a limiting luminosity for increasing total energy released by accretion $GM\dot{M}/R$? To address this question, Basko & Sunyaev (1976) assume that gas flows down the channel of a hollow-cone accretion column onto the surface of a neutron star. This assumption follows from the flow of matter outside the Alfvén surface as originating from disk accretion. They assume (i) the accretion funnel is azimuthally symmetric with a width less than the funnel circumference; (ii) matter flows along the field lines of a magnetic dipole; (iii) radiation pressure $\ll$ gas pressure; and (iv) the flow is steady state. With these assumptions they are able to simplify the equations of mass, momentum and energy conservation to a coupled set of time-independent PDE's. To simplify the equations further to a coupled set of ODE's, they take the values of the mass density, velocity and energy density of radiation in the center of the accretion column as representative of the entire accretion column and average the transverse radiation diffusion flux over the transverse thickness of the accretion column. The equations now depend upon the vertical length scale only and can be solved numerically. This key assumption of Basko & Sunyaev (1976) and the solution to the resulting equations leads to a total amount of energy L_X, radiated per unit time by the side walls of the two sinking columns, that increases with the accretion rate $L_t \equiv GM\dot{M}/R$. They find, however, that at $L_t \sim L^{**}$ [defined by equation (3) below], the rate of this increase diminishes and a kind of saturation regime occurs. In particular, their solutions show that the X-ray luminosity attains its peak value $\sim (2\text{--}4)L^{**}$. Thus they find that there is a limiting luminosity. The luminosity L_X, can be represented over wide range of parameter values by

$$L_X = L^{**} \exp(L^{**}/L_t) E_1(L^{**}/L_t), \qquad (2)$$

where $E_1(x)$ is the exponential integral, and the limiting luminosity L^{**} is given by

$$L^{**} = 2(l_0/d_0)(c/\kappa)GM = 8 \times 10^{38}(l_0/40d_0)(\sigma_T/\sigma)(M/M_\odot). \qquad (3)$$

As we shall see, this conclusion is indeed incorrect and is a result of their one-dimensionalization approximation. This incorrect assumption has also led to the difficulties of Braun & Yahel (1984) in finding high-luminosity, steady-flow solutions.

The next important work on accretion occurred with Wang & Frank (1981) who were the first to obtain full 2-D steady-state numerical solutions. Their solutions clearly demonstrate the formation of an accretion mound bounded by a radiatively mediated shock resulting in a fan beam formation of radiation emitted from the side of the accretion column.

Restricted analytic solutions to the steady-state accretion problem taking into account variations in directions both parallel and transverse to the field lines were obtained by Kirk (1985) and Arons (1992). Arons (1992) assumed steady flow along the field lines, prompt emission (flow time along a scale height $\gg$ characteristic emission/absorption times), strong radiation pressure, LTE, magnetic modifications of the Rosseland averaged scattering opacity independent of density and temperature, supersonic free-fall flow above the radiation-mediated shock, and subsonic settling in the region below. These assumptions allowed an exact solution which demonstrated that the transverse scale of the plasma determining the rate of diffusive flow across the magnetic field is small compared to the width of the accretion column for most of the known X-ray pulsars. Basko & Sunyaev (1976) and Braun & Yahel (1984) assumed that the transverse diffusion length scale is equal to the width of the column. This incorrect assumption fixed the sideways diffusion velocity of radiative energy to a value inadequate to carry away the gravitational energy released at a given height, thus causing the energy released at the shock to be dragged down into the subsonically settling mound and released by sideways diffusion only well below the shock. This leads to the erroneous conclusion that there is an upper limit to the X-ray luminosity of an AXP.

3. Time-Dependent Radiation-Hydrodynamics of Accretion on Magnetized Neutron stars

Past observations of X-ray pulsars suggest the flow may be strongly fluctuating. Voges *et al.* (1988) reported EXOSAT results indicating strong variability in the light curves of Hercules X-1, and suggested this might be a result of low-dimensional chaos in the flow, rather than from random fluctuations. Recently, several RXTE observations of both X-ray pulsars and bright non-pulsing LMXB's have discovered evidence for remarkable variability at oscillation frequencies from 40 Hz up to the kHz range. As we shall see, this higher frequency regime, only accessible by RXTE, is of direct relevance to the dynamics of the plasma flow in the low-altitude region of X-ray emission, at and just above the magnetic poles. We believe that oscillations found in these higher frequency regimes finds its explanation as phenomena associated with photon bubble instabilities.

3.1. Photon bubble instability and linear stability analysis

The radiative support of the plasma in a mound model is a delicate balance. Because the radiation diffuses slowly out of the mound in a sideways direction, the energy density greatly exceeds $F_\parallel/c$, where $F_\parallel$ is the component of the radiation flux parallel to the magnetic field and the flow. The flux automatically adjusts itself to be just slightly less than the standard Eddington value, so that gravity is almost exactly balanced. This delicate balance is ripe for instability. Imagine a disturbance in which a small hole appears in the plasma, with both density and radiation pressure lower than the surroundings. On the diffusion time scale, fresh photons diffuse in from the surroundings of the optically-thick medium, which cause the hole to become buoyant and rise faster than fresh plasma can flow in and restore the equilibrium. Having risen into the lower density overlying regions, the hole or bubble rises further. The possibility of photon bubble formation was first suggested many years ago by Prendergast & Spiegel (1973).

The situation in polar cap accretion is locally super-Eddington, however. The investigation of such phenomena requires a self-consistent solution of the full two-dimensional, time-dependent equations of radiation hydrodynamics governing the accretion of matter onto the polar caps. In previous studies (Klein & Arons, 1989; 1991), preliminary results suggested that indeed the settling mound of a neutron star develops a new form of "turbulence" in which "photon bubbles" form in the medium and transport energy to the surface of the accretion column.

The fully non-linear development of these flows is not accessible to analytic solutions, especially if one attempts to include enough physics to describe also the light curve and spectrum that would emerge as the time-dependent flow develops. On the dynamical side alone, one expects the photon bubbles to coalesce as they rise, a phenomenon far beyond the reach of analytic theory and yet of fundamental importance to the time scale and magnitude of the luminosity fluctuations created by this phenomenon. Therefore, a two-dimensional, axisymmetric radiation hydrodynamics code has been written to model the spatially inhomogeneous time-dependent structure likely to be present in polar cap

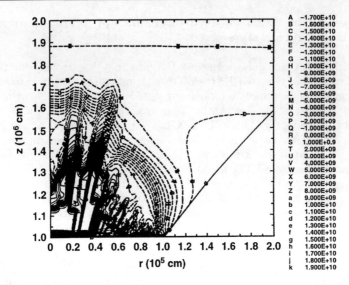

Figure 1.

models incorporating the full effect of super-stong magnetic fields on all the relevant physics. The code explicitly treats the radiative processes of primary importance. Bremsstrahlung, including the effects of the magnetic field on the photon production rate, is included and is the dominant source of photons for weakly-magnetized objects. Cyclotron line radiation is included as a continuum source. Comptonization including magnetic field effects is also included as an energy transfer process between the photons and electrons. The code keeps track of the relativistic inertia of the plasma. Other features include the ability to calculate separate electron and ion temperatures, which can be of importance in shocked regions. A new algorithm for flux limiters has been developed that takes into account the transition from optically-thick to optically-thin regions, including both the atmosphere of the accretion column (where radiation escapes into black sky outside the column) and the enclosed optically-thin pockets that form within the optically-thick accretion mound (photon bubbles). The plasma flow occurs along the field lines of a star-centered dipolar magnetic field. The physical model has been described by Arons *et al.* (1987). The specific form of the equations and their solution has been described briefly in Klein & Arons (1989; 1991). Klein *et al.* (in preparation) describe the full details of the equations and the numerical methods.

3.2. Non-linear evolution of photon bubbles in an accretion column

We have used our code to discover the presence of photon bubbles in numerical models of polar cap accretion flow (Klein & Arons, 1989; 1991; Klein *et al.*, 1996a). Figure 1 shows isocontours of the velocity in the column after 3 ms for a neutron star mass of $1.4 M_\odot$, $R_* = 10$ km, a polar cap opening angle of 0.1 rad corresponding to a circular polar cap area of 3 km^2, an accretion luminosity of 3×10^{37} erg s^{-1}, and a surface dipole magnetic field of 3×10^{12} G. The dashed contours represent infall, while the solid lines correspond to upflowing plasma; these are photon bubble regions. An accretion mound develops reaching 3–4 km above the surface, partially filled with low-density photon bubbles where the inflow velocity is $\sim 0.1c \sim v_{\text{ff}}(R_*)/3$. The accretion mound is surrounded by a strong radiation-dominated accretion shock that decelerates the infalling matter to subsonic velocities. Comptonization in this shock transfers most of the infall energy to the photons created in the dense mound. The photon bubbles are regions with density 2–3 orders of magnitude lower than that of surrouding regions. Because the column has large magnetically modified Thomson optical depth, the radiation emerges in a fan beam with flutuations in outgoing transverse flux from the sides of the column caused by temporal variations in the photon bubbles.

3.3. Photon bubble oscillations in accretion-powered pulsars

With time-dependent, two-dimensional axisymmetric radiation hydrodynamic calculations of locally super-Eddington accretion onto highly magnetized neutron stars appropriate to the flow onto the polar caps of high luminosity X-ray pulsars, Klein et al. (1996a) provide the first theoretical predictions of the existence of Photon Bubble Oscillations (PBO). These are a consequence of the dynamical evolution of the photon bubbles that form in the accretion mound below the shock as a result of the photon bubble instability. They show that small radiation pockets created within the mound merge into larger structures on the photon diffusion time scale within the mound ($t_{\text{diff,mound}} \sim 0.1$–$1.0$ ms). These lose photons through diffusion into the shock and fragment as the decreasing bubble bouyancy allows the elevated plasma to fall to the stellar surface.

After the photon bubbles develop non-linear growth, the resulting limit cycle creates relatively well-defined, moderately high Q power peaks in the power density spectrum of the luminosity time series (PBO) on the $t_{\text{diff,mound}}$ time scale of the light curve. They predict that photon bubble instabilities in the accretion mound result in an observable signal in the power spectrum of the luminosity with a time scale of tenths of a ms over a range of accretion luminosities. Klein et al. (1996a) find that as a larger number of small photon bubbles form, they coelesce to form a smaller number of bubbles with larger total volume. Radiation fills these low-opacity, large bubbles, decreasing the emergent radiative flux and therefore the luminosity from the accretion column. This process appears to occur on a combination of the diffusion time scale within the mound and the bubble coelescence time scale. Large bubbles release their photons through diffusion close to the accretion shock, allowing the plasma pushed up by the bubbles to fall down and break up the larger bubbles, winning the competition between plasma weight and bubble bouyancy. Diffusion of the released photons into the shock, where they are energized by Comptonization, and out the side column results in an increase in the luminosity. Repetition of this cycle causes the statistically steady-state fluctuation in the pulsar's radiative output.

3.4. Evidence for PBO and photon bubble turbulence in accreting neutron stars (GRO J1744-28 and Sco X-1)

Recent Rossi XTE high time resolution observations of a newly discovered, remarkably bright X-ray pulsar, GRO J1744-28, has uncovered moderately high frequency quasi-periodic oscillations at 20, 40 and 60 Hz, with a striking high frequency red-noise component with a −5/3 power-law index extending from 40–600 Hz (Zhang *et al.*, 1996; Klein *et al.*, 1996b). In addition, van der Klis *et al.* (1996) have made RXTE observations of Sco X-1, the brightest bulge LMXB source, and have discovered the highest frequency QPO at 1,100 and 830 Hz. The discovery of X-ray high frequency variability and high frequency red-noise would not be possible without the extraordinary high time resolution capability of the Rossi XTE.

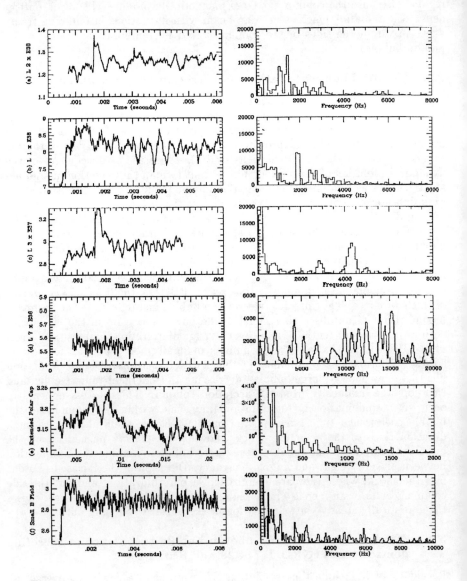

Figure 2.

Klein et al. (1996b) made recent multi-dimensional radiation-hydrodynamics calculations of accretion onto magnetized stars in an attempt to explore the possiblility that these recent observations find explanation in photon bubble instabilities formed in the accretion mound of a neutron star. In their earlier work, Klein et al. (1996a) show that for narrow polar caps, $\theta_{cap} \sim 0.1$ radians, sub-millisecond PBO would be produced corresponding to quasi-periodic frequencies of a few $\times\ 10^3$ Hz. In a recent series of calculations they have computed several models at different accretion luminosities, L_a, ranging from 2×10^{38} erg s^{-1} to 7×10^{36} erg s^{-1} with the same field strength ($B = 3 \times 10^{12}$ G) and narrow polar cap size to ascertain the variation of PBO frequency with L_a. Figures 2a–2d show the time series of the bolometric luminosity emerging from the side of the accretion column and the associated power density spectrum for four models. The power density spectrum for these models reveals power (PBO) concentrated at peak frequencies of 1,592, 1,953, 4,102 and 12,000 Hz respectively. The diffusion time for a large polar cap can be substantially longer than for the $\theta_{cap} \sim 0.1$ models. Thus, if GRO J1744-28 has a polar cap radius $\theta_{cap} \gg 0.1$, it is plausible to find PBO at frequencies substanially lower than 2,000 Hz. An initial calculation was performed with $\theta_{cap} = 0.4$ and $L_a = 3 \times 10^{37}$ erg s^{-1}. The luminosity time series and the power density spectrum are shown in Figure 2e. The power density spectrum reveals PBO at 170 Hz and 260 Hz interpreted as the first and second harmonic with the implied fundamental PBO at 85 Hz lost in the red-noise of the underlying spectrum.

In Figure 3, we show the averaged power density of the RXTE data of GRO J1744-28 from 0.1 Hz to 1,000 Hz (upper half of the figure). The pulsations of the neutron star at 2 Hz is clearly seen as well as the 40 Hz QPO. The 20 Hz and 60 Hz features are not seen in this 12 minute exposure. A dashed reference line of slope $-5/3$ is included to show the striking $-5/3$ power law in the data from 40 Hz out to 600 Hz. In the lower half of the figure, we show the results of their recent calculation (Klein et al., 1996b) with an extended polar cap.

They slide the spectrum so that the 170 Hz PBO feature in the simulation lines up with the observed 40 Hz QPO. The first and second harmonics of the fundamental PBO are clearly evident. What is equally noticeable is the striking high-frequency power-law spectrum of the noise upon which the PBO is superimposed. This power law is $-5/3$ and extends from 100 Hz to several $\times\ 10^4$ Hz. The power law indicates that the coalescence and break-up of photon bubbles in the accretion column may result in an energy cascade from large bubble structures down to smaller bubbles as the bubbles release their radiation from the interior of the column. They have identified this as photon bubble turbulence; a key signature of photon bubble instabilities that may be present in the power-law spectra of most, if not all, highly luminous X-ray pulsars. Although the correspondence between these first models and the observations for GRO J1744-28 is not perfect, the calculations have shown that with large polar caps, PBO can exist at frequencies approaching 40 Hz.

Recent RXTE observations of Sco X-1 (van der Klis et al., 1996) have found sub-millisecond QPO at 1,080–1,130 Hz with a second high frequency QPO found at 800 Hz. The QPO fractional rms amplitudes were observed to be between 0.5% and 0.75% of the total count rate. Klein et al. (1996b) carried out a simulation with a weak magnetic field ($B = 5 \times 10^9$ G) which

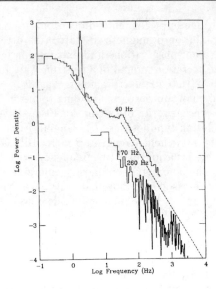

Figure 3.

nevertheless is strong enough to cause plasma to fall along the field lines, and an accretion luminosity appropriate to Sco X-1 ($L_a = 3 \times 10^{37}$ erg s^{-1}). We show the time series of the bolometric luminosity and the corresponding power density spectrum in Figure 2f. Clear evidence of PBO at 700 Hz and 1,100 Hz has been found in the calculations in striking agreement with the RXTE observations. Furthermore, the calculations predict that additional high frequency PBO at 2,000 Hz and 2,600 Hz may exist. Klein *et al.* (1996b) showed that as with GRO J1744-28, the power spectrum of the simulation of Sco X-1 yields a remarkable $-5/3$ power law extending from 3,000 Hz out to 5×10^4 Hz. They predict that a power law with this slope (and thus indicating photon bubble turbulence) should be seen with further observations. They identify the QPO at 800 Hz and 1,100 Hz and the predicted power law as the first evidence of PBO and photon bubble turbulence in a non pulsing weak field bulge source. Furthermore, they have calculated the rms amplitude of the PBO features found in their calculations at 700 Hz and 1,100 Hz. They found rms amplitudes between 0.2% and 0.8%, completely consistent with the recent observations.

Arons (1992) has shown that photon bubble instabilities can form with magnetic fields as small as 10^8 G. Classical X-ray pulsars easily fullfil this criterion, as do all reasonable models for the bursting pulsar GRO J1744-28. That Sco X-1 might indeed be showing the PBO and photon bubble turbulence suggests that this neutron star, and by extension other neutron stars in LMXBs, may be weakly magnetized objects, as has been suggested by the evolutionary scenarios connecting these objects to the millisecond rotation-powered pulsars. If our identification of the QPO in Sco X-1 with PBO is corrrect, study of this phenomenon will yield new insight into the general problem of magnetic field evolution in accreting neutron stars. The PBO phenomena will give us a new probe for the magnetic field, accretion flow, field geometry and polar cap extension near the surface of accreting neutron stars.

Major computational challenges remain in the study of photon bubble instabilities in the accretion columns of neutron stars. Perhaps the greatest challenge awaits the generalization of these studies to three spatial dimensions. A key question that must be addressed eventually is the three-dimensional coherency of the collective phenomena giving rise to photon bubble oscillations. A modestly resolved 3-D calculation will take 1,600 hours at a sustained 10 Gflops rate to evolve the flow to 100 ms. Potential computational breakthroughs could occur with the development of adaptive time-stepping techniques. Another area that will have major impact in time-dependent studies of accretion flow onto neutron stars will be the development of new approaches to treat the transport of radiation accurately in a dynamically changing 3-D medium. A major breakthorough in this area would be the development of adaptive angle transport.

Acknowledgments. I am indebted to Jon Arons and Garrett Jernigan for many stimulating discussions. This research was partially supported under the auspices of the US Department of Energy at the Lawrence Livermore National Laboratory under contract W-7405-Eng-48.

References

Arons, J. 1992, ApJ, 388, 561.
Arons, J., Klein, R. I., & Lea, S. M. 1987, ApJ, 312, 666.
Basko, M. M., & Sunyaev, R. A. 1976, MNRAS, 175, 395.
Braun, A., & Yahel, R. Z. 1984, ApJ, 278, 349.
Davidson, K. 1973, Nature, 246, 1.
Hameury, J. M., Bonnazola, S., & Heyvaerts, J. 1980, A&A, 90, 359.
Kirk, J. G. 1985, A&A, 142, 430.
Klein, R. I., & Arons, J. 1989, in *Proceedings of the 23rd ESLAB Symposium on Two Topics in X-ray Astronomy*, eds. N. E. White & T. D. Guyenne, ESA SP-296 (Paris: European Space Agency), 1, 89.
Klein, R. I., & Arons, J. 1991, in *Stellar Atmospheres: Beyond Classical Models*, ed. L. Crivillari, (Boston: Kluwer), 205.
Klein, R. I., Arons, J., Jernigan, G., & Hsu, J. 1996a, ApJ, 457, L85.
Klein, R. I., Jernigan, G., Arons, J., Morgan, E. H., & Zhang, W. 1996b, ApJ, 469, L119.
Prendergast, K. H., & Spiegel, E. A. 1973, Comments Ap. Space Phys., 5, 43.
Pringle, J. E., & Rees, M. J. 1972, A&A, 21, 1.
van der Klis, M., et al. 1996, ApJ, 469, L1.
Wang, Y.-M. 1982, A&A, 112, 24.
Wang, Y.-M., & Frank, J. 1981, A&A, 93, 255.
Zhang, W., et al. 1996, ApJ, 469, L29.

On the Existence of Intermediate Shocks

S. A. E. G. Falle & S. S. Komissarov[1]
Department of Applied Mathematics, The University of Leeds, Leeds LS2 9JT, U.K.

Abstract. In recent years, a number of conservative upwind schemes for magnetohydrodynamics have been developed (*e.g.*, Brio & Wu, 1988; Zachary *et al.*, 1994; Ryu & Jones, 1995). Although these work reasonably well, they all appear to produce intermediate shocks for certain Riemann problems. Since it is well known that intermediate shocks are unphysical (*e.g.*, Jeffreys & Taniuti, 1964), this is obviously worrying. In fact, it has even led Wu (1987) to claim that such shocks really exist when the dissipative terms are included in the equations. We show that this is not true and that numerical schemes only generate such shocks in problems with certain kinds of symmetry. However, since many numerical simulations do indeed have such symmetries, it may be necessary to take steps to eliminate intermediate shocks. It is, as yet, not clear whether they are present in any of the magnetohydrodynamic calculations in the literature, but if they are, then this must cast serious doubt on the validity of such calculations.

1. Ideal Magnetohydrodynamics

We start by discussing the mathematical structure of the equations of ideal magnetohydrodynamics. For our purposes it is sufficient to consider the one-dimensional equations, which can be written in the conservation form (*e.g.*, Brio & Wu, 1988)

$$\frac{\partial \mathbf{U}}{\partial t} + \frac{\partial \mathbf{F}}{\partial x} = 0.$$

The conserved quantities $\mathbf{U}$ and the corresponding fluxes $\mathbf{F}$ are

$$\mathbf{U} = \begin{bmatrix} \rho \\ \rho v_x \\ \rho v_y \\ \rho v_z \\ e \\ B_x \\ B_y \\ B_z \end{bmatrix}, \quad \mathbf{F} = \begin{bmatrix} \rho v_x \\ \rho v_x^2 + p_g + B^2/2 - B_x^2 \\ \rho v_x v_y - B_x B_y \\ \rho v_x v_z - B_x B_z \\ \{e + p_g + B^2/2\} v_x - B_x (\mathbf{v} \cdot \mathbf{B}) \\ v_x B_y - v_y B_x \\ v_x B_z - v_z B_x \end{bmatrix}.$$

---

[1] Astrospace Centre, Lebedev Physical Institute, Leninsky Prospect 53, Moscow B-333, 177924, Russia

Intermediate Shocks

Here p_g is the gas pressure and

$$e = \frac{p_g}{(\gamma - 1)} + \frac{1}{2}B^2 + \frac{1}{2}\rho v^2$$

is the total energy per unit volume. The units in these equations are such that the velocity of light and the factor 4π do not appear.

1.1. Characteristic wave speeds

There are only seven variables in this system since the condition $\nabla \cdot \mathbf{B} = 0$ requires the x-component of the magnetic field to be constant if there is no y- or z-dependence. There are, therefore, seven waves whose speeds are given by the eigenvalues, λ_i ($i = 1 \ldots 7$), of the Jacobian matrix

$$\underline{A} \equiv \frac{\partial \mathbf{F}}{\partial \mathbf{U}}.$$

These are

Fast Waves	$\lambda_{1,7}$	$=$	$v_x \mp c_f,$
Alfvén Waves	$\lambda_{2,6}$	$=$	$v_x \mp c_a,$
Slow Waves	$\lambda_{3,5}$	$=$	$v_x \mp c_s,$
Entropy Waves	λ_4	$=$	$v_x,$

where the Alfvén speed, c_a, and the slow and fast speeds, c_s and c_f, respectively, are given by

$$c_a = B_x/\sqrt{\rho}, \quad c_{s,f} = \frac{1}{2}\left[a^2 + \frac{B^2}{\rho} \mp \left\{\left(a^2 + \frac{B^2}{\rho}\right)^2 - \frac{4a^2 B_x^2}{\rho}\right\}^{1/2}\right].$$

Here $a = \sqrt{\gamma p/\rho}$ is the sound speed. Note that $0 \leq c_s \leq c_a \leq c_f$.

1.2. Shock relations

The states $\mathbf{U}_L$ and $\mathbf{U}_R$ on the left and right of a discontinuity traveling with speed s are related by

$$s(\mathbf{U}_L - \mathbf{U}_R) = \mathbf{F}_L - \mathbf{F}_R.$$

These require that

$$[\mathbf{B}_t(c_a^2 - v_x^2)] = 0, \tag{1}$$

where v_x is the velocity in the shock frame ($s = 0$) and $\mathbf{B}_t$ is the transverse component of the field. As usual, $[\cdot]$ denotes the jump across the shock.

The following kinds of non-linear discontinuity are permitted:

1. **Slow/Fast Shocks** $sign\{(c_a^2 - v_x^2)_L\} = sign\{(c_a^2 - v_x^2)_R\}$

   Equation (1) implies that there is no rotation of the transverse magnetic field.

2. **Intermediate Shocks** $sign\{(c_a^2 - v_x^2)_L\} \neq sign\{(c_a^2 - v_x^2)_R\}$

Equation (1) implies that the transverse field changes sign (*i.e.*, it rotates by π).

1.3. Uniqueness

There are 15 unknowns at a discontinuity, $\mathbf{U}_L$, $\mathbf{U}_R$ and the shock speed. Since there are seven shock relations, this means that we need eight incoming characteristics to determine the solution uniquely. We can most conveniently discuss this by dividing the states in the shock frame into four categories

1) $|v_x| \geq c_f > c_a > c_s$,   2) $c_f \geq |v_x| \geq c_a > c_s$,

3) $c_f > c_a \geq |v_x| \geq c_s$,   4) $c_f > c_a > c_s \geq |v_x|$.

We say that a shock is of type $m \to n$ if the upstream and downstream states are of types m and n respectively. The number of incoming characteristic for the various types of non-linear discontinuity are then:

Fast Shock $(1 \to 2)$: The number of incoming waves is

Upstream: 2 Fast, 2 Alfvén, 2 Slow, 1 Entropy,
Downstream: 1 Fast.

Since this gives a total of eight, these shocks are uniquely determined.

Slow Shock $(3 \to 4)$: The number of incoming waves is

Upstream: 1 Fast, 1 Alfvén, 2 Slow, 1 Entropy,
Downstream: 1 Fast, 1 Alfvén, 1 Slow.

Again we have a total of eight, so these shocks are also uniquely determined.

Intermediate Shock $(1 \to 3)$: The number of incoming waves is

Upstream: 2 Fast, 2 Alfvén, 2 Slow, 1 Entropy,
Downstream: 1 Fast, 1 Alfvén.

The total is now nine, which means that these shocks are overdetermined.

Intermediate Shock $(1 \to 4)$: The number of incoming waves is

Upstream: 2 Fast, 2 Alfvén, 2 Slow, 1 Entropy,
Downstream: 1 Fast, 1 Alfvén, 1 Slow.

The total is ten, so these shocks are also overdetermined.

Intermediate Shock $(2 \to 3)$: The number of incoming waves is

Upstream: 1 Fast, 2 Alfvén, 2 Slow, 1 Entropy,
Downstream: 1 Fast, 1 Alfvén.

Intermediate Shocks

Although these shocks have eight incoming characteristics, they are overdetermined because they have too many Alfvén waves. The shock relations requires the transverse fields on either side of the shock to be parallel, but this is not compatible with the presence of three Alfvén waves.

Intermediate Shock ($2 \to 4$): The number of incoming waves is

Upstream: 1 Fast, 2 Alfvén, 2 Slow, 1 Entropy,

Downstream: 1 Fast, 1 Alfvén, 1 Slow.

The total is nine, so these shocks are overdetermined.

This tells us that, as far as the hyperbolic equations are concerned, the solution is overdetermined if it contains intermediate shocks. They are therefore not generic in the sense that they can only arise for very special initial conditions. Shocks which have too few incoming waves, such as rarefaction shocks, are allowed by the hyperbolic equations, but they must be excluded since one can show that no steady dissipative shock structure exists if there are too few incoming waves.

Things are somewhat different for co-planar problems, that is those in which the plane defined by the velocity and magnetic field remains invariant. There are then only five variables and we must exclude the Alfvén waves. The $1 \to 4$ intermediate shock then has too many waves and the $2 \to 3$ has too few, but $1 \to 3$ and $2 \to 4$ have the correct number of waves. Furthermore, since there are no Alfvén waves, the sign of the transverse field can only be changed by an intermediate shock.

2. Numerical Calculations

Brio & Wu (1988) considered a co-planar Riemann problem in which the transverse field changes sign across the initial discontinuity. This problem therefore admits a solution containing an intermediate shock and indeed all numerical schemes generate a solution containing a compound wave consisting of $2 \to 3/4$ intermediate shock with a slow rarefaction attached to its downstream end (Brio & Wu, 1988; Ryu & Jones, 1995; Barmin et al., 1996; Falle et al., 1997). Here 3/4 means a state in which the normal velocity in the shock frame is equal to the slow speed. However, Falle et al. (1997) showed that there exists another solution in which the compound wave is replaced by a slow shock and an Alfvén shock that changes the sign of the transverse field.

This problem is clearly one of those for which the solution is not unique if intermediate shocks are allowed. However, the solution with the intermediate shock is not generic in the sense that it only exists because the direction of the transverse field rotates by a multiple of π across the initial discontinuity. This solution is therefore unphysical and Barmin et al. (1996) have shown that it does not arise in numerical calculations if the problem is slightly altered so that the initial magnetic field does not rotate by exactly π.

Figure 1 shows that intermediate shocks also do not appear if the initial discontinuity is replaced by a small, but finite region in which the transverse

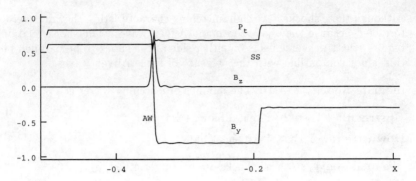

Figure 1. Breakup of a 2 → 4 intermediate shock into an Alfvén wave (AW) and a slow shock (SS) caused by a field rotation in the initial conditions. P_t is the total pressure.

field rotates smoothly through an angle π. With $\gamma = 5/3$, the initial conditions are:

$$\begin{aligned} & x \leq -0.01 & \mathbf{U} = \mathbf{U}_\mathrm{L}, \\ -0.01 < & x < 0.01 & \mathbf{U} = \mathbf{U}_\mathrm{I}(x), \\ & x \geq 0.01 & \mathbf{U} = \mathbf{U}_\mathrm{R}, \end{aligned}$$

where

$\mathbf{U}_\mathrm{L}$: $\rho = 0.4262$, $p_\mathrm{g} = 0.0306$, $\mathbf{B} = (-0.7, 0.7454, 0)$, $\mathbf{v} = (0.9232, -1.4636, 0)$,
$\mathbf{U}_\mathrm{R}$: $\rho = 1$, $p_\mathrm{g} = 0.6$, $\mathbf{B} = (-0.7, -0.3, 0)$, $\mathbf{v} = (0.25, 0, 0)$

and $\mathbf{U}_\mathrm{I}(x)$ is the same as $\mathbf{U}_\mathrm{L}$ except that the field is

$$\mathbf{B}(x) = (-0.7, 0.7454\cos\phi, 0.7454\sin\phi), \quad \phi = 50\pi(x + 0.01).$$

Without the intermediate state, $\mathbf{U}_\mathrm{I}$, these initial conditions would generate a 2 → 4 intermediate shock. However, we can see from Figure 1 that the presence of the field rotation ensures that this splits into an Alfvén wave and a slow shock. We find that the intermediate shock in the Brio & Wu problem also disappears if we do the same thing to the initial conditions. A numerical scheme can therefore be forced to generate the physically correct solution simply by telling it that the world is not really that symmetric.

It is also interesting to ask what the dissipation in a numerical scheme does when the initial conditions should generate an intermediate shock that has no dissipative shock structure, such as a 2 → 3 intermediate shock in the co-planar case. As before $\gamma = 5/3$ and the initial data are

$x \leq 0$ $\rho = 0.4262$, $p_\mathrm{g} = 0.0306$, $\mathbf{B} = (0.7, 0.7454, 0)$, $\mathbf{v} = (0.1732, 1.4636, 0)$,
$x > 0$ $\rho = 0.6509$, $p_\mathrm{g} = 0.2797$, $\mathbf{B} = (0.7, -0.68, 0)$, $\mathbf{v} = (-0.2318, -0.5319, 0)$.

Figure 2 shows that in this case, the numerical scheme turns the shock into a compound wave consisting of a 2 → 3/4 shock with a slow rarefaction attached to its downstream end, which is exactly what should happen.

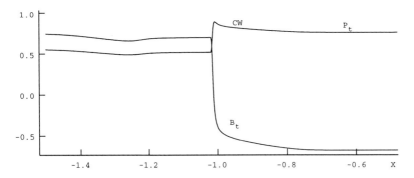

Figure 2. Decay of a 2 → 3 intermediate shock into a compound wave (CW). P_t is the total pressure and B_t is the transverse magnetic field.

3. Conclusions

Both these results and those of others show that upwind numerical schemes for magnetohydrodynamics are really pretty good. For sensible problems they give sensible results and if we give them a stupid problem, that is one with a special symmetry, then they give the correct stupid answer. There is therefore no reason to suppose that there is anything wrong with the theory, but it is obviously important to avoid using numerical calculations to solve problems with excessive symmetry.

References

Barmin, A. A., Kulikovskiy, A. G., & Pogorelov, N. V. 1996, J. Comput. Phys., 126, 77.
Brio, M., & Wu, C. C. 1988, J. Comput. Phys., 75, 400.
Falle, S. A. E. G., Komissarov, S. S., & Joarder, P. 1997, MNRAS, submitted.
Jeffreys, A., & Taniuti, T. 1964, *Nonlinear Wave Propagation*, (New York: Academic Press).
Ryu, D., & Jones, T. W. 1995, ApJ, 442, 228.
Wu, C. C. 1987, Geophys. Res. Lett., 14, 668.
Zachary, A. L., Malagoli, A., & Colella, P. 1994, SIAM J. Sci. Comput., 15, 263.

Numerical Simulations Can Lead to New Insights

Robert F. Stein

Physics and Astronomy Department, Michigan State University, East Lansing, MI 48824, U.S.A.

Mats Carlsson

Institute of Theoretical Astrophysics, University of Oslo, Blindern, N-0315 Oslo, Norway

Åke Nordlund

Theoretical Astrophysics Center and Astronomical Observatory NBIfAFG, Juliane Maries vej 30, DK-2100, Copenhagen Ø, Denmark

Abstract. Numerical simulations can lead to new insights into astrophysical processes that would be difficult to achieve otherwise. We present a few examples from our 1.5-D radiation-magnetohydrodynamic calculations and 3-D convection calculations:

1) Magnetic-field-free regions of the chromosphere have enhanced emission without any increase in the average gas temperature;

2) Solar convection is driven non-locally by cool, turbulent, filamentary, downdrafts produced by radiative cooling in the surface boundary layer;

3) The lower part of the solar convection zone is slightly stably stratified.

1. Introduction

Numerical simulations can lead to new insights into astrophysical processes that would be difficult to achieve otherwise. We present a few examples from one-dimensional radiation hydrodynamic and three-dimensional magnetohydrodynamic simulations of the solar atmosphere and convection zone.

2. Solar Chromosphere

To model dynamics of the solar atmosphere, we solve the 1.5-dimensional conservation equations, the induction equation, the non-LTE radiative transfer equation, and the equations for atomic level populations. All quantities are a function of height and time only. We solve the equations implicitly on an adaptive mesh (Dorfi & Drury, 1987) using van Leer's (1977) second-order upwind advection scheme and Scharmer's method for calculating the radiative transfer (Scharmer & Carlsson, 1985).

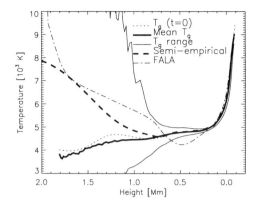

Figure 1. Time averaged (mean) gas temperature in the simulation and the semi-empirical temperature that gives the best fit to the time average of the emergent intensity from the simulation. Also shown are: the minimum and maximum (range) temperatures in the simulation, the starting model temperature, and the semi-empirical model of Fontenala et al., 1993. The semi-empirical model giving the same intensities as the dynamical simulation shows a chromospheric temperature rise while the mean temperature in the simulation does not rise.

These radiation hydrodynamic simulations of the solar atmosphere have radically changed our view of the solar chromosphere. They have shown that enhanced chromospheric emission, which corresponds to an outwardly increasing semi-empirical temperature structure, can be produced by acoustic waves without any increase in the mean gas temperature (Figure 1). Localized, brief periods of high temperature behind shocks lead to enhanced emission in the ultraviolet, where the emission is an exponential function of temperature. Integrated over time, the viscous shock dissipation is balanced by the radiative cooling and there is no increase in temperature. There is, however, an increase in radiation (radiative cooling). The semi-empirical chromospheric temperature rise is an artifact of temporal averaging of the highly non-linear UV emission. The Sun does not have a classical chromosphere in magnetic-field-free internetwork regions (Carlsson & Stein, 1995)!

3. Solar Surface Convection

To model convection, we solve the three-dimensional equations of mass, momentum and energy conservation and the induction equation in non-conservative form on a non-staggered grid. We use a tabular realistic equation of state, including ionization energy. Radiative energy exchange is critical in determining the structure of the upper convection zone. Since the top of the convection zone occurs near the level where the continuum optical depth is unity, neither the optically-thin nor the diffusion approximations give reasonable results. We therefore include 3-D, LTE radiation transfer in our model (Nordlund, 1982). We use a third-order leapfrog predictor-corrector in time (Hyman, 1979) and cal-

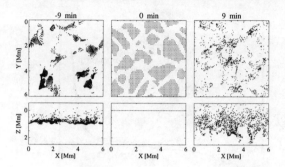

Figure 2. Motion of fluid parcels. The location of fluid parcels which are moving upward through the visible surface at a given time, are shown nine minutes earlier and later. These form the interiors of the granules. Nine minutes earlier, most of this fluid has come from about the same depth below the surface, but from a much smaller region of the horizontal plane, because the upflows diverge in order to conserve mass. Only a small fraction of fluid reaches the surface where it can cool by radiating to space. Nine minutes later, these fluid parcels are heading downward with large velocity and have converged into filaments.

culate spatial derivatives using compact third- and sixth-order fits to the functions (Lele, 1992). The code is stabilized by a hyper-viscosity which removes short-wavelength noise without damping the longer wavelengths. Horizontal boundaries are periodic, the top boundary is transmitting, and at the bottom boundary we specify the entropy of the incoming fluid, zero net mass flux and make pressure uniform over the bottom boundary (Nordlund & Stein, 1990).

Such realistic simulations of convection near the solar surface have led to a paradigm shift in our perception of convection. Convection is inherently nonlocal. It is driven from the surface boundary layer, on the intermediate scale of granulation, by radiative cooling. This produces low-entropy fluid that descends in the intergranule lanes and merges into filamentary downdrafts that penetrate through the convection zone (Figures 2 and 3). Most of the buoyancy work that

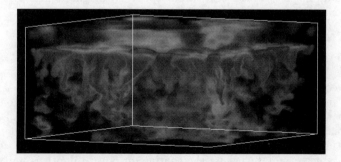

Figure 3. Entropy fluctuations occur primarily as low-entropy, filamentary, downdrafts, descending through nearly uniform, entropy-neutral ascending fluid and are produced by radiative cooling of fluid that reaches the surface.

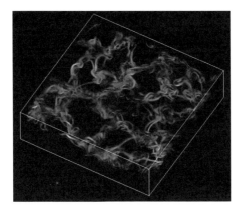

Figure 4. Vorticity is produced by the shear of overturning flows and occurs primarily in the intergranule lanes and the filamentary downdrafts. Upflows, in contrast, are very laminar and free of vorticity because the upflow diverges and all fluctuations get smoothed out quickly.

drives the convection occurs in these overdense downdrafts. They drive both larger-scale laminar cellular flows and smaller-scale turbulent motions. This new paradigm replaces the older one of energy cascade from large-scale driven eddies to smaller eddies.

Shear at the edges of the downflows creates vorticity and turbulence in the downdrafts (Figure 4). Ascending fluid, in contrast, has a very low level of fluctuations because it diverges and overturns as it rises in order to conserve mass as it enters lower-density layers. Hence, the upflows are laminar and entropy neutral (Nordlund & Stein, 1995).

4. Solar Interior Convection

To model the deep layers of the solar convection zone, we use a three-dimensional magnetohydrodynamic code that solves the equations for conservation of mass, momentum, and internal energy in conservative form together with the induction equation for the magnetic field, on a staggered mesh. An ideal equation of state is used (Stein et al., 1993). Radiation transfer is treated in the diffusion approximations using a Kramer's-type opacity. In order to reduce the thermal relaxation time, we increase the energy flux to 10^7 times the solar value, so this is a "toy model".

Simulations of such highly idealized models of the deep layers of the solar convection zone have revealed that the entropy structure in the bulk of the convection zone is slightly stable, caused by the gradual transition from radiative to convective energy transport (Figure 5). The radiative flux decreases with increasing height, because the Kramer's opacity increases with decreasing temperature. This heats both the ascending and descending gas. As a result, the entropy of the ascending fluid increases with height and the entropy of the descending fluid increases with depth (the entropy of descending fluid also in-

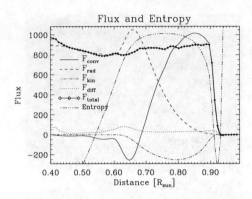

Figure 5. The total energy flux and the fluxes that contribute to it (enthalpy, radiative, kinetic energy and viscous) are shown for a toy solar model along with the mean entropy. Entropy decreases rapidly with height near the top boundary of the convective layer (very unstable) because of radiative cooling. In the bulk of the convection zone, the mean entropy is nearly constant, but increases slightly with height (slightly stable) because radiation heats the fluid and increases its entropy.

creases with depth because of entrainment). Given the intermittency of cool, descending fluid, the median and average entropy increases slightly with height, producing a slightly stable entropy gradient (Figures 5 and 6). It is also clear from Figure 6 that entropy fluctuations injected into the ascending fluid at the bottom boundary layer of the convection zone are much smaller than those injected into descending fluid by radiative cooling in the top boundary layer of the convection zone (Nordlund & Stein, 1995).

5. Conclusion

We have called attention to three significant new insights into the structure of the solar chromosphere and convection zone arising from the analysis of numerical simulations: the lack of a chromospheric gas temperature increase in magnetic-field-free regions, the non-local driving of convection on intermediate scales arising from radiative cooling at the top boundary layer of the convection zone, and the slightly stable stratification of the bulk of the convection zone.

Acknowledgments. This work was supported by NASA grants NAGW 1695 and NAG 5-2489, by NSF grant AST 95-21785, by a grant from the Norwegian Research Council and by a grant from the Danish Research Foundation. The computations were performed at the National Center for Supercomputer Applications and with support from the Norwegian Research Council, Tungregneprogrammet and the Danish Natural Science Research Council.

References

Carlsson, M., & Stein, R. F. 1995, ApJ, 440, L29.

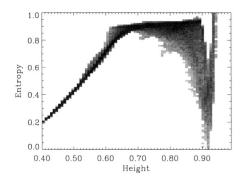

Figure 6. Histogram of entropy fluctuations at each depth. The minimum entropy, corresponding to the coolest downdrafts, increases toward the bottom of the convection zone. The entropy contrast decreases with depth because ascending, entropy-neutral, fluid overturns and mixes with the entropy-deficient descending fluid. The median entropy (the darkest band) decreases slightly with depth in the bulk of the convection zone, because radiation heats the fluid.

Dorfi, E. A., & Drury, L. O. 1987, J. Comput. Phys., 69, 175.
Fontenala, J. M., Avrett, E. H., & Loeser, R. 1993, ApJ, 406, 319.
Hyman, J. 1979, Adv. in Comput. Meth. for PDE's – III, eds. R. Vichnevetsky & R. S. Stepleman, 313.
Lele, S. K. 1992, J. Comput. Phys., 103, 16.
Nordlund, Å. 1982, A&A, 107, 1.
Nordlund, Å., & Stein, R. F. 1990, Comput. Phys. Comm., 59, 119.
Nordlund, Å., & Stein, R. F. 1995, in *Stellar Evolution: What Should be Done?*, 32nd Liege Int. Astroph. Coll., eds. A. Noels, D. Fraipont-Caro, M. Gabriel, N. Grevesse, & P. Demarque, 32, 75.
Scharmer, G. B., & Carlsson, M. 1985, J. Comput. Phys., 59, 56.
Stein, R. F., Galsgaard, K., & Nordlund, Å. 1993, in *Proceedings of the Cornelius Lanczos International Centenary Conference*, eds. J. D. Brown, M. T. Chu, D. C. Ellison & R. J. Plemmons, (Philadelphia: SIAM), 440.
van Leer 1977, J. Comput. Phys., 23, 276.

Angular Momentum Transport in Simulations of Accretion Disks

James Rhys Murray

CITA, University of Toronto, 60 St. George St., Toronto, ON M5S 3H8, Canada

Abstract. In this paper, we briefly discuss the ways in which angular momentum transport is included in simulations of non-self-gravitating accretion disks, concentrating on disks in close binaries. Numerical approaches fall in two basic categories; particle-based Lagrangian schemes, and grid-based Eulerian techniques. Underlying the choice of numerical technique are assumptions that are made about disk physics, and in particular, about the angular momentum transport mechanism. Grid-based simulations have generally been of hot, relatively inviscid disks whereas particle-based simulations are more commonly of cool, viscous disks. Calculations of the latter type have been instrumental in developing a model for the superhump phenomenon. We describe how we use an artificial viscosity term to introduce angular momentum transport into our smoothed particle hydrodynamics (SPH) disk code.

1. Introduction

The principal difficulty in modelling accretion disks stems from our ignorance of the mechanism by which angular momentum is transported through the disk. This process (or processes) must transfer angular momentum from the inner disk to the outer disk, and also convert kinetic energy to thermal energy in order to allow a mass flux inwards through the disk. Our ignorance is usually reduced to a single shear viscosity term in the Navier-Stokes equation, where the kinematic shear viscosity is given by:

$$\nu = \alpha c H. \qquad (1)$$

In writing equation (1), Shakura & Sunyaev (1973) assumed the unknown mechanism to be turbulent viscosity, in which case the scale length of the largest turbulent eddies had to be less than the pressure scale height of the disk H, and the eddy velocity had to be less than the sound speed c. Thus the dimensionless parameter α that measures the efficiency of the angular momentum transport is expected to be less than unity. The accretion disks in outbursting dwarf novae for example, are observed to have $\alpha \simeq 0.2$. The corresponding shear viscosity is several orders of magnitude too large to be molecular.

A requirement of our simulations then, is that they reproduce observed rates of angular momentum transport. But does the transport occur via disk scale structures such as non-linear density waves that need to be resolved in the simulations, or via very small scale, sub-grid, phenomena (*e.g.*, turbulence) that we cannot explicitly include in the simulations but whose effects must be allowed for in the equations of motion? In the past, authors investigating the former possibility have used high-order, "inviscid", grid-based hydrodynamics algorithms, whilst the latter has been investigated with "artificially-viscous" particle schemes.

Sawada et al. (1986) simulated an accretion disk in a close binary system using the second-order Osher upwind scheme, a grid-based technique that does not require an artificial viscosity to resolve shocks. In a two-dimensional calculation set on the equatorial plane of the binary, they found that the secondary star launched spiral density waves at the outer edge of the disk. These waves propagate inwards, carrying a deficit of angular momentum which they transmit to the general flow when they damp. As a result, there is a net outwards flux of angular momentum and a mass flux inwards. A self-similar analytic solution by Spruit (1987) showed that a disk in which spiral shocks provided the only mechanism for angular momentum transport would have an effective $\alpha \lesssim 0.01$. Calculations by Savonije et al. (1994) found that spiral shocks were most efficient for hot disks with Mach numbers $\mathcal{M} = v_\phi/c \leq 10$, where v_ϕ is the azimuthal velocity of the gas. Disks in cataclysmic variables have Mach numbers closer to 30. Spiral shocks then are not the most important means for angular momentum transport in these objects.

From the above discussion we conclude that there is another mechanism besides spiral shocks at work. At this stage the most likely candidate is a weak-field magnetohydrodynamic shear instability discussed by Chandrasekhar (1960), but applied to the accretion disk problem by Balbus & Hawley (1991). Three-dimensional simulations of small regions of the disk (in the shearing sheet approximation) by Stone et al. (1996) showed that magnetohydrodynamic turbulence generated by the instability transported angular momentum with an estimated efficiency $\alpha \lesssim 0.01$, similar to the maximum efficiency of spiral shocks. This is disturbing as it is an order of magnitude less than observational estimates of α, which are obtained by fitting axisymmetric viscous disk solutions to the decay in luminosity of dwarf novae after outburst (Cannizzo, 1993). Dwarf novae in quiescence though, are thought to have an effective α more in line with the Stone et al. (1996) result (although in this case, the observational results are less well constrained).

MHD turbulence is obviously a phenomenon that we cannot hope to capture in a global disk simulation. However, we can simply introduce a shear viscosity term into our numerical scheme to simulate the action of the weak field instability. Stone et al. (1996) suggest that the shear stress caused by the instability is consistent with the Shakura & Sunyaev parameterisation [equation (1)].

2. Particle Methods and the Explanation for Superhump

Lin & Pringle (1976) developed a particle-based scheme that included an artificial method for "viscously" dissipating energy, in order to study the evolution of disks in close binaries. In their "sticky particle method", gas pressure is neglected and particles move as test particles in the restricted three-body problem. Then, at the end of each time step, a grid is placed over the computational domain. Particle velocities are adjusted in order to minimise the kinetic energy in

each cell, whilst conserving linear and angular momentum. Lucy (1977) cited
the sticky particle method as being a progenitor of smoothed particle hydrodynamics. Having neglected pressure forces, Lin & Pringle (1976) excluded the
possibility of angular momentum transport via density waves. They simulated
the evolution of accretion disks under steady mass transfer from the companion
star for three different binary mass ratios, and in each obtained a steady-state
disk that was comparable in size to the accreting star's Roche lobe.

Whitehurst (1988a; 1988b) developed a completely Lagrangian scheme that
included pressure forces, and subsequently found that in close binaries with extreme mass ratios, the secondary star is able to excite a large eccentricity in
the disk. This was particularly exciting as some extreme mass ratio cataclysmic
variables exhibited a periodic luminosity variation known as "superhump" that
was thought to be the signature of an eccentric disk. The superhump period was
always a few percent larger than the orbital period. In Whitehurst's simulation,
the eccentric disk (as seen in the inertial frame) exhibited a slow prograde precession. In his model, the superhump signal was caused by the periodic stressing
of the eccentric disk by the tidal field, and the superhump period was the beat
period between the disk precession period and the binary period.

Lubow (1991) showed analytically that eccentricity growth occurred when
the disk was large enough to encompass an eccentric inner Lindblad resonance.
This could only happen when the mass ratio $q = M_s/M_p \lesssim 0.25$, where M_p is
the mass of the accreting star and M_s is the mass of the companion.

Whitehurst's result was verified and expanded upon by Hirose & Osaki
(1990) using the original sticky particle method, and later by Murray (1996) using SPH. The three particle techniques introduced viscous dissipation in different
ways, and the resulting kinematic viscosities had different forms. In the sticky
particle and SPH simulations, ν was constant, whereas Whitehurst's default
choice was $\nu \propto r^{-3/2}$ (Whitehurst also ran a simulation with $\nu \propto r^{1/2}$). Thus,
the detailed functional dependence of the viscous dissipation has not proven to
be critical. What is important, and common to the three sets of simulations, is
a large ν at large radii which allowed the disk to spread out to the 3 : 1 resonance. Heemskerk (1994) studied the phenomenon using an Eulerian, grid-based
scheme for solving the inviscid equations of hydrodynamics. Heemskerk (1994)
performed some simulations using only the $m = 3$ component of the tidal field
(that being the term responsible for the Lindblad resonance), and found that
the disk became eccentric. However, when he used the full tidal potential, the
accretion disk was unable to maintain contact with the resonance, even though
they had initially overlapped, and significant eccentricity growth did not occur.

3. Artificial Viscosity in our SPH Algorithm

Smoothed particle hydrodynamics is a completely Lagrangian numerical scheme,
which uses an ensemble of particles to model a fluid (see Monaghan, 1992). The
fluid equations are replaced by a set of equations for the evolution of the particles.
For example, the fluid momentum equation becomes, in the SPH scheme, a set
of equations for the forces on each particle with spatial derivatives estimated by
interpolating quantities from neighbouring particles. The "standard" SPH form

of the momentum equation for particle a, neglecting gravitational forces, is

$$\frac{d\mathbf{v}_a}{dt} = -\sum_b m_b \left(\frac{P_a}{\rho_a^2} + \frac{P_b}{\rho_b^2} + \frac{\beta \mu_{ab}^2 - \zeta \bar{c}_{ab} \mu_{ab}}{\bar{\rho}_{ab}} \right) \nabla_a W(\mathbf{r}_a - \mathbf{r}_b, h). \quad (2)$$

Here $\mathbf{r}_a$, $\mathbf{v}_a$ and m_a are the position, velocity and mass of particle a. P_a, ρ_a and c_a are the pressure, density and sound speed of the fluid evaluated at $\mathbf{r}_a$. The density ρ is obtained by interpolation, and then P and c are obtained using the chosen equation of state. The $\bar{X}_{ab}$ notation indicates the arithmetic mean of quantity X evaluated at $\mathbf{r}_a$ and $\mathbf{r}_b$. ζ and β are the coefficients of the linear and non-linear artificial viscosity terms. $W(\mathbf{r}, h)$ is the interpolation kernel and μ_{ab} is defined as

$$\mu_{ab} = \frac{(\mathbf{v}_a - \mathbf{v}_b) \cdot (\mathbf{r}_a - \mathbf{r}_b)}{(\mathbf{r}_a - \mathbf{r}_b)^2 + \eta^2}, \quad (3)$$

with η being a softening parameter usually set equal to one-tenth the smoothing length h. Artificial viscosity was originally added to the SPH equations in order to improve the resolution of shocks, and is usually used with a switch that sets it to zero when $(\mathbf{v}_a - \mathbf{v}_b) \cdot (\mathbf{r}_a - \mathbf{r}_b) > 0$. This ensures that dissipation only occurs for compressive flows. For disk simulations however, we want viscous dissipation to occur wherever there is velocity shear, so we use only the linear viscosity term with the switch disabled. By letting the number of particles $n \to \infty$, the smoothing length $h \to 0$, and approximating the summation with an integral, we can obtain the continuum equivalent of the linear artificial viscosity term. Following Pongracic (1988) and Meglicki et al. (1993), we find that the "viscous" force per unit mass is

$$\mathbf{a}_{\text{visc}} = \frac{\zeta h \kappa}{2\rho} \left[\nabla \cdot (c\rho \mathbf{S}) + \nabla(c\rho \nabla \cdot \mathbf{v}) \right], \quad (4)$$

where the deformation tensor is

$$S_{ab} = \frac{\partial v_a}{\partial x_b} + \frac{\partial v_b}{\partial x_a}. \quad (5)$$

Here κ, a constant dependent only upon the smoothing kernel used, is $1/4$ for the standard cubic spline kernel (in two dimensions). If we assume the density and sound speed vary on much longer length-scales than the velocity, then we have

$$\mathbf{a}_v = \frac{\zeta hc}{8} \left[\nabla^2 \mathbf{v} + 2\nabla(\nabla \cdot \mathbf{v}) \right]. \quad (6)$$

In other words, the linear artificial viscosity term generates both shear and bulk viscosity *in a fixed ratio*. In the interior of the disk where $\nabla \cdot \mathbf{v} \simeq 0$, we simply have a kinematic viscosity

$$\nu = \frac{1}{8} \zeta hc. \quad (7)$$

In order to check the accuracy of equation (7), we simulated the viscous spreading of an axisymmetric disk with a Gaussian initial surface density profile

$$\Sigma(r, t=0) = \Sigma_0 \, e^{\frac{(r-r_0)^2}{l^2}}. \quad (8)$$

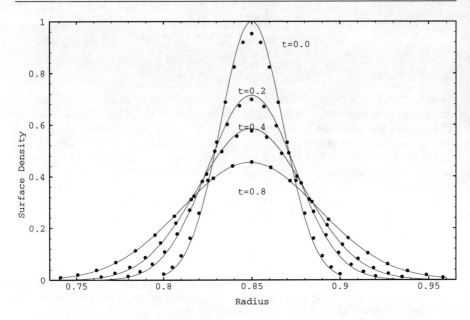

Figure 1. Evolution of an axisymmetric ring with a Gaussian initial density profile. The solid lines denote the analytic solution at the times shown, and the heavy points show the SPH solution. In this calculation we use $r_0 = 0.85$, $l = 0.025$, $\Sigma_0 = 1.0$ and $\nu = 2.5 \times 10^{-4}$ [see equation (8)]. Units have been scaled so $\mathcal{G}M = 1$, where M is the mass of the central star. We used a fixed smoothing length $h = 0.01$.

For this (two-dimensional) calculation, we laid down 22,400 particles in 21 equally spaced concentric rings, varying the particle mass to generate the Gaussian density profile. Gas pressure forces were switched off, and a constant smoothing length was used. We find that the actual shear viscosity generated by the linear artificial viscosity term was within ±10% of the value given by equation (7). Figure 1 shows the close agreement between the analytic solution for a Gaussian ring with viscosity given by equation (7), and our SPH simulation. The test is described in more detail in Maddison et al. (1996) along with a study of the stability of a narrow ring of viscously interacting particles.

It is possible to derive SPH interpolant estimates of $\nabla^2 \mathbf{v}$, and thus to include shear viscosity directly in the SPH equations. Flebbe et al. (1994) and Watkins et al. (1996) introduced formulations and tested them with ring spreading calculations similar to the one above, obtaining reasonable agreement between predicted and actual values for ν. However, their results are clearly more noisy than those shown in Figure 1. Both Flebbe et al. (1994) and Watkins et al. (1996) must interpolate twice in order to obtain $\nabla^2 \mathbf{v}$ which may increase the susceptibility of the calculations to the single ring instability discussed in Maddison et al. (1996). Most importantly, in the Flebbe et al. (1994) and Watkins et al. (1996) formulations, the forces between two particles are not anti-symmetric and along the particles' line of centres. Therefore, they do not exactly conserve linear and angular momentum. With our term, linear and angular momentum are conserved to machine accuracy.

4. Conclusions

We have described superhump simulations completed with a low-viscosity code that gives qualitatively different results from more viscous calculations. Clearly, we must quantify the effective shear viscosity of accretion disk simulations if we are to make meaningful comparisons with observations. The ring-spreading calculation is a simple way to do this. There are several ways to introduce a known shear viscosity into SPH disk calculations. We described in detail a method which conserves linear and angular momentum. Momentum conservation is important in disk calculations that may run for thousands of dynamical time scales. We showed that the effective ν introduced by our term could be accurately estimated with equation (7).

References

Balbus, S. A., & Hawley, J. F. 1991, ApJ, 376, 214.
Cannizzo, J. K. 1993, ApJ, 419, 318.
Chandrasekhar, S. 1960, Proc. Nat. Acad. Sci., 46, 53.
Flebbe, O., Münzel, S., Herold, H., Riffert, H., & Ruder, H. 1994, ApJ, 431, 754.
Heemskerk, M. H. M. 1994, A&A, 288, 807.
Hirose, M., & Osaki, Y. 1990, PASJ, 42, 135.
Lin, D. N. C., & Pringle, J. E. 1976, in *Structure and Evolution of Close Binary Systems*, ed. P. Eggleton (Dordrecht: Reidel), 237.
Lubow, S. H. 1991, ApJ, 381, 259.
Lucy, L. 1977, AJ, 82, 1013.
Maddison, S., Murray, J., & Monaghan, J. J. 1996, Publ. Astron. Soc. Australia, 13, 66.
Meglicki, Z., Wickramasinghe, D., & Bicknell, G. V. 1993, MNRAS, 264, 691.
Monaghan, J. J. 1992, ARA&A, 30, 543.
Murray, J. 1996, MNRAS, 279, 402.
Pongracic, H. 1988, Ph.D. thesis, (Monash University).
Savonije, G. J., Papaloizou, J. C. B., & Lin, D. N. C. 1994, MNRAS, 268, 13.
Sawada, K., Matsuda, T., & Hachisu, I. 1986, MNRAS, 219, 75.
Shakura, N. I., & Sunyaev, R. A. 1973, A&A, 24, 337.
Spruit, H 1987, A&A, 184, 173.
Stone, J. M., Hawley, J. F., Gammie, C. F., & Balbus, S. A. 1996, ApJ, 463, 656.
Watkins, S. J., Bhattal, A. S., Francis, N., Turner, J. A., & Whitworth, A. P. 1996, A&AS, 119, 177.
Whitehurst, R. 1988a, MNRAS, 232, 35.
Whitehurst, R. 1988b, MNRAS, 233, 529.

On Radiative Transfer in SPH

Alexei Razoumov

Department of Physics and Astronomy, University of British Columbia, Vancouver, BC V6T 1Z4, Canada

1. Introduction

In recent years, smoothed particle hydrodynamics (SPH) has been extensively used for solving the equations of fluid flow in a variety of astrophysical problems. There have been several attempts to extend the gridless nature of SPH to nonhydrodynamical equations (*e.g.*, Laguna, 1995). The aim of the present study is to explore the possibility of including radiative transfer into particle hydrodynamics.

The major drawback of smoothed interpolation is its inability to conserve variables properly, along with its weak approximation to discontinuities. This leads to poor performance in the propagation of steep features, compared to modern finite-difference methods (*e.g.*, Balsara, 1995). However, the robustness and complete Lagrangian nature of particle-based schemes make them very attractive for astrophysical applications. Besides that, the interaction of radiation with matter usually occurs in dense regions, which are naturally populated with particles.

Time-dependent radiative transfer has recently been included in a number of astrophysical hydrodynamical codes. The most sophisticated approach to the problem as yet is the grid-based ZEUS-2D solver of two-dimensional magnetic RHD conservation equations (Stone *et al.*, 1992). At the same time, there have been very few attempts to put radiative transport into SPH. Lucy (1977) modeled the diffusion of radiation in stars finding that the calculated flux diverged from the exact solution by several tens of percent. Brookshaw (1985), being concerned with the accuracy in the diffusion limit, proposed another formulation for second-order spatial derivatives and modified the SPH diffusion equation to include free-radiating boundaries. In all of these problems, it is assumed that radiative transport is described by the diffusion equation. The difficulties arise when we try to model radiative transfer in SPH with a variable optical depth.

2. Treatment of Radiative Transport

A comprehensive review of the equations of radiative transport in comoving frames can be found in Mihalas & Mihalas (1984) or in a series of ZEUS-2D papers (Stone *et al.*, 1992, and references therein). The transfer problem can be written as a system of two moment equations plus a closure scheme relating the second moment to the first. The last operation is often performed by *ad hoc* flux limiting which assumes an analytical dependence between the moments. More sophisticated codes are based on the iterative reconstruction of intensity as a function of direction at each point in space at a fixed time step until the scheme converges.

Rather than solving a three-dimensional problem, we confine our discussion to the one-dimensional transfer problem in a static medium. For our test purposes, we use the combined moment equation (Mihalas & Mihalas, 1984)

$$\frac{1}{\rho}\frac{\partial E}{\partial t} = \epsilon_h + \frac{1}{\rho}\frac{\partial}{\partial x}\left[c\lambda\frac{\partial}{\partial x}(fE)\right] - \frac{1}{\rho}\frac{\partial}{\partial x}\left(\frac{\lambda}{\lambda_t}F^n\right), \qquad (1)$$

where E is the radiation energy density and F^n is the flux at the beginning of the nth time step. The radiative energy source and sink terms are given by ϵ_h, and $\lambda^{-1} \equiv \lambda_t^{-1} + \lambda_p^{-1}$ is defined by the mean free path of a photon, λ_p, and its free travel time, $\lambda_t \equiv c\Delta t$, during a time step Δt. The flux at a new time step is given by

$$F^{n+1} = -c\lambda^{n+1}\frac{\partial}{\partial x}\left(f^{n+1}E^{n+1}\right) + \frac{\lambda}{\lambda_t}F^n. \qquad (2)$$

To simplify the problem even further, we consider pure scattering so that ϵ_h in equation (1) gives the net non-radiative heating. Since the properties of the scattering medium do not change in time, the problem will be linear in the diffusion and free-streaming modes. In optically-intermediate regions, the equations become non-linear because of the dependence of the transport mechanism on local properties.

3. Tests in the Diffusion Limit and for Free Streaming

With reflective boundary conditions in the optically-thick regime, the system should quickly come to equilibrium, with the radiation energy density at each point determined by the total energy of the initial configuration. For all tests with 100 interpolation points, the relative error is smaller than 10^{-6} after about ~ 30 diffusion times t_d, an accuracy which is normally achieved only with finite difference schemes (Stone et al., 1992). When we relax the particle distribution to a completely random configuration, the relative error after $30 \times t_d$ becomes as large as 10^{-3}.

In optically-thin layers, the equation of transfer becomes an advection equation with first-order spatial differencing. In this case, we solve the Eulerian equation since particles represent the fluid, while in standard SPH we deal with advection of particles. The problem is equivalent to the Eulerian wave solver in Laguna (1995). We find the solution is always unstable with oscillations first appearing behind the wave front. These oscillations become smaller but do not disappear with fully-implicit time differencing and tend to weaken when larger numbers of interpolation points are used. The growing error is mostly caused by the second-order spatial accuracy of the smoothing error. The runs with higher-order kernels (*e.g.*, a super-Gaussian) produce somewhat better results. In practice, this instability is controlled by the change of the mean intensity within the smoothing length and breaks the solution on discontinuities if one does not introduce artificial dissipation. The next alternative is to construct kernels with smaller discretization errors or to adjust the number of close neighbors to account for nonuniformity of the point set.

4. Temporal Adaptability for Radiative Transport

To follow the evolution of a real system with radiating flows, we should be able to evolve our equations on a hydrodynamical time scale (Mihalas & Mihalas, 1984). The argument for this is that in the optically-thin regime, the radiation field instantaneously adjusts to slow changes in the density distribution. Thus, at any given time step (determined by the hydrodynamical time scale), we should solve time-independent equations for radiative transfer. In static diffusion, radiation is essentially frozen into material and the mixture evolves on a fluid flow time scale again. For opaque and fast but still non-relativistic flows ($v/c \geq \lambda_p/l$), the hydrodynamical time scale becomes comparable to radiation time scales and so the equations should be solved on a fluid time scale.

The tests described above were calculated on much smaller radiative time scales. Choice of the global time step is determined by the type of the problem being investigated. For instance, if we are interested in diffusion of radiation, we have to find the solution on a diffusion time scale everywhere including optically-thin regions in between dense clumps. One way to impose the same time step is to modify our equations in such a way that the time derivatives of the radiation field disappear when we do not need them. We saw that the accuracy of the solution in the streaming (and intermediate) regime on a radiation time scale is not comparable to the solution of the transport problem with modern grid methods. For instance, we cannot propagate a square wave in a similar way to total variation diminishing schemes (Stone & Mihalas, 1992). Solution on a bigger time scale (either diffusion or hydrodynamical) at once could help to avoid these instabilities. In between dense clumps, one then solves time-independent equations. Therefore, the information relating radiating surfaces of optically dense regions propagates immediately. The major difficulty is to come up with a formulation which preserves radiation variables accurately near the surfaces where the type of equations changes.

In attempts to increase spatial accuracy, one must pay special attention to the nature of the leading error. While it is possible to construct kernel functions which give both smaller smoothing and discretization errors, the change in the total resolution may need to be accompanied by proper adjustment in the number of close neighbors if the discretization error prevails. In conclusion, both particle and grid methods have their own problems. It is possible that SPH will give better performance than grids in connecting the radiating surfaces.

References

Balsara, D. 1995, J. Comput. Phys., 121, 357.
Brookshaw, L. 1985, Astron. Soc. of Australia, Proc., 6, 207.
Laguna, P. 1995, ApJ, 439, 814.
Lucy, L. 1977, AJ, 82, 1013.
Mihalas, D., & Mihalas, B. 1984, *Foundations of Radiation Hydrodynamics* (New York: Oxford University Press).
Stone, J. M., & Mihalas, D. 1992, J. Comput. Phys., 100, 402.
Stone, J. M., Mihalas, D., & Norman, M. L. 1992, ApJS, 80, 819.

Semi-Empirical Wind Models for λ Velorum (K4 Ib–II)

Philip D. Bennett & Graham M. Harper[1]

Center for Astrophysics and Space Astronomy, University of Colorado, Boulder, CO 80309-0389, U.S.A.

Abstract. We present techniques used to construct semi-empirical wind models of evolved late-type stars constrained by HST/GHRS spectra of ultraviolet Fe II emission lines. High quality UV spectroscopic observations are now available for selected K and M (super)giants. We focus on the computational challenges faced in modeling the wind of the K supergiant λ Velorum (K4 Ib–II), using GHRS spectra of lines of Fe II, Mg II h and k, and C II] 2325 Å. Fe II line photons are scattered throughout the wind acceleration region; these lines represent the best available diagnostics for the construction of semi-empirical wind models. However, their modeling requires solving the combined radiative transfer and level population problem simultaneously in a moving spherical wind, and represents a state-of-the-art computational problem because of the complexity of the Fe II ion. For Fe II, the high degree of line interlocking requires that a minimum of 10^2 energy levels and 10^4 transitions be included in a realistic calculation. This work represents the first fully self-consistent treatment of the Fe II multi-level problem in a cool stellar wind.

1. Introduction

The winds of evolved, late-type stars display two very different morphologies depending upon location in the H-R diagram. Stars with spectral types earlier than about K2 generally have hot ($T_e \sim 10^6$ K) solar-like coronal winds, while later spectral-types possess cool ($T_e \leq 2 \times 10^4$ K), low velocity (≤ 200 km s^{-1}) winds. The nature and mass-loss rates of these non-coronal winds remain poorly determined, and detailed empirical descriptions of the winds are lacking. There is no shortage of theoretical models: acoustic waves, radiation pressure on molecules and dust grains, and Alfvén waves have all been proposed as driving mechanisms, but there are substantial difficulties with all existing theoretical descriptions. The purpose of the present investigation is to obtain a reliable empirical model of a non-coronal stellar wind, including the velocity law and temperature and density structure, using high quality UV spectroscopic observations obtained with the Hubble Space Telescope. These empirical models will tightly constrain future theoretical investigations.

---

[1] Based on observations made with the NASA/ESA Hubble Space Telescope, obtained at the Space Telescope Science Institute, which is operated by the Association of Universities for Research in Astronomy, Inc., under NASA contract NAS 5-26555.

We are analyzing the ultraviolet spectrum of the non-coronal supergiant λ Vel, the brightest K supergiant in the sky, obtained with the Goddard High Resolution Spectrograph (GHRS) on board the Hubble Space Telescope. A very extensive set of GHRS spectral data exists for this star, obtained originally by R. D. Robinson (HST GO-5307). The principal diagnostics for analyzing the wind acceleration region consist of the rich UV emission line spectrum of Fe II, along with the Mg II h and k, S I UV1, C II] 2325 Å and Al II] 2669 (Judge & Jordan, 1991). The formation depths of these diagnostics span the range from the lower chromosphere for optically-thin lines such as C II], to the regime of the terminal velocity wind for the very opaque Mg II resonance lines, which have line center optical depths $\tau \sim 10^5$. The shape of the emergent line profiles for these diagnostics depends critically upon the local velocity field and gradient in the line-forming region. Therefore, it is possible to use the information in the emergent Fe II line profiles to reconstruct the velocity, density and electron temperature in the line-forming region. Our goal is to map the velocity law, density and temperature structure throughout the wind acceleration region by using lines with a suitable range of optical depths. The numerous Fe II lines, plus Mg II h and k, and C II] 2325 Å, available in the HST/GHRS archival spectra of λ Vel span line center optical depths from $\tau \sim 10^5$ down to $\tau < 1$, permitting the wind to be mapped from the lower chromosphere out to $\sim 1,000$ stellar radii. The variation in line profile appearance with optical depth is shown in Figure 1 (after Judge & Jordan, 1991).

Although the Fe II (and Mg II) wind lines sample large optical depths, these lines remain effectively thin for late-type giants and supergiants (see Judge, 1990), and the radiative transfer problem lies close to the pure scattering limit. The richness of the Fe II spectrum is caused by the presence of a large number (63) of low-lying even parity levels. Electric dipole radiative transitions are strictly forbidden among these levels and thus radiative decays of excited levels can occur only via much weaker magnetic dipole or electric quadrupole radiation. Therefore, at typical temperatures ($T_e \sim 7,000$ K) and densities ($n_e \sim 10^6$ cm^{-3}) of stellar winds, collisional rates dominate over radiative decays, and *an approximately Boltzmann Fe II population is maintained*. This behavior allows both the density and electron temperature of the gas to be estimated. However, the presence of the large number of populated levels greatly complicates the radiative transfer problem because of *line interlocking*. Photons no longer resonantly scatter in one strong line (as in the two-level atom approximation) but instead are redistributed among several transitions sharing a common upper level. To model individual line profiles, it is necessary to find the self-consistent solution of the radiative transfer equation for each of the radiative transitions and the atomic statistical equilibrium (rate) equations. The complexity of the Fe II energy level structure leads to the demanding computational problem that we seek to address.

2. The Spherical Radiative Transfer Problem

In order to recover chromosphere and wind models from the emergent line profiles, we must make some assumptions about the geometry. The simplest reasonable geometry is that of a steady, spherically-symmetric wind; we adopt this

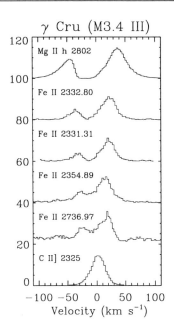 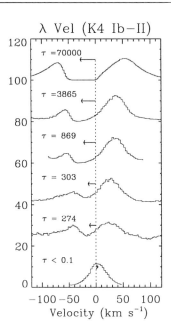

Figure 1. The variation in observed wind line profiles with optical depth for γ Cru and λ Vel as observed with the HST/GHRS. An arrow marks the apparent line center, which is blue-shifted with increasing optical thickness as the formation depth moves further out in the accelerating wind.

model for our analysis. The equation of radiative transfer in the fluid or comoving frame and neglecting radiative advection and aberration terms of order v/c is given by (Mihalas, 1978; Baron et al., 1996)

$$\mu \frac{\partial I_\nu(r,\mu)}{\partial r} + \frac{1-\mu^2}{r}\frac{\partial I_\nu(r,\mu)}{\partial \mu} - \left(\frac{\nu v}{cr}\right)\left[1 - \mu^2 + \mu^2 \left(\frac{d\ln v}{d\ln r}\right)\right]\frac{\partial I_\nu(r,\mu)}{\partial \nu}$$

$$= \eta_\nu(r) - \chi_\nu(r)I_\nu(r,\mu), \quad (1)$$

where I_ν is the specific intensity of the radiation field, $\mu = \cos\theta$, θ is the radiation propagation angle relative to the radial direction, ν denotes the frequency of the radiation, v is the (radial) flow velocity, χ_ν is the opacity and η_ν the emissivity of the gas. Thus, the spherical transfer problem alone involves the solution of a 3-D $(N_r \times N_\mu \times N_\nu)$ hyperbolic partial differential equation. Since each grid size is of the order of 10^2, the total 3-D problem requires $\sim 10^6$ grid points to obtain solutions of $I_\nu(r,\mu)$ for each radiative transition. Assuming we need $\sim 10^2$ operations per grid point, and allowing an extra factor of 10 per grid point for various overheads (computation of opacities, etc.), then the number of floating point operations, to solve the radiative transfer problem for a single line is of order 10^9. The minimum requirement to be able to study the Fe II lines in the GHRS spectra of λ Vel is an atomic model containing $\sim 10^2$ fine-structure levels, which results in about 10^4 spectral lines. This implies a computational effort of $\sim 10^{13}$ operations. This estimate is for the direct (formal) solution of the transfer equation, where the opacity and emissivity are assumed known. For multi-level non-LTE problems, this is not the case *a priori* because η_ν and χ_ν depend upon the non-LTE level populations.

3. The Fe II Multilevel Statistical Equilibrium

For a given radiation field, the non-LTE Fe II level populations are given by the simultaneous solution of statistical equilibrium (rate) equations,

$$n_i \sum_{j \neq i}^{n_l} P_{ij} - \sum_{j \neq i}^{n_l} n_j P_{ji} = \nabla \cdot (n_i \vec{v}), \qquad (2)$$

where n_i is the population of level i and $\vec{v}$ is the flow velocity. For a spherical geometry, the right hand side of equation (2) becomes $r^{-2} d(n_i r^2 v)/dr$. This equation expresses that the total rates P in and out of level i must be zero for a time-independent problem. Here, P_{ij} gives the total rate (*i.e.*, probability per unit time) that an atom in level i will undergo a transition to a different level j. The total rate can be expressed as

$$P_{ij} = R_{ij} + C_{ij}, \qquad (3)$$

where R_{ij} is the radiative rate from $i \to j$ while C_{ij} denotes the collision rate (usually dominated by electrons) for this transition. Although C_{ij} depends only upon the local electron density n_e and temperature T_e and is thus locally determined, R_{ij} depends upon the radiation field and introduces a nonlocal (and nonlinear) coupling into the rate equations. A very useful approximation for R_{ij} is obtained by assuming that the frequency of the photon emitted after a radiative transition is not correlated with the frequency of the photon absorbed during the transition. This assumption of *complete frequency redistribution* (CRD) leads to the expressions

$$R_{ij} = \begin{cases} A_{ij} + B_{ij} \bar{J}_{ij} & \text{if } i > j; \\ B_{ij} \bar{J}_{ij} & \text{if } i < j, \end{cases} \qquad (4)$$

where $\bar{J}_{ij}$ is I_ν integrated first over propagation angle, and then averaged in frequency over the line profile. Thus, in CRD, the interaction of the radiation field in a single spectral line with the gas at a single point in space is completely described by $\bar{J}$. It is important to note that $\bar{J}$ depends only upon position.

To solve the Fe II atomic model, we must simultaneously solve the transfer and statistical equilibrium equations [equations (1) and (2), hereafter, the RT+SE problem] for the solution $\{I_\nu(r,\mu), \vec{n}(r)\}$. The solution of the RT+SE problem cannot be solved directly, but must instead be iterated to convergence. Traditionally, solutions have been obtained using linearization methods, but in recent years the efficient and robust approximate lambda operator (ALO) methods have become increasingly popular (*cf.*, Scharmer & Carlsson, 1985; Olson *et al.*, 1986; Rybicki & Hummer, 1991; Hauschildt, 1993). Experience shows that typically $\sim 10^2$ iterations are required for convergence so that the total computational effort to solve a minimal 100-level Fe II RT+SE problem is of order 10^{15} operations.

4. The J-MULTI Project

To make the solution of the Fe II problem more tractable, we are developing a modified version, J-MULTI, of the spherical multi-level code S-MULTI (Harper, 1994), itself an extension of Carlsson's (1985) MULTI code. The key difference is that for most of the iterative procedure, J-MULTI solves only for the angle-averaged moments of I_ν, unlike (S-)MULTI which computes the entire angle-dependent intensity at each iteration. The radiation moments are

$$\{J_\nu(r), H_\nu(r), K_\nu(r), N_\nu(r) \cdots\} = \tfrac{1}{2} \int_{-1}^{1} I_\nu(r,\mu)\{1, \mu, \mu^2, \mu^3 \cdots\} d\mu.$$

Of these, only J_ν is needed to compute the radiative rates. The moment approach avoids wasting computational effort computing the full angle-dependent radiation intensity at every iteration when the radiative rates (and therefore level populations) depend only upon the angle-averaged mean intensity.

The moment solution is achieved by simultaneously solving the set of equations obtained by forming moments of equation (1) over μ. However, this procedure yields an infinite hierarchy of moment equations, and each finite subset of equations always possesses one more moment variable than there are equations. This is the famous closure problem, and to solve the set of equations requires adopting additional relations between the moments. Radiative transfer provides just such a relation in the limit of large optical depth where the diffusion limit holds (this forms the inner boundary condition for the wind problem). Here, the Eddington factor $f_\nu = K_\nu/J_\nu = 1/3$. Another limit is reached in the optically-thin wind far from the star where the solid angle subtended by the stellar disk $\Omega_* \ll 4\pi$. In this regime, $J_\nu = K_\nu$ and $f_\nu = 1$. Thus, f_ν varies by only a factor of 3 over the entire range of the wind problem, suggesting that the Eddington factor is insensitive to the model solution. Numerical studies confirm this behavior, and demonstrate that f_ν almost always lies between $1/3$ and 1.

Auer & Mihalas (1970) utilized the behavior of f_ν to develop a practical solution relying on an approximate closure of the moment equations that was updated with successive iterations. The final solution is exact as the f_ν's are also iterated to convergence. We have adapted this Eddington factor scheme for use in the spherical J-MULTI code. More details of the method applied to spherical geometry can be found in Mihalas & Hummer (1974). For the static problem, retaining the first two moment equations gives a system with three moment variables present, which is conveniently closed using the approximate value of f_ν. In the case of the spherical, moving atmosphere, *four* moments appear in the first two equations. Closure of this problem thus requires the use of two Eddington factors, f_ν as before, and $g_\nu = N_\nu/H_\nu$, which is defined analogously to f_ν and which is similarly well-behaved. With these definitions, the spherical wind moment equations become (Mihalas et al., 1976)

$$q_\nu \frac{\partial (r^2 H_\nu)}{\partial X_\nu} + \frac{\nu v(r)}{cr\chi_\nu} \left[\frac{\partial (1-f_\nu) r^2 J_\nu}{\partial \nu} + \frac{d \ln v}{d \ln r} \frac{\partial (f_\nu r^2 J_\nu)}{\partial \nu} \right] = r^2 (J_\nu - S),$$

$$\frac{\partial (f_\nu q_\nu r^2 J_\nu)}{\partial X_\nu} + \frac{\nu v(r)}{cr\chi_\nu} \left[\frac{\partial (1-g_\nu) r^2 H_\nu}{\partial \nu} + \frac{d \ln v}{d \ln r} \frac{\partial (g_\nu r^2 H_\nu)}{\partial \nu} \right] = r^2 H_\nu,$$

where $dX_\nu = -\chi_\nu q_\nu dr$.

Auer's (1971) sphericality function $q_\nu(r)$ is introduced to remove analytically terms in $1/r$ which are numerically ill-conditioned, and is defined by

$$\frac{\partial \ln(r^2 q_\nu)}{\partial r} = \frac{3f_\nu - 1}{f_\nu r}.$$

Note that the Eddington factor approach still requires a solution of the full angle-dependent intensities (referred to as the formal solution) from time to time. The values of $I_\nu(r,\mu)$ found from the formal solution are then used to compute the various moments and from these, values of the Eddington factors f_ν and g_ν are updated for use in another round of moment equation iterations. Therefore, the main iteration loop of the Fe II RT+SE problem becomes $\sim 10^2$ times faster, since $J_\nu(r)$ need only be solved on a 2-D grid (instead of 3-D for I_ν). For a problem requiring $\sim 10^2$ iterations to converge, we still need to evaluate ~ 10 formal solutions, and these will continue to dominate the timings. The $\sim 10^2$ moment solutions will altogether be about as computationally demanding as just 1 formal solution. The net result in moving from S-MULTI to J-MULTI is that $\sim 10^2$ formal solutions is reduced to $\sim 10^2$ moment solutions ($= 1$ formal solution in timing) plus ~ 10 formal solutions. This corresponds to a factor of 10 reduction in effort, for a total of 10^{14} operations for the Fe II problem. We are solving this problem on a dedicated Sun workstation, rated at $\sim 10^8$ floating point operations s^{-1} (100 Mflops). Therefore, the computation of a single Fe II wind model will require $\sim 10^6$ s or 10 days on this machine. We are also investigating optimizing the angle quadratures in the formal solution to minimize the number of angles required. We may be able to save another factor of 3–5 here also, which would bring the solution timing of the Fe II wind problem close to 10^5 s.

5. Construction of the Wind Model

The previous discussion concerned the direct solution, in which the emergent Fe II line profiles are computed in a self-consistent manner for a specified wind model (*i.e.*, a given velocity, density, and temperature structure). However, to produce a wind model which reproduces the observed λ Vel Fe II profiles requires solving the inverse problem. To do this, we assume the velocity structure of λ Vel is given by a parametrized wind law similar to that found for the primary of the eclipsing binary ζ Aur (Figure 2). This result was obtained from analysis of the wind *absorption* spectrum seen superimposed upon the UV continuum of the hot secondary near eclipse (Bennett et al., in preparation). We will also adopt a parametrized temperature structure based on our ζ Aur results. The λ Vel wind density then follows from the velocity law, assuming a steady, spherical flow, given the mass loss rate. This procedure fully describes the wind model in terms of a small (~ 10) number of parameters, and makes the solution of the inverse problem tractable. To restrict the parameter space of viable solutions initially, we construct approximate solutions using the escape probability formalism to replace the formal solution (*e.g.*, see reviews in Kalkofen, 1984).

Acknowledgments. This work was supported by NASA ATP grant NAG5-3033 and STScI grant GO-5307.02-93A to the University of Colorado.

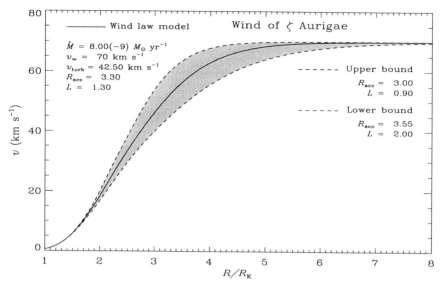

Figure 2. Wind velocity law of the K4 Ib primary of the eclipsing binary ζ Aur. The wind is described by five parameters in addition to the stellar mass and radius: $\dot{M}$, the mass loss rate; v_∞, the terminal velocity; $v_{\text{turb}} = (2gH)^{1/2}$, the turbulent velocity needed for pressure support (H is the pressure scale height); R_{acc}, the radius where deposition of mechanical energy into the wind ceases; and L, the effective length over which the deposition ceases. The shaded area indicates the uncertainty in the wind law. From Bennett et al. (in preparation).

References

Auer, L. H. 1971, JQSRT, 11, 573.
Auer, L. H., & Mihalas, D. 1970, MNRAS, 149, 65.
Baron, E., Hauschildt, P. H., & Mezzacappa, A. 1996, MNRAS, 278, 763.
Carlsson, M. 1985, Uppsala Astr. Obs. Report No. 33.
Harper, G. M. 1994, MNRAS, 268, 894.
Hauschildt, P. H. 1993, JQSRT, 50, 301.
Judge, P. G. 1990, ApJ, 348, 279.
Judge, P. G., & Jordan, C. 1991, ApJS, 77, 75.
Kalkofen, W. 1984, *Methods in Radiative Transfer*, (Cambridge: Cambridge University Press).
Mihalas, D. 1978, *Stellar Atmospheres, 2nd ed.*, (San Francisco: Freeman).
Mihalas, D., & Hummer, D. G. 1974, ApJS, 28, 343.
Mihalas, D., Kunasz, P. B., & Hummer, D. G. 1976, ApJ, 210, 419.
Olson, G. L., Auer, L. H., & Buchler, J. R. 1986, JQSRT, 35, 431.
Rybicki, G. B., & Hummer, D. G. 1991, A&A, 245, 171.
Scharmer, G. B., & Carlsson, M. 1985, J. Comput. Phys., 59, 56.

Interaction of the Pulsar-Driven Wind with Supernova Ejecta

Byung-Il Jun[1]

Department of Astronomy, University of Minnesota, 116 Church Street S.E., Minneapolis, MN 55455, U.S.A.

Abstract. Recent HST (Hubble Space Telescope) observations of the Crab Nebula show filamentary structures that appear to originate from the Rayleigh-Taylor (R-T) instability operating on the supernova ejecta accelerated by the pulsar-driven wind. In order to understand the origin and formation of the filaments in the Crab Nebula, we study the dynamics of a normal Type-II supernova remnant model for the Crab Nebula numerically. In this model, the pulsar-driven wind pushes into the uniformly expanding supernova ejecta. Therefore, the model includes four shocks from the outside inward: a supernova blast wave, a reverse shock moving into the supernova ejecta, a shock wave driven by expanding pulsar bubble, and a wind termination shock. Numerically, we find three independent instabilities in the interaction region. The first weak instability occurs in the very beginning, and is caused by the impulsive acceleration of supernova ejecta by the pulsar wind. The second instability occurs in the post-shock (shock wave driven by pulsar bubble) flow during the intermediate stage. This second instability develops briefly while the gradients of density and pressure are of opposite sign (satisfying the criterion of the R-T instability). The third and most important instability develops as the shock driven by the pulsar bubble becomes accelerated ($r \sim t^{6/5}$). This is the strongest instability and produces pronounced filamentary structures which resemble the observed filaments in the Crab Nebula.

1. Introduction

The Crab nebula is believed to be the remnant of the supernova of 1054 AD and it is the nearest pulsar-powered nebula. The "Crab" has a number of remarkable and puzzling properties: the observed mass and kinetic energy are much smaller than expected from a typical supernova (Davidson & Fesen, 1985). In addition, the emission-line filaments are helium-rich, suggesting the core material of the progenitor for their origins.

The interaction between a pulsar-driven wind and the enveloping supernova ejecta has provided a particularly challenging and important problem since de-

---

[1] also, Minnesota Supercomputer Institute, 1200 Washington Avenue South, Minneapolis, MN 55415, U.S.A.

tailed images of the Crab were obtained by the HST. The complex emission-line filaments would appear to originate from the Rayleigh-Taylor instability operating on the ejecta accelerated by the pulsar-driven wind (Hester et al., 1996). These finger-like structures seem to grow inward (toward the center of the remnant) and to be connected to each other at their origins by a faint, thin, tangential structure (the "skin"). Some of the long finger-like structures terminate in dense "head" regions. The interface between the synchrotron nebula and an ejecta shell has been thought to be R-T unstable since motions of the filaments (Trimble, 1968) and the synchrotron nebula itself (Bietenholtz et al., 1991) revealed their outward post-explosion acceleration. The magnetic field in the Crab has been inferred to be strong (a few hundred μG) and seems to play an important role in the formation of the filaments. Polarimetric VLA observations of the Crab show a radially-aligned magnetic field orientation (Bietenholtz & Kronberg, 1990), although the rotating pulsar is likely to produce toroidal (tangential) magnetic fields (Rees & Gunn, 1974). The radial magnetic field can be produced as R-T fingers stretch the existing field lines.

Several authors have proposed that the Crab formed from a normal Type-II supernova remnant (Chevalier, 1977; Kennel & Coroniti, 1984), despite the absence of convincing detection of a surrounding fast-moving shell (e.g., Frail et al., 1995). In this model for the Crab, the pulsar-driven wind pushes into the uniformly expanding supernova ejecta. Therefore, the model includes four shocks from the outside inward: a supernova blast wave, a reverse shock moving into the supernova ejecta, a shock wave driven by expanding pulsar bubble, and a wind termination shock (Figure 1). This model is particularly attractive because there is no need for a peculiar low-energy supernova event and it can be applied to other pulsar-powered remnants. Also, the observed acceleration is naturally explained, because the pulsar's wind shock expands with the law $r \sim t^{6/5}$ in the self-similar stage if the pulsar's luminosity is assumed to be constant (Chevalier, 1977). In this paper, we will present the result of our numerical investigation of a Type-II supernova remnant model on the origin of the filamentary structures in the Crab Nebula.

2. Numerical Methods and Initial Condition

Understanding the dynamical interaction between the pulsar-driven wind and the ejecta requires multidimensional numerical simulations with high resolution, because the development of the instability is highly non-linear. The high-resolution simulation is particularly necessary to resolve small regions containing filaments that would give us the density profiles needed to calculate ionization models and compare with the HST images. For this problem, we utilize the ZEUS-3D code, developed and tested at the National Center for Supercomputing Applications (Clarke & Norman, 1994). ZEUS-3D is a three-dimensional, Eulerian, finite-difference code for solving the equations of astrophysical fluid dynamics, including the effects of magnetic fields. The grid velocity is also allowed to change every time step so that the grid can follow the expanding system. This is necessary in modeling the Crab in order to treat the expanding supernova remnant accurately and keep adequately high resolution at the interaction region between the pulsar wind and supernova ejecta.

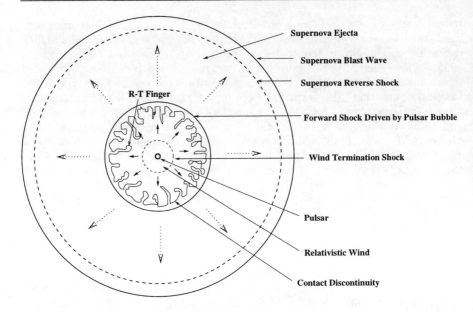

Figure 1.   Normal Type-II Supernova Remnant Model

The model includes the pulsar wind, moving supernova ejecta, and supernova shell confined by the supernova blast wave and reverse shock. Since we are interested in simulating the interaction region between the pulsar wind and the supernova ejecta, our simulation excludes the supernova shell. Readers are referred to Jun & Norman (1996a; 1996b) for the detailed study on the dynamics of supernova shell. The pulsar wind is generated by the constant mechanical luminosity, $L = 10^{40}$ erg s^{-1}. This luminosity is much higher than the current luminosity from the Crab pulsar, $L \sim 10^{38}$ erg s^{-1}, and is determined by accounting for the observed rate of decrease in the luminosity. We are currently investigating the effect of a time-varying luminosity and the detailed results will be presented in Jun (in preparation). The initial radius of the wind is chosen to be 0.1 pc. The supernova ejecta is initialized to be uniformly moving with $v = 500$ km s^{-1} at 0.1 pc and the velocity at larger radius is linearly proportional to the radius. The constant density of supernova ejecta is assumed to be $\rho = 1.67 \times 10^{-22}$ g cm^{-3} over the entire ejecta. Note that the ejecta density decreases with time because of ejecta expansion. The simulation is carried out in a spherical coordinate system with a resolution of 300×400 zones in the $r - \phi$ plane. The computational plane is divided into two blocks of zones: The outer quarter of the grid where the instability develops is resolved with 200×400 uniform zones and the inner three quarters of the grid is resolved with 100×400 ratioed zones. We assume the entire fluid to be nonrelativistic gas with $\gamma = 5/3$, for simplicity. We update the outer boundary condition in the r-direction to take into account the condition of the moving supernova ejecta every time step. The boundary condition in the ϕ-direction is assumed to be periodic. The density field is perturbed with an amplitude of 1%–10% in the entire region.

3. Hydrodynamic Evolution and Instabilities

The global evolution of the flow is studied by one-dimensional numerical simulations. In the early stage, the shock wave driven by the expanding pulsar bubble propagates much faster than the moving supernova ejecta and the density of supernova ejecta can be considered roughly constant (stationary medium stage). Therefore, the flow structure resembles the wind solution in a stationary medium as described by Weaver et al. (1977). The shock expansion can be approximated as $r_{shock} \sim t^{3/5}$. As the shock velocity decreases and becomes comparable to the ejecta velocity, the ejecta can no longer be considered as a stationary medium and the effect of decreasing density of the ejecta becomes important (moving medium stage). Finally, the shock accelerates and enters the self-similar stage which is described by $\rho_{ejecta} \sim t^{-3}$ and $r_{shock} \sim t^{6/5}$ (Chevalier, 1977; 1984; Jun, in preparation). The thickness of the region between the shock and the contact discontinuity becomes considerably thinner, about $0.02\, r_{shock}$.

Numerically, we find three independent instabilities in the interaction region between the pulsar-driven wind and the supernova ejecta. The first weak instability occurs in the very beginning, and is caused by the impulsive acceleration of dense ejecta by the low-density pulsar wind (Figure 2a). The second instability occurs in the post-shock flow during the intermediate stage (Figure 2b). This second instability develops briefly while the gradients of density and pressure are of opposite sign (satisfying the criterion of the R-T instability). This unstable flow develops in the transition between the stationary medium stage (Weaver's self-similar stage) and the moving medium stage (Chevalier's self-similar stage). In the former, both pressure and density in the post-shock region increase with increasing radius (Weaver et al., 1977). On the other hand, in the latter, both pressure and density in the post-shock region decrease with increasing radius (Chevalier, 1984). In each stage, the post-shock flow is stable. However, while the stationary medium stage evolves into the moving medium stage, the post-shock pressure and density evolve with $P \sim v^2 t^{-3}$ and $\rho \sim t^{-3}$, respectively. Therefore, the pressure decreases more rapidly with time than the density if the power-law index of the shock expansion ($r \sim t^a$) is smaller than unity. The different evolutions of the density and pressure result in an unstable flow that shows a positive gradient in density profile but a negative gradient in pressure profile. This unstable flow disappears when the shock expansion is accelerated. The second instability is also found to exist in the self-similar solution when the expansion index (a) is much smaller than 1 and the ejecta density decreases with $\rho \sim t^{-3}$ (Jun, in preparation).

The third and most important instability develops in the contact discontinuity between the pulsar wind and the shocked supernova ejecta as the wind becomes accelerated ($r \sim t^{6/5}$). This is the strongest instability and produces pronounced filamentary structures (Figures 2c and 2d). The thin layer between the shock and the contact discontinuity is severely deformed because of the instability. (Note that the shock front is not affected by the first and second instabilities.) Figure 2c shows a number of thin fingers pointing toward the center. At later time (Figure 2d), some of these thin fingers become unstable while some fingers maintain stability and develop into long thin structures. The formation of these stable long fingers is possible because of the large density dif-

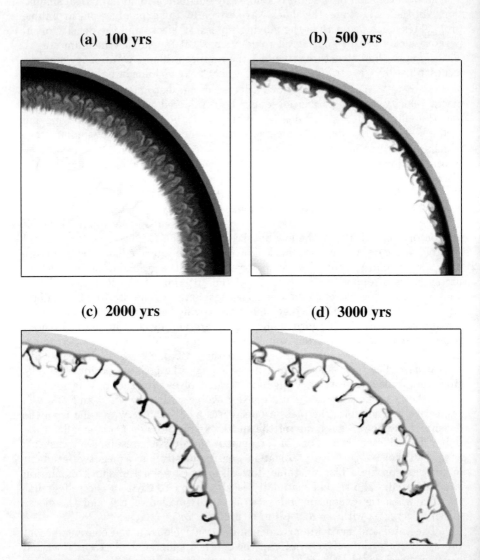

Figure 2. Instabilities in the interacting pulsar bubble with expanding supernova ejecta.

ference between the shocked thin layer and the low density pulsar bubble. What is noticeable is the formation of dense heads on the R-T fingers, which can be compared to regions F or G in the HST image of Hester et al. (1996). These dense finger heads are attributed to the compressibility of the gas. The density in dense finger heads is found to be about 10 times higher than in other regions of the fingers. The general morphology of our numerical results is compared to the HST observation of the filaments in the Crab Nebula (see Hester et al., 1996) and found to look remarkably similar to the observations.

4. Conclusions

Our numerical simulation of an interacting pulsar bubble with expanding supernova ejecta produces well-developed filamentary structures which resemble the filaments in the Crab Nebula. Three instabilities are found to develop independently in different evolutionary stages. The Rayleigh-Taylor instability driven by the acceleration of the thin shell which develops in the later stage is found to be the main mechanism for the origin of the filaments.

Our model can explain important observational features of the Crab Nebula. They are the accelerations of the filaments and the synchrotron nebula, formation of a thin skin connecting the filaments, formation of the filaments pointing toward the center, and several thin stable fingers and high-density finger heads.

Acknowledgments. I am grateful to Roger Chevalier, Jeff Hester, Tom Jones, Mike Norman, and Jim Stone for useful discussion and encouragement. The work reported here is supported by NSF grant AST-9318959 and by the Minnesota Supercomputer Institute.

References

Bietenholtz, M. F., & Kronberg, P. P. 1990, ApJ, 357, L13.
Bietenholtz, M. F., Kronberg, P. P., Hogg, D. E., & Wilson, A. S. 1991, ApJ, 373, L59.
Chevalier, R. A. 1977, in *Supernovae*, ed. D. N. Schramm (Dordrecht: Reidel), 53.
Chevalier, R. A. 1984, ApJ, 280, 797.
Clarke, D. A., & Norman, M. L. 1994, NCSA Tech. Report 15,
    http://zeus.ncsa.uiuc.edu:8080/ lca/zeus3d/zeus32.ps.
Davidson, K., & Fesen, R. A. 1985, ARA&A, 23, 119.
Frail, D. A., Kassim, N. E., Cornwell, T. J., & Goss, W. M. 1995, ApJ, 454, L129.
Hester, J. J. et al. 1996, ApJ, 456, 225.
Jun, B.-I., & Norman, M. L. 1996a, ApJ, 465, 800.
Jun, B.-I., & Norman, M. L. 1996b, ApJ, 472, 245.
Kennel, C. F., & Coroniti, F. V. 1984, ApJ, 283, 694.
Rees, M. J., & Gunn, J. E. 1974, MNRAS, 167, 1.
Trimble, V. L. 1968, AJ, 73, 535.
Weaver, R., McCray, R., Castor, P. R., & Moore, R. T. 1977, ApJ, 218, 377.

The Evolution of Blue Stragglers: Merging SPH and Stellar Evolution Codes

Alison Sills
Department of Astronomy, Yale University, P.O. Box 208101, New Haven, CT, 06520-8101, U.S.A.

Abstract. We have been investigating the creation of blue stragglers in stellar collisions by combining the results of Smoothed Particle Hydrodynamics (SPH) simulations of the stellar collision with stellar evolution calculations. The difficulties of converting the results of a hydrodynamic calculation into a starting model for a stellar evolution code which assumes hydrostatic equilibrium will be discussed, and preliminary results will be presented.

1. Introduction

Recently, SPH calculations of stellar collisions relevant to stars in globular clusters have been performed (Benz & Hills, 1987; Lombardi et al., 1996; Sandquist et al., 1997). A number of groups (Bailyn & Pinsonneault, 1995; Sills et al., 1995; Ouellette & Prichet, 1996; Sandquist et al., 1997) attempted to make some approximation to the starting model for their evolution calculations using the results of these SPH calculations. However, no one has used the full results of the SPH code as the starting model for their stellar evolution calculations. Instead, only the chemical composition profile was imposed on an otherwise normal stellar model. Sandquist et al. (1997) attempted to model the internal energy distribution of a collision product by adding an arbitrary amount of energy to their normal stellar models until the stars were on the Hayashi track, and then began their evolutionary runs.

The interpretation of the results of these studies has been inconsistent. The later SPH results show that the collision products are not fully mixed, as Benz & Hills (1987) proposed. However, Sills et al. (1995) and others discovered that some of the observations can only be explained if the blue stragglers are fully mixed. In an attempt to solve this dilemma, Leonard & Livio (1995) proposed that the stars are mixed during a fully convective pre-main sequence phase directly following the collision. It was this effect that Sandquist et al. (1997) attempted to model in an *ad hoc* way. However, it was not clear that collision products actually evolve in this way.

2. Procedure

We are using the results of the SPH calculations by Lombardi et al. (1996) as the starting models for use in the Yale Rotating stellar Evolution Code (YREC,

Guenther et al., 1992) in its non-rotating mode. We used the entropy and chemical composition profiles from the SPH results as inputs to the equations of stellar structure to determine the profiles of pressure, temperature, density, radius, hydrogen abundance, and luminosity as a function of mass for our starting model.

Using the results of one code directly as the initial conditions for another is tricky. Since many assumptions about the physical situation are very different between codes, consistency is difficult. The kinds of assumptions which cause the most problems include the typical time scales, the geometry of the situation, and which physical effects are important and which can be neglected.

3. Assumptions for Each Code and How to Deal with Them

SPH codes are inherently hydrodynamic, while stellar evolution codes depend on the star being in hydrostatic equilibrium. Therefore, the collision must be complete in the sense that the SPH merger remnant must be in hydrostatic equilibrium before the results can be used for the evolution code. We calculate the pressure gradient from the density gradient using the equation of hydrostatic equilibrium to ensure that the star is in equilibrium.

Stellar evolution codes are one-dimensional since the star is modelled as a set of spherical shells. SPH codes follow the motions of particles, and so are three-dimensional. To convert from SPH to the evolution code, we average all quantities in spherical shells. If the merger remnant is not spherical, we assign an average radius to each shell of constant density.

The time scales of the two codes are significantly different. In SPH, a time step is much less than one dynamical time scale (~ 1 hour), with total elapsed time on the order of hours. In stellar evolution, a time step is a thermal ($\sim 10^6$ yr) to nuclear ($\sim 10^8$ yr) time scale, with total elapsed time on the order of Gyr. We use the shortest possible time step for the initial evolution calculations. In YREC, the shortest numerically stable time step is $\sim 10^2$ yr.

SPH and stellar evolution codes are quite different in both their time scales and treatment of energy transport. This drastically affects their treatment of convection, which can cause problems for the merging of the two codes. In SPH simulations, the motion of convective elements is treated explicitly. Energy is not transported between elements and their surroudings, so mixing can be modelled, but not energy transport. This is valid, since the amount of energy transported by convection in a few days is insignificant. YREC uses the mixing length theory of convection. We replace the complicated situation in convective regions with average quantities. The time step between models is longer than the convective turnover time scale so convective regions are mixed "instantaneously". The energy transported by convection is based on average motion of convective elements. Therefore, we must model the treatment of convection more carefully than usual in stellar evolution calculations.

4. Treatment of Convection

The usual criterion for convective stability is the Schwarzschild criterion: $\nabla_{\rm rad} < \nabla_{\rm ad}$. However, gradients in chemical composition can stabilize fluid against convection. The more general condition for convective stability is the Ledoux criterion:

$$\nabla_{\text{rad}} < \nabla_{\text{ad}} + \frac{\beta}{4 - 3\beta}\nabla_\mu.$$

When a region is stable according to the Schwarzshild criterion and unstable according to the Ledoux criterion, it is called semi-convective. In these regions, fluid elements oscillate slowly and mix the region over a thermal time scale. The mixing and energy transport can be treated using a diffusion approximation (Langer et al., 1983). The Schwarzschild criterion in commonly used in stellar evolution codes since the time step is longer than the thermal time, and so any semi-convective motions can be treated as convective. However, when evolving stellar collision products, we wish to follow the initial thermal relaxation carefully, and so a full treatment of semi-convection is a necessary element in the evolution code used to model stellar collision products.

5. Results

Using the results of SPH calculations of stellar collisions directly as starting models for stellar evolution calculations involves careful consideration of the different assumptions of each code and their effects on the evolution. Convection is the most problematic process to model correctly. Semi-convection must be treated explicitly. The collision products never become fully convective, and do not even develop large convective envelopes. Therefore, these stars do not spend any time on the Hayashi track, contrary to the speculation of Leonard & Livio (1995) and the assumption of Sandquist et al. (1997).

References

Bailyn, C., & Pinsonneault, M. 1995, ApJ, 439, 705.
Benz, W., & Hills, J. 1987, ApJ, 323, 614.
Guenther, D., Demarque, P., Kim, Y. C., & Pinsonneault, M. 1992, ApJ, 387, 372.
Langer, N., Sugimoto, D., & Fricke, K. 1983, A&A, 126, 207.
Leonard, P., & Livio, M. 1995, ApJ, 447, L121.
Lombardi, J., Rasio, F., & Shapiro, S. 1996, ApJ, 468, 797.
Ouellette, J., & Prichet, C. 1996, in *The Origins, Evolution and Destinies of Binary Stars in Clusters*, ed. E. F. Milone, (San Francisco: ASP), 356.
Sandquist, E., Bolte, M., & Hernquist, L. 1997, ApJ, 477, 335.
Sills, A., Bailyn, C., & Demarque, P. 1995, ApJ, 455, L163.

Detonation Waves in Type Ia Supernovae

D. J. R. Wiggins & S. A. E. G. Falle

Department of Applied Mathematics, The University of Leeds, Leeds LS2 9JT, U.K.

Abstract. The propagation of 2-D nuclear detonation waves in Type Ia supernovae is calculated from Huygens' principle and the resulting flow is then computed using an adaptive hydrodynamic code. This has the advantage that one can study axisymmetric explosions using wave speeds determined from one-dimensional calculations with very detailed nuclear chemistry. The method is applied to two cases, the one-dimensional versions of which have previously been considered by Nomoto (1982). We find that the detonation wave forms a pair of cusps which are likely to have a significant effect on both the abundances and the final asymmetry of the expansion.

1. Introduction

Type Ia supernovae result from explosive nuclear burning. Whether the mechanism is a detonation, a deflagration, or some combination of the two is the cause of current debate. Any model needs to fulfill certain criteria: enough ^{56}Ni must be produced to power the light curve, excessive neutronisation needs to be avoided and there need to be sufficient amounts of intermediate mass elements created and travelling with the observed range of velocities.

The off-centre helium detonation of a sub-Chandrasekhar mass star seems a reasonable scenario for Type Ia explosions (Munari & Renzini, 1992; Kenyon et al., 1993). This allows a wider range of progenitor mass which is consistent with the considerable dispersion among observed peak absolute magnitudes (Hamuy et al., 1994). Many previous models have assumed spherical symmetry (Nomoto, 1982; Khokhlov, 1991; Woolsley & Weaver, 1994). The presence of a companion along with the highly sensitive temperature dependence of nuclear reaction rates means it is more likely that helium ignition occurs at a point rather than simultaneously over the whole layer. The asymmetries of such explosions are considered here.

2. Wave Front Calculation

For the densities encountered in Type Ia supernova progenitors, Khokhlov (1989) finds that detonation waves travel at the Chapman-Jouguet (CJ) speed in helium and less than 1% faster than the CJ speed in CO. In a CJ detonation (CJD), the wave speed, which is sonic with respect to the downstream gas, is uniquely determined by the initial state of the gas and the energy released by the nuclear burning. As it is reasonable to assume an off-centre detonation generates a CJD, the propagation of the wave can be determined independently of any dynamical effects it produces.

It is reasonable to assume a spherically-symmetric density distribution as any asymmetry required for a point explosion need not be so large as to have a significant effect on the wave speed (which is not particularly sensitive to the density). The wave speed can thus be determined as a function of position. This is analogous to that of the propagation of light in a medium with a variable refractive index. Huygens' construction can be used to determine the position of the wave front as a function of time.

The path of an ignition ray (which is the analogue of a light ray) satisfies Fermat's principle. If $D(R)$ is the wave speed then,

$$\delta \int_{\theta_A}^{\theta_B} L d\theta = 0, \quad L = \frac{(R^2 + R'^2)^{1/2}}{D(R)}, \quad R' = dR/d\theta. \tag{1}$$

Solving the Euler equation and parameterising for R and θ in terms of the travel time t, one finds

$$\frac{d\theta}{dt} = -\frac{CD^2}{R^2} \quad \frac{dR}{dt} = \pm \frac{D(R^2 - C^2 D^2)^{1/2}}{R}, \tag{2}$$

where C is a constant.

The locus of points on all the ignition rays at time t gives the wave front. If there are sharp changes in the wave speed, then it is possible for rays with different initial directions to intersect. Such rays need to be destroyed whereas new ones need to be created when the distance between two neighbours becomes too large.

The wave front at time t not only consists of points for which the rays can be traced back to the ignition point, but also points that can be traced back to the position of the wave front at some time $t^* < t$, *i.e.*, the wave front is the source of secondary waves. This is just Huygens' construction.

We shall consider approximations to Nomoto's cases (A & B). The initial parameters for these cases are given by Wiggins & Falle (1997). Speed functions for both cases can be found by approximating their density distribution and using wave speed data given in Buchler *et al.* (1972) and Mazurek (1973). Figures 1 and 2 show the propagation of the detonation waves for both cases.

The wave front can be split into three different parts: one propagating towards the centre of the star, a part that propagates around the outer envelope, and a refraction region joining the two. The different parts are joined by cusps. The outer cusp exhibits a constant angle (caused by refraction at the CO/He interface). The time scale for complete burning here is comparable to the hydrodynamic time. Such effects, which are two-dimensional in nature, can be expected to alter the nucleosynthesis (from the spherically-symmetric case). The outer cusp in case B is smaller than the one in A, as the distance of the explosion point from the interface (along with the ratio of wave speeds across it) determines the cusp angle. The dynamics and nucleosynthesis will be influenced by its location. The angle of the inner cusp sharpens as part of the wave in the envelope runs away from the part propagating towards the centre. The outer part of the wave becomes less convex as t increases. If the detonation is initiated

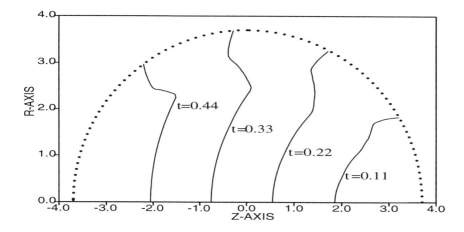

Figure 1. The propagation of the wave front for case A. The dotted semicircle is the outer radius of the star. Distances are in units of 10^8 cm.

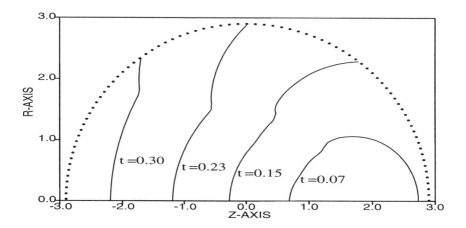

Figure 2. The propagation of the wave front for case B. The dotted semicircle is the outer radius of the star. Distances are in units of 10^8 cm.

at a reasonable distance from the interface, this would result in some focusing of the detonation at the axis of the interface. Perhaps this would make it easier for the He detonation to ignite the core.

3. Hydrodynamics

The numerical integration of the gas dynamic equations is done using a second-order upwind scheme as described in Falle (1991). The detonation wave thickness is assumed to extend over a small region (5–10 cells). Nearly all the energy in

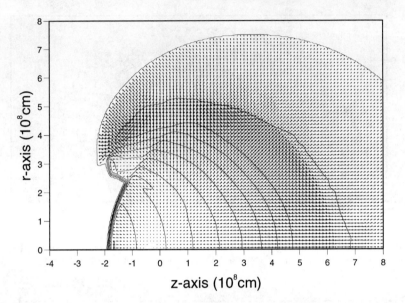

Figure 3. Case A: logarithmic contours of pressure with velocity vectors at 0.421s. There are nine equally-spaced contours between $\log p = 23$ and $\log p = 25.4$ (cgs). The maximum velocity vector is 1.79×10^9 cm s^{-1}.

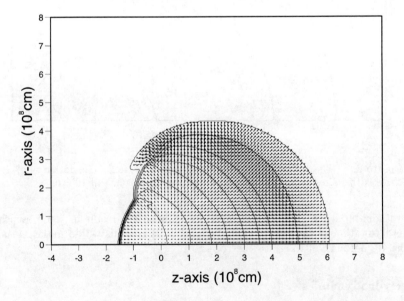

Figure 4. Case B: logarithmic contours of pressure with velocity vectors at 0.244s. There are seven equally-spaced contours between $\log p = 23$ and $\log p = 25.4$ (cgs). The maximum velocity vector is 2.43×10^9 cm s^{-1}.

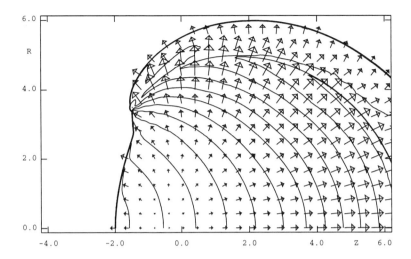

Figure 5. Case A: logarithmic contours of pressure and velocity vectors for the adaptive grid calculation at $t = 0.421$. There are 12 equally-spaced contours between $\log p = 22$ and $\log p = 25.45$ (cgs). The maximum velocity vector is 1.98×10^9 cm s^{-1}.

a detonation wave is released in burning up to NSQE (Khokhlov 1989), *i.e.*, over a comparatively thin region. The reaction rate is taken to be Gaussian in form.

Figures 3 and 4 show logarithmic contours of pressure and velocity vectors for cases A and B on a grid of 200 cells in the r-direction and 300 cells in the z-direction, for a simple equation of state, no gravity, and a negligible initial pressure. It is interesting to note the presence of a rarefaction wave caused by the flow divergence behind the outer cusp, and a shock, generated at the inner concave cusp, which travels into the refraction region. Differences between cases A and B show how a subtle difference (cusp angle) can have a marked effect on the asymmetry of the expansion.

The method outlined above enables us to identify, quite cheaply, the more "interesting" cases. However, a uniform grid is computationally expensive if we wish to explore these with the correct degree of numerical resolution. In order to overcome this, we used a hierarchical adaptive code μCobra. Since μCobra uses the estimated truncation error to refine the grid, its performance does not depend on the geometry of the flow. It is particularly useful for problems in which high resolution is only required in a small fraction of the computational domain, *i.e.*, near the detonation front in our case. The code has been applied to a number of astrophysical problems (*e.g.*, Falle & Raga, 1993; Falle & Komissarov, 1996). Figure 5 shows an exploratory calculation for Case A using μCobra. The propagation of the wave front is handled by a scalar variable [varying from 0 (unburnt) to 1 (burnt)] which reacts at a rate W given by,

$$W = K(1-\alpha)^{\frac{1}{2}}, T > T_i$$
$$= 0, \text{otherwise.}$$

K determines the width of the reaction front, T_i is set in each cell to be twice the initial temperature in that cell, *i.e.*, the cells only start to react when they are hit by the shock. The absence of a well-defined kink in this calculation is the result of the lower He-to-CO-speed ratio at the interface which results from the detonation having to accelerate up to the correct speed rather than obtaining it instantaneously. This is only an exploratory calculation and need not concern us much. However, it shows the importance of using our wave-front method as a guide.

In the calculation, we use six grids $G^0 - G^5$, with $\Delta x = 0.2$ on G^0 and $\Delta x = 6.25 \times 10^{-3}$ on G^5. This calculation provides a 35-fold increase in speed over a uniform grid (which would require $960 \times 2{,}560$ cells to get the same resolution).

4. Conclusion

The hydrodynamics of our two-dimensional computations show that the asymmetry of the Type Ia explosion is fairly sensitive to the initial state. Since there are a large number of possible initial conditions, we need a cheap computational method to explore them all. Once we have found some viable models, we can then use more detailed nuclear chemistry to determine the resulting abundances. It should be possible to do this with our adaptive scheme.

Acknowledgments. DJRW acknowledges support from PPARC during the course of this work. μCobra has been generously supplied by Mantis Numerics Ltd.

References

Buchler, J. R., Wheeler, J. C., & Barkat Z. 1972, ApJ, 172, 713.
Falle, S. A. E. G. 1991, MNRAS, 250, 581.
Falle, S. A. E. G., & Komissarov, S. S. 1996, MNRAS, 278, 586.
Falle, S. A. E. G., & Raga, A. C. 1993, MNRAS, 261, 573.
Hamuy, H., *et al.* 1994, AJ, 108, 2226.
Khokhlov, A. M. 1989, MNRAS, 239, 785.
Khokhlov, A. M. 1991, A&A, 245, 114.
Kenyon, S. J., Livio, M., Mikołajewska, J., & Tout, C. A. 1993, ApJ, 407, 81.
Mazurek, T. J. 1973, Ap&SS, 23, 365.
Munari, U., & Renzini, A. 1992, ApJ, 397, 87.
Nomoto, K. 1982, ApJ, 257, 780.
Wiggins, D. J. R., & Falle, S. A. E. G. 1997, MNRAS, in press.
Woolsley, S. E., & Weaver, T. A. 1994, ApJ, 371, 62.

Monte-Carlo Simulations of X-Ray Binary Spectra

D. A. Leahy

Department of Physics and Astronomy, University of Calgary, Calgary, AB T2N 1N4, Canada

Abstract. Monte Carlo methods are used to calculate the spectrum of radiation from accreting neutron stars. The spectrum is reprocessed by several physical processes in matter with various properties and geometries in the binary system.

1. Introduction

Several studies already have been done on reprocessing and iron line fluorescence. For planar matter geometry, George & Fabian (1991) calculated the spectrum for reflection of X-rays from a disk of matter. Basko (1978) used an analytic first-scattering approximation to calculate the profile of the singly scattered iron K line wing, and pointed out the dependence on the exit angle from the plane. Hatchett & Weaver (1977) calculated the iron line profile in the isotropic scattering approximation. For uniform spherical matter, Makino et al. (1985) calculated equivalent widths (EW) *vs.* column density. Inoue (1985) compared EW for planar and spherical geometries for an unabsorbed central source. Makashima (1986) compared EW for the cases of unabsorbed, absorbed, and blocked central source. Leahy et al. (1989) calculated EW and spectra for uniform and thin-shell spherical matter. EW for a beamed X-ray source inside a spherical cloud of matter were given by Leahy & Creighton (1993).

2. The models

The basic description of the method of calculation is given in Leahy & Creighton (1993). The source has a power-law energy spectrum, and can be either isotropic or beamed. The matter which reprocesses the source radiation can be arbitrarily specified in spatial distribution. The total cross-section included absorption and electron scattering. For electron scattering, the Klein-Nishina formula was used. Iron and nickel K-shell fluorescence was included. Atomic data for the various line energies and cross-sections were taken from CRC Tables. Energy losses for Compton-scattered line photons were taken into account; this determines the fluorescent line shapes.

The photon propagation is carried out in three dimensions. Specific geometries considered include:

i) a sphere with constant density (from $r = 0$ to $r = R$);

ii) a $1/r^2$ density profile for $r > R$;

iii) a thin shell of constant density, zero density inside, with $\delta r \ll R$;

iv) a spherical cloud of uniform density illuminated externally by a point source;

v) disk geometry, both static and rotating.

3. Results and Discussion

The results for the isotropic source case have been summarized elsewhere (Leahy & Creighton, 1993). Uniform, $1/r^2$ and exponential density distributions were considered. Spectra and EW have been determined for the different density distributions and for different power-law indices. Line profiles were also calculated. Single-scatter and double-scatter peaks can be identified in the line profile. The ratios of line photons in these peaks to the unscattered line depends strongly on optical depth, and very weakly on geometry. The line profile depends on both the optical depth and the geometry of the gas distribution. A distinct difference in shape is found between line profiles for the uniform density and the $1/r^2$ cases, allowing the line shape to be a diagnostic of the density distribution.

Since X-ray pulsars are beamed, we also considered the effects of beaming on spectrum and EW. Much higher EW and more apparent absorption of the continuum results when looking off-axis to the beam, except at the highest optical depths. Iron line profiles have also been calculated, and show a strong dependence on the off-axis angle. The line intensity along the beam axis is strongest, that antiparallel to the beam axis is weakest. Because of the higher optical depth, most of the photons reaching the observer at any angle have been forward scattered, more so for directions nearer antiparallel to the beam axis.

The simulations of externally illuminated clouds are of interest, since clumps of matter are known to orbit then accrete onto the neutron stars in some systems and possibly AGN as well. The clumps are also expected on theoretical grounds: irradiated matter in many X-ray binaries is unstable to breaking up into a two-phase system: cold dense clumps in a hot coronal medium. The spectra reprocessed by an external cloud are very strongly angle-dependent because of both electron scattering and absorption. The iron line profile is a strong function of viewing angle and thus, this is a powerful diagnostic technique. One can also derive the radial velocity of the cloud from the overall shift of the iron line energy. This opens up the possibility of determining orbits of clouds near the compact X-ray source. Current satellite resolution is just inadequate to resolve the line shapes. The EW relations were calculated, and also the spectra summed over many clouds distributed isotropically around the source. The many-cloud spectrum is a good candidate to explain the low-energy excess observed in many X-ray binary sources.

Several calculations of reprocessing of X-ray spectra by external matter have been carried out. Simple measures such as EW are no longer seen to be very useful on their own, since they are now known to be functions of both optical depth and radiation pattern. This means a combination of iron line shape, EW, and continuum spectral shapes must be used to constrain the reprocessing geometry.

A practical approach to the problem of determining source and cloud properties from an observed X-ray spectrum is as follows. The optical depth can be determined to first order from the observed continuum spectrum (*e.g.*, from the iron K-edge depth). To extract information from the EW and profile data, one needs to assume a realistic beaming pattern for the source and a geometry. One must also calculate a set of line profiles for a two-dimensional grid of optical depth and angle values. Then the observed line profile and EW determine viewing angle and optical depth. One should then compute an expected continuum spectrum and compare this to the observed continuum spectrum. If the comparison is good, then some analysis of the uncertainties can be done. If not, then the assumed reprocessing geometry or beaming pattern need to be changed and the process repeated.

In practise, one cannot utilize line profile information simply because it is not available; the resolution of past and present detectors has been too limited. However, improvements in detector energy resolution at the energy of the fluorescent iron line will soon allow determination of iron line profiles, and allow these new techniques for determining source properties to be applied for the first time.

Acknowledgments. This work supported by a research grant to D.A.L. from the Natural Sciences and Engineering Research Council of Canada.

References

Basko, M. 1978, ApJ, 223, 268.
George, I., & Fabian, A. 1991, MNRAS, 249, 352.
Hatchett, S., & Weaver, R. 1977, ApJ, 215, 285.
Inoue, H. 1985, in *Proceedings of the Japan-US Seminar on Galactic and Extragalactic Compact X-ray Sources*, eds. Y. Tanaka & W. Lewin (Tokyo: ISAS), 283.
Leahy, D., Matsuoka, M., Kawai, N., & Makino, F. 1989, MNRAS, 236, 603.
Leahy, D., & Creighton, J. 1993, MNRAS, 263, 314.
Makashima, K. 1986, in *The Physics of Accretion onto Compact Objects*, eds. K. Mason, M. Watson, & N. White, (New York: Springer-Verlag), 249.
Makino, F., Leahy, D., & Kawai, N. 1985, Space Sci. Rev., 40, 421.

Effects of Scattering on Pulse Shapes from Accreting Neutron Stars

D. A. Leahy

Department of Physics and Astronomy, University of Calgary, Calgary, AB T2N 1N4, Canada

Abstract. Time delays of scattered radiation by circumstellar matter are calculated for some basic geometries. The results show that significant changes to the pulse shapes observed from neutron stars can occur.

1. Introduction

The effects of scattering X-rays in matter surrounding a central X-ray source can be quite important. Several studies have been done on the effects by reprocessing on X-ray spectra in matter of planar geometry (George & Fabian, 1991; Matt et al., 1991) or spherical geometry (Leahy & Creighton, 1993) or even a random set of blobs of matter (Bond et al., 1993; Sivron & Tsuruta, 1993).

However, here the effects on the pulse shapes is the topic of study. The importance of such effects has been realized as more data on pulse shapes from X-ray pulsars become available from satellite instruments. A model of a spherical shell of matter around the central source is used to calculate what kind of effects from scattering might be expected to occur for an X-ray pulsar.

2. The Model and Calculations

The basic model assumes a central rotating X-ray source inside a spherical shell of scattering material. Other geometries considered include a linear stream of matter directed along constant spherical polar angles, and a parabolic stream. A single-scatter approximation is used, which is valid in the optically-thin limit for the shell. All time delays from source to observer are calculated, so the correct distribution of photon arrival times is derived.

The calculations are done for the case where the angular distribution of radiation from the central source is a delta function. The scattering effects for any more complicated (and realistic) distribution of radiation can be calculated by a convolution of the results for a delta function beam with the actual angular distribution. Here we calculate only the scattered radiation. To calculate what the actual pulse shape looks like, one needs to add the appropriate amount of direct radiation from the central source to the scattered radiation.

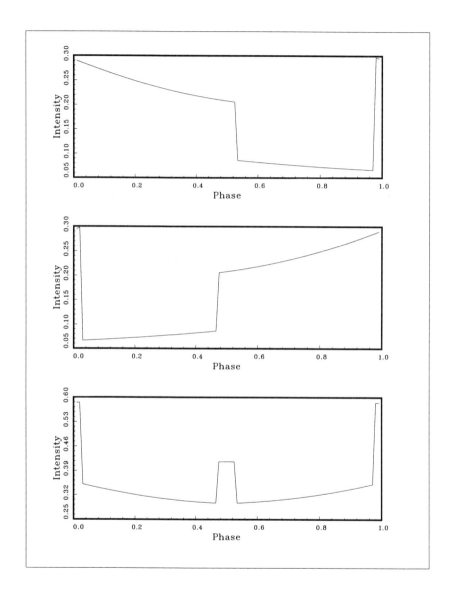

Figure 1. Pulse phase dependence of scattered radiation from a thin shell of radius $R = cP$ for the case of $\theta_r = \pi/4$, $\theta_m = \pi/4$. Top panel: radiation from primary beam, middle panel: radiation from the antipodal beam, bottom panel: the sum from both beams.

3. Results

The pulse shapes depend critically on i) the size of the shell (in units of cP, where P is the rotation period of the pulsar), ii) the orientation of the primary beam with respect to the rotation axis (angle θ_m), and iii) the orientation of the observer with respect to the rotation axis (angle θ_r). Generally, for small shell radii ($R \ll cP$), the time delays become negligible and the only important factors are the distribution of the scattering matter convolved with the angular dependence of the Compton scattering cross-section. For larger R, the effects of time delays become significant and can dominate the observed pulse profile.

Here, only some sample results are given. Figure 1 gives the observed pulse shapes of the scattered radiation (only) for a rotating beam at 45° to the rotation axis and for the observer also at 45° to the rotation axis. The top plot is for radiation from the primary beam, the middle plot is for the secondary (antipodal) beam (directed 180° away from the primary beam), and the lower plot is for summed radiation from both beams. Rotation phase is defined so that phase 0° corresponds to the primary beam lying in the rotation axis-observer axis plane. Figure 1 is for $R = cP$. Generally, for $R < R_{\text{crit}}$, the pulse shapes are smooth and continuous. For $R = R_{\text{crit}}$ or greater, cusps in the pulse shapes occur, since scattered radiation from the back of the shell will lag radiation from the front of the shell which was emitted from the source half a period later. R_{crit} is determined by setting the photon propagation time between the back limit and front limit of the scattering shell equal to one half of a rotation period. Thus, for the example above, $R = cP/2$ is the minimum size shell that will give cusps in scattered pulse shapes.

4. Discussion

At the current stage, the calculations of scattered radiation have been done for sources with different angles θ_r and θ_m and for a thin shell with different dimensionless shell radii R/cP. Some calculations have been done for different scatterer geometries, but the results from these calculations remain to be explored in a systematic way. However, it already appears that a large variety of pulse shapes for scattered radiation can occur, including cusp-like features which are caused by forward-scattered radiation from the pulsar overtaking back-scattered radiation emitted at some time earlier.

Acknowledgments. This work is supported by a research grant to D.A.L. from the Natural Sciences and Engineering Research Council of Canada.

References

Bond, I., Matsuoka, M., & Yamauchi, M. 1993, ApJ, 405, 179.
George, I., & Fabian, A. 1991, MNRAS, 249, 352.
Leahy, D., & Creighton, J. 1993, MNRAS, 263, 314.
Matt, G., Perola, G., & Piro, L. 1991, A&A, 247, 25.
Sivron, R., & Tsuruta, S. 1993, ApJ, 402, 420.

Part III
YSO/ACCRETION/ OUTFLOW/ISM

Collapse and Outflow: Towards an Integrated Theory of Star Formation

Ralph E. Pudritz[1]

CITA, University of Toronto, 60 St. George St., Toronto, ON M5S 1A1, Canada

Dean E. McLaughlin & Rachid Ouyed

Department of Physics and Astronomy, McMaster University, Hamilton, ON L8S 4M1, Canada

Abstract. Observational advances over the last decade reveal that star formation is associated with the simultaneous presence of gravitationally collapsing gas, bipolar outflow, and an accretion disk. Two theoretical views of star formation suppose that either stellar mass is determined from the outset by gravitational instability, or by the outflow which sweeps away the collapsing envelope of initially singular density distributions. Neither picture appears to explain all of the facts. This contribution examines some of the key issues facing star formation theory.

1. Introduction

How stars form is one of the most important unsolved problems in astrophysics. It has turned out that the process is surprisingly rich, involving the formation of dense cores in magnetized molecular clouds, gravitational collapse, the ubiquitous presence of accretion disks around young stellar objects, and most surprisingly perhaps, the presence of high speed bipolar outflows and jets. While gravitational collapse and the formation of disks might have been expected in any model of star and planet formation, the role of bipolar outflows has yet to be fully integrated into our thinking.

Star formation theory has scored some notable successes, among them being the elucidation of the importance of magnetic fields, and the physics of gravitational collapse and of magnetized outflows from the central engine. In spite of these advances however, there is still no generally accepted answer to the most basic question of all: What determines the mass of a star? In this review, we discuss some of the main ideas of star formation with a view to addressing this question. Its solution will no doubt require sophisticated mathematical and numerical tools.

---

[1] on leave from Department of Physics and Astronomy, McMaster University, Hamilton, ON L8S 4M1, Canada

1.1. Basic facts

Star formation occurs within very specific, over-dense regions within molecular clouds known as molecular cloud cores (Benson & Myers, 1989). The physical conditions within these cores presumably provide the initial conditions for star formation. One of the most basic properties of cores is that their mass distribution is very well defined. Measurements indicate that the clumps and cores within molecular clouds obey a well-defined relation in which the number of cores per unit mass scales as

$$\frac{dN(m)}{dm} \propto m^{-\alpha},$$

where the index $\alpha = 1.6 \pm 0.2$ (*e.g.*, Blitz, 1991).

The internal structure of cores has been intensively studied in the last decade. The observations suggest that molecular cloud cores are prolate structures (Myers *et al.*, 1991) in which rotation is insignificant in comparison with the self-gravity of the core. The line widths in molecular cloud cores *increase* as one moves outwards to larger radii; thermal gas pressure can dominate only in the innermost regions (0.01 pc scales) in cores (Fuller & Myers, 1992; Caselli & Myers, 1995). Nonthermal motions dominate on larger scales in cloud cores, with the nonthermal velocity dispersion in low mass cores scaling as

$$\sigma_{\rm NT} \propto r^{0.5},$$

while for higher mass cores

$$\sigma_{\rm NT} \propto r^{0.25}.$$

These relations *appear to hold for both starless and star-containing cores*. We therefore appear to be seeing the initial conditions for star formation. Several theorists argue that this is misleading and that all cores are already affected by outflows from newly formed stars. If the nonthermal line width in cores is produced by the interaction of bipolar outflows with their surrounding core gas, then it has been argued that one could construct a model for the stellar initial mass function, or IMF (*e.g.*, Silk, 1995).

The source of energy sufficient to balance gravity within molecular clouds and their cores is the magnetic field that threads them. Zeeman measurements show that molecular cloud and core fields have energy densities comparable to gravity (Myers & Goodman, 1988, Heiles *et al.*, 1993). The nonthermal line widths would presumably reflect a generic MHD turbulence, or perhaps superposition of MHD waves in clouds. No general theory for how such MHD turbulence might be excited yet exists, although outflows have been suggested as a possible source of excitation.

Of central importance for an integrated theory of star formation is an understanding of why outflow and collapse are operative at the same time as a star forms. The so-called Class 0 sources, which are objects having virtually no emission at wavelengths below 10 μm and spectral energy distributions characterized by single blackbodies at $T \simeq 15$–30 K, are important in this regard. There is some evidence to suggest that these are protostars whose collapsing envelopes may exceed the central protostar in mass suggesting an age of $t \simeq 2 \times 10^4$ yr in some models (*e.g.*, André *et al.*, 1993; but see Pudritz *et al.*, 1996). The key point here is that such objects have particularly strong and well-collimated outflows (*e.g.*, Bontemps *et al.*, 1996).

Finally, infrared camera observations of embedded young stellar objects within molecular cloud cores indicate that stars don't form individually, but as members of groups and clusters. Almost of necessity, the most detailed available calculations of star formation focus on the formation of individual stars. However, this theoretical focus may be blinding us to the solution to our basic question.

1.2. Basic ideas

Theoretical thinking about the star formation process stems from two fundamental, but quite different aspects of the physics of self-gravitating gas clouds. The first view is that stellar mass is determined by gravitational instability, while the second is that it is determined by shutting off accretion in collapsing cores. There are persuasive arguments for and against both of these pictures.

Gravitational instability: One of the classic calculations is Ebert (1955) and Bonnor's (1956) analysis of the stability of an isothermal sphere of self-gravitating gas of mass M that is embedded in an external medium with a pressure P_s. The critical mass for a cloud at temperature $T = 10$ K and supported purely by thermal gas pressure is

$$M_J = \frac{1.2(T/10 \text{ K})^2}{(P_s/10^5 k_B \text{ cm}^{-3} \text{ K})^{1/2}} M_\odot .$$

It is impressive that this argument picks out a solar mass so that one can say that self-gravitating gas at 10 K naturally forms solar mass objects (*e.g.*, Larson, 1992). The 10 K temperature arises from the balance of cosmic ray heating and cloud cooling by millimetre radiation from collisionally excited CO molecules (Goldsmith & Langer, 1978). The Jeans mass is significantly larger in turbulent media, where turbulent rather than purely thermal pressure enters into the above expression (*e.g.*, Mckee *et al.*, 1993).

There are several problems with this view however. There doesn't seem to be an obvious way of explaining the initial mass function of stars, which ranges over two decades in mass. Why wouldn't the clump mass spectrum be the same as the IMF in this theory? The measured mass spectral index for the IMF is $\alpha_\star = 2.35$ (Salpeter, 1955). Thus, while the total gas mass in the CMF is dominated by its most massive core, the stellar mass in the IMF is dominated by the low mass end. This fact suggests that many low mass objects prefer to form in the more massive clumps; *i.e.*, cluster formation is required (Patel & Pudritz, 1994). Secondly, the role of outflows seems incidental to the process except insofar as it removes the angular momentum of collapsed core gas allowing the star to form via accretion through the disk.

Truncating the collapse: An equally fundamental view of a self-gravitating cloud is that the accretion rate in the collapse of singular isothermal spheres is fixed by molecular cloud core conditions (*e.g.*, Shu, 1977). For an isothermal equation of state, the accretion rate is a constant,

$$\dot{M} = 0.975 \frac{a_{\text{eff}}^3}{G} = 1.0 \times 10^{-5} \left(\frac{a_{\text{eff}}}{0.35 \text{ km s}^{-1}}\right)^3 M_\odot \text{ yr}^{-1},$$

where a_{eff} is an effective sound speed in the core. In these self-similar theories, gravitational collapse and accretion onto a central protostar can go on indefinitely. Mass must therefore be fixed by the mechanism that truncates the accretion phase such as jets and outflows. This view is interesting because it incorporates outflow into the basic mechanism of star formation.

This view also has its problems. As with the first theory, there seems to be no obvious way in which an IMF could be produced. The role of the CMF is equally mysterious. Secondly, while jets do pack considerable power, they also appear to be highly collimated. This is especially true of outflows associated with the so-called class 0 sources. An outflow that doesn't cover a fair fraction of 4π is unlikely to be able to eject the remains of a collapsing envelope. Such material could still end up on the disk, and accrete from there onto the central star.

2. Initial States

The basis for the gravitational instability picture arises most simply in the model worked out by Bonnor and Ebert. Consider applying an external surface pressure P_s on an isothermal cloud of mass M. Calculate the structure of the resulting clump that is in hydrostatic balance with its own gravity and internal pressure P and the external pressure. For such pressure bounded equilibria, we may ask the question, at what radius R will one find pressure balance between clump pressure and P_s? This problem is solved by finding solutions to the Lane-Emden equation for these truncated configurations. If one now asks for the characteristics for our isothermal, non-magnetic cloud, that is *critically stable* ($dP_s/dR = 0$), Ebert and Bonnor found that

$$M_{\text{crit}} = 1.18 \frac{\sigma_{\text{ave}}^4}{(G^3 P_s)^{1/2}},$$

$$\Sigma_{\text{crit}} = 1.60 (P_s/G)^{1/2}.$$

These models have a finite central density and attain a $\rho \propto r^{-2}$ structure at large radii.

The generalization of this analysis for arbitrary equations of state may be found in McLaughlin & Pudritz (1996), where one finds that the expressions for M_{crit} and Σ_{crit} for clouds with any equation of state will differ from the isothermal, Bonnor-Ebert ones at the < 10% level; the physical scalings remain the same. [The effect of the equation of state is to modify the numerical value of the line width σ_{ave}; see McLaughlin & Pudritz (1996).] The gradual loss of magnetic flux by ambipolar diffusion implies that cores are supported against gravitational collapse in their innermost regions for about ten free fall times. As long as the central regions of cores are magnetically supported, their central densities continue to grow slowly. Detailed calculations show that once a critical value of the ratio of the gas mass to magnetic flux in the central region is surpassed, then the central density profile of magnetized cores begins to approach a singular solution more rapidly. During this more dynamic phase, collapse probably begins before a singular state is actually achieved (*e.g.*, Basu & Mouschovias, 1994).

Shu (1977) and Li & Shu (1996) argue that this steepening of the density profile continues until the density actually becomes singular. In this event, which occurs say at time $t = 0$, its structure is simply described by the relation;

$$\rho(r) = \frac{\sigma^2}{2\pi G} r^{-2} .$$

This singular distribution has a finite mass at the centre (related to a numerical constant, m_0). Once it is achieved (in finite time), it is impossible for gas in the vicinity of the newly-formed protostar to evade gravitational collapse. Thus, the main accretion phase in singular isothermal sphere (SIS) models consists of an "inside-out" collapse of the envelope onto this protostar. The evolution during this phase is then best discussed in terms of an outwards-moving collapse front—the so-called "expansion wave."

2.1. Equations of state

A theoretical model of star formation in more massive cores must incorporate some means of describing the nonthermal motions. One way of proceeding is to model the gas with an effective equation of state (EOS). If one resorts to polytropic models as an example, then the total line width scales as $\sigma^2 \propto P/\rho \propto \rho^{\gamma-1}$. Now since the lines are observed to broaden as one goes to larger physical scales (where the density is decreasing), then this simple result requires models with $\gamma < 1$, which brings up the idea of negative polytropic indices (Maloney, 1988). Lizano & Shu (1989) modeled the general structure of cores by breaking the pressure up into isothermal and turbulent contributions; $P = P_{\text{iso}} + P_{\text{turb}}$. In order to handle the turbulent motion, they suggested a so-called "logotropic" relation (their words) between turbulent gas pressure and density; $P_{\text{turb}} = \kappa \ln(\rho/\rho_{\text{ref}})$. On the other hand, McKee & Zweibel (1995; see also Pudritz 1990) later noted that a gas dominated by the pressure of Alfvén waves would have an effective EOS of the form $P_{\text{wave}} \propto \rho^{1/2}$.

The data of Caselli & Myers (1995) provide a way of testing possible EOS. The challenge is to fit the trends in both the low and higher mass cores using a single EOS. McLaughlin & Pudritz (1996) found that the models mentioned above did not fit the data. Their best fit is achieved by the so-called pure logotrope,

$$P_{\text{total}}/P_c = 1 + A \ln(\rho/\rho_c),$$
$$A \simeq 0.2 ,$$

in which the *total* gas pressure $P_{\text{iso}} + P_{\text{turb}}$ has a logarithmic dependence on density. In the singular limit, these models have density profiles that are much shallower than SIS models;

$$\rho(r) = \left(\frac{AP_c}{2\pi G}\right)^{1/2} r^{-1} .$$

Figure 1 shows the fit of the pure logotrope to observed line widths inside cores, and illustrates how well constrained the value of the coefficient A is. The vertical lines mark the radius of the critically stable model. Unmagnetized logotropes have critical masses of $92 M_\odot$ while magnetized ones have critical

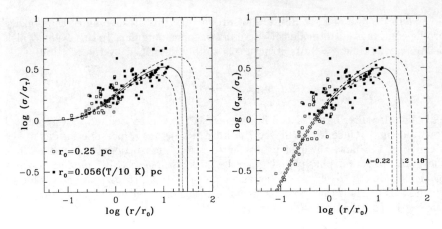

Figure 1. Total line width σ, and turbulent component σ_{NT}, relative to the thermal contribution σ_T. Data on low mass cores (*open symbols*) are referred to a larger scale radius than high mass cores (*filled symbols*); the former are less centrally concentrated. Curves are the predictions of the logotropic EOS, for three values of the parameter A; vertical lines mark the radius of a critical mass, $A = 0.2$ logotrope (from McLaughlin & Pudritz 1996).

masses of $250 M_\odot$. Such cores are obviously far more massive than $1 M_\odot$, and must therefore be the objects in which multiple star formation occurs. Since solar mass clumps fall far below the critical mass for a logotrope, their internal structures are dominated more by gas pressure than by gravity. Thus, they are less centrally concentrated than higher mass clumps, and this is reflected in their larger scale radius r_0 in Figure 1. Note that star formation appears to be occurring in a wide variety of these clumps which suggests perhaps that the gravitational instability analysis may have less to do with the issue of stellar mass determination than does the accretion picture.

3. Gravitational Collapse

Insight into gravitational collapse in molecular cores has been gained by considering special cases that are analytically tractable such as the collapse of SIS models. Non self-similar models, such as the Bonnor-Ebert solutions, require a detailed numerical solution. Thus, Foster & Chevalier (1993) investigated the collapse of Bonnor-Ebert spheres and found that their numerical solutions produced supersonic velocities during initial stages of the collapse. Central inflow speeds reached -3.3 times the sound speed, and a central density distribution $\rho \propto r^{-2}$ developed. Their work generalizes the collapse calculations of Larson (1969) and Penston (1969) who began with uniform spheres. The difference in the results is that Foster & Chevalier find that supersonic speeds develop only in a small region in the centre, and not throughout the model. Also, the mass accretion rate is constant only if the initial configurations are very highly centrally concentrated.

Following Larson (1969) and Shu (1977), it is convenient to define similarity variables; if a_t (in general, time dependent) is the sound speed, then the self-similar dimensionless length is $x = r/a_t t$. The equations of motion then imply an accretion rate

$$\dot{M}(0,t) \propto \frac{a_t^3}{G}.$$

In what follows, we take the extreme cases of the SIS (Shu, 1977) and the singular, pure logotrope models (McLaughlin & Pudritz, 1997). The different character of the self-similar collapse solutions for these two different models illustrates well the effect that the EOS has upon the physics of the collapse. For the SIS model, the position of the expansion wave (see § 1.2) in the self-similar variable x is always located at $x_{\exp} = 1$. The expansion wave moves outwards through the undisturbed envelope at constant speed $a_t = a$; its position at any time is then $r_{\exp} = at$, and the mass of the central protostar grows as

$$M = m_0 \frac{a^3 t}{G}, \qquad m_0 = 0.975.$$

For the logotrope on the other hand, the expansion wave is located at $x_{\exp} = 1/4\sqrt{2}$ and the sound speed is no longer a constant; $a_t = (AP_c 4\pi G)^{1/2} t \propto t$. The position of the expansion wave in space is $r_{\exp} = a_t t/4\sqrt{2} \propto t^2$ and the mass of the protostar grows as

$$M = m_0 \frac{(AP_c 4\pi G)^{3/2} t^4}{G}, \qquad m_0 = 0.667 \times 10^{-3}.$$

The infall speed at any time is lower for the logotrope than for the SIS model because the latter has a more centrally concentrated density profile. Note also that the numerical constant m_0, which scales the mass that has already fallen into the centre, is much smaller for the logotrope. This has the consequence that it takes much longer to grow low mass protostars.

One of the main results of McLaughlin & Pudritz (1997) is that the time required to accumulate a solar mass star is of order 2×10^6 yr, which is much longer than for an SIS model. On the other hand, all stars in the logotrope picture accumulate in roughly the same time which is not true of SIS models. This has a major impact on our ideas of IMF formation. It suggests that star formation in the logotropic picture must really be sequential in time. Indeed, any theory for the formation of star clusters must guarantee that low mass star formation gets started first, since when massive stars turn up, the molecular cores will be obliterated. While SIS models could only pertain to low mass cores, high mass star formation necessarily takes place in more turbulent conditions so that star formation time scales are much shorter (*e.g.*, Myers & Fuller, 1992). Thus, here too, sequential star formation needs to be invoked.

4. Outflows

Episodic jets are observed in AGNs, regions of star formation (*e.g.*, Edwards *et al.*, 1993), and binary systems with black holes. Whenever one observes a jet, there is good evidence that an accretion disk is also present; a fact that is probably not fortuitous. Young stellar objects have associated outflows that last a long time, at least $1-2 \times 10^5$ yr according to Parker *et al.* (1991). The outflows in Class 0 submm sources have mechanical luminosities that rival the accretion luminosity of the central object with $L_{\mathrm{mech}} \simeq L_{\mathrm{bol}}$. In all outflows, radiation pressure fails by several orders of magnitude to provide the observed thrusts in winds so that mechanisms involving magnetic drives seem to be suggested.

Current models for outflows invoke magnetic fields that thread Keplerian disks. They are of two types; (*i*) *hydromagnetic disk winds* wherein the engine consists of a Keplerian disk threaded by a magnetic field that is either generated *in situ*, or advected in from larger scales (*e.g.*, Blandford & Payne, 1982; Camenzind, 1987; Lovelace *et al.*, 1987; Heyvaerts & Norman, 1989; Pelletier & Pudritz, 1992; Li, 1995; Appl & Camenzind, 1993; Königl & Ruden, 1993); or (*ii*) *X-winds*, which are magnetized stellar winds where the interaction of a protostar's magnetosphere with a surrounding disk results in the opening of some of the magnetospheric field lines (Shu *et al.*, 1987; 1994).

Perhaps the most important difference between these two classes of models lies in the role of the central object. For disk winds, only the depth of the gravitational well created by the central object is of any importance. The energy source for the flow is the gravitational energy release of material in the Kepler disk as the wind torque extracts its angular momentum. This view implies that the physics of jets from the environs of protostars or black holes is essentially the same. For X-wind models on the other hand, the magnetization and structure of the central object is critical. Its magnetic field strength must be sufficient to carve a magnetosphere inside the disk and outflow requires that the magnetopause and corotation radii of the star are virtually identical; $R_{\mathrm{m}} \simeq R_{\mathrm{co}}$.

These two different wind mechanisms make different predictions about the possibility of truncating the collapse of the surrounding envelope. The X-wind model has a low density, radial component to the wind that could possibly clear out the envelope. Disk wind simulations, such as those of Ouyed *et al.* (1997) and Ouyed & Pudritz (1997a; 1997b—see below) find that a finite fraction of the disk is involved in outflow and that outflows are rapidly collimated towards the outflow axis. This implies that they may not be able to clear out the envelope. As far as we are aware, extensive calculations of this type have never been done in either of these theoretical models so the jury is still out.

Numerical simulations of disk winds by Ouyed *et al.* (1997; see also Ouyed, these proceedings) were run in order to test the predictions of steady-state theory and to see whether or not time-dependent calculations would yield jets that are truly episodic. The simulations have an initial state consisting of a central point mass, the surface of a surrounding Keplerian accretion disk (inner radius r_i), and a disk corona that is in exact (analytical and numerical) hydrostatic balance in the gravitational field of the central object and in pressure balance with the accretion disk below. The disk and corona is threaded by a magnetic field configuration chosen to have initial current $\mathbf{J} = 0$ so that no magnetic force is exerted initially upon the corona. Two different magnetic configurations were investigated; the first was a vacuum solution for a field with a conducting plate at its base (called a potential distribution), and the second was a constant uniform magnetic field that is parallel to the z-axis and perpendicular to the disk. This second configuration was chosen because no outflow is expected in steady state theory. The models depend on five parameters; three prescribe the initial corona (ratio of gas to magnetic pressure, thermal to rotational energy density, and the ratio of the density of the base of the corona to disk density; all these measured at r_i), one gives the ratio of the toroidal to poloidal field strength in the disk, and a final parameter measures the speed at which mass is injected from the disk into the base of the corona. This latter speed is taken to be a thousandth of the local Kepler speed, or a hundredth of the disk sound speed. All lengths in our simulations are in units of r_i, and all times (τ) are in units of the Kepler time $t_i = r_i/v_{K,i}$ at the inner edge of the disk.

Simulations using the potential field configuration (see Ouyed et al., 1997; Ouyed, these proceedings) clearly show a bow shock separating the outflow that has started from the disk from the undisturbed corona. The field lines and flow behind the bow shock are collimated towards the z-axis into a jet-like, cylindrical outflow. The result shows that a cylindrically collimated, stationary outflow is achieved. Cylindrical collimation is predicted to be a generic feature of magnetized outflows in which the dominant toroidal field of the outflow together with its associated current (which flows up the jet) together exert a pinching Lorentz force towards the outflow axis (e.g., Heyvaerts & Norman, 1989). Ouyed et al. (1997) also show the position of the Alfvén and fast magnetosonic (FM) surfaces where the outflow speed achieves the propagation speeds of two of the three important wave speeds in magnetized gas. The data are compared with the position of the Alfvén point on each field line in the simulation as predicted by steady state theory (e.g., Blandford & Payne, 1982). The agreement is very good. This and many other diagnostics (see Ouyed et al., 1997) show that there is good agreement with steady state disk wind theory and our simulations. While this is important and interesting, nature prefers to produce highly time-dependent, episodic outflows. Why is this?

Figure 2 shows that outflow occurs even for our initially uniform magnetic field configuration. The highly collimated, jet-like outflow is, in this case, dominated by a series of dense knots that are produced periodically on a time scale of $t_{knot} \simeq 11 t_i$. Knots are produced in a generating region close to the central source; at a distance $z_{knot} \simeq 6$–$7 r_i$. Knots continue to be produced for as long as we have run our simulations, up to 1,000 time units, and so they are truly generic and are not transients. Figure 2 shows three snapshots of a highly zoomed-in simulation [$(z \times r) = (20 \times 10) r_i$] designed to show the details of the knot generating region. The left panels show the poloidal magnetic field structure of the flow at three times during which a new knot is formed. The right panels show the opening angle of the magnetic field lines as a function of their footpoint radius r_0 on the surface of the accretion disk. One sees from these graphs that the field lines have been pushed open in a small region of the disk, making an angle of 50°–60° with respect to the disk surface. These field lines have been opened up by the toroidal magnetic field pressure arising from the Keplerian rotation of each field line. Since Kepler rotation is faster at smaller radii, one expects that torsional waves introduce stronger toroidal field into the corona at smaller radii. This creates the radial gradient in toroidal field pressure that opens the field lines to less than the critical angle of 60° from the axis (Blandford & Payne, 1982).

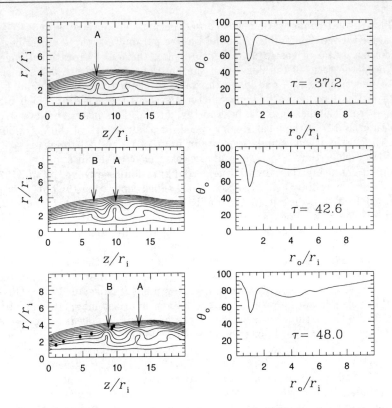

Figure 2. The left panels show the magnetic field structure of the knot generating region at the three times; 37.2, 42.6 and 48.0 inner time units. The right panels show the angle θ_o of field lines at the base of the flow, with the disk surface, at these times. Note the narrow band of field lines which is sufficiently opened ($\theta_o \leq 60°$) so as to drive the outflow. Only field lines involved in the knot generation process are shown; field lines at larger disk radius stay reasonably vertical as seen in the right panels (from Ouyed et al., 1997).

We found that knots are produced whenever the toroidal field in this inner region is sufficient to recollimate the newly accelerated gas back towards the outflow axis (see Ouyed et al., 1997). The gas necessarily speeds up. Because the gas is rotating however, it encounters a centrifugal barrier at $r \geq r_0$. As it reflects off of this barrier, it collides with the slower gas around it and shocks. The shocked gas regions move away from this generator region, and are kept coherent by strong enhancements of the toroidal field both ahead of it, and behind it. The knots, which are the overdense regions, have low toroidal field strengths and conversely the space between the knots is dominated by high toroidal field strength. The time scale for the passage of an Alfvén wave (in the toroidal field) from the jet radius towards the axis and back again, turns out to be precisely the knot generation time. This episodic behaviour of jets may reflect on the nature of the accretion disk that is feeding gas into the corona. If the entry ram pressure of newly injected material in the corona exceeds the toroidal field at the base of the corona, then we found that the outflow develops into a stationary flow. Thus, the general time-dependent behaviour of episodic jets may be intimately related to conditions in the underlying accretion disks. One must ultimately remove the constraint of keeping the disk as a fixed boundary condition in the problem if one hopes to explore this idea by numerical simulation.

5. An Integrated Model?

What general points about an integrated star formation theory arise from these considerations? Perhaps the least controversial point is that accretion disks may be the glue that binds outflow and infall together. Outflows may commence as soon as the collapse has been sufficient to create even a tiny, centrifugally supported region in a disk (*e.g.*, Pudritz *et al.*, 1996). Accretion of infalling material onto and through the disk will drive the outflow. Thus, the continuous feeding of the disk by the infalling envelop should help to sustain a vigorous outflow. It is completely unclear as to whether or not the details of the gravitational collapse are important for the formation of an outflow (*e.g.*, singular logotropic *vs.* isothermal collapse; or non-self-similar collapse). It is sobering to note that no self-consistent numerical simulation of collapse that we are aware of has shown that outflows are produced. If outflows are disk winds, then their efficient removal of disk angular momentum would help to drive an accretion flow through the disk. Of course, significant turbulent disk viscosity, such as could be produced by MHD turbulence (Balbus & Hawley, 1991), could also transport disk angular momentum (radially). The high collimation of hydromagnetic disk winds makes it unlikely that they will clear out the infalling envelope. X-winds, if they occur, may have less of a problem in this regard.

While many details need to be checked before any useful predictions can be made, we suggest from all of this that the physical processes of collapse and outflow at the level of the formation of an individual star, have no obvious means of dictating the mass of a star. If this is correct, then the answer to our basic question must take place in more general, larger scale processes. Thus, the idea that stars form as members of groups and hence must somehow compete for their gas supply, may be of central importance to the theory of star formation.

Acknowledgments. REP thanks the organizers of this most stimulating conference for the invitation to give this talk. REP acknowledges the financial support and stimulating environment of CITA where he worked on this paper during a sabbatical leave. He also thanks S. Basu and D. Johnstone at CITA for stimulating conversations and a critical reading of this manuscript. The research of DEM and RO was supported by McMaster University, and that of REP through an operating grant from the Natural Science and Engineering Research Council of Canada.

References

André, P., Ward-Thompson, D., & Barsony, M. 1993, ApJ, 406, 122.
Appl, S., & Camenzind, M. 1992, A&A, 256, 354.

Balbus, S. A., & Hawley, J. F. 1991, ApJ, 376, 214.
Basu, S., & Mouschovias, T. Ch. 1994, ApJ, 432, 720.
Benson, P. J., & Myers, P. C. 1989, ApJS, 71, 89.
Blandford, R. D., & Payne, D. R. 1982, MNRAS, 199, 883.
Blitz, L. 1991, in *The Physics of Star Formation and Early Stellar Evolution*, (NATO ASI Series), eds. C. J. Lada & N. D. Kylafis (Dordrecht: Kluwer), 3.
Bonnor, W. B. 1956, MNRAS, 116, 351.
Bontemps, S., André, P., Tereby, S., & Cabrit, S. 1996, A&A, 311, 858.
Camenzind, M. 1987, A&A, 184, 341.
Caselli, P., & Myers, P. C. 1995, ApJ, 446, 665.
Ebert, R. 1955, Z. Astrophys., 37, 217.
Edwards, S., Ray, T. P., & Mundt, R. 1993, in *Protostars and Planets III*, eds. E. Levy & J. Lunine (Tucson: University of Arizona Press), 567.
Foster, P. N., & Chevalier, R. A. 1993, ApJ, 416, 303.
Fuller, G. A., & Myers, P. C. 1992, ApJ, 384, 523.
Goldsmith, P. F., & Langer, W. D. 1978, ApJ, 222, 881.
Heiles, C., Goodman, A., McKee, C. F., & Zweibel, E. G. 1993, in *Protostars and Planets III*, eds. E. H. Levy & J. I. Lunine (Tucson: University of Arizona Press), 279.
Heyvaerts, J., & Norman, C. 1989, ApJ, 347, 1055.
Königl, A., & Ruden, S. P. 1993, in *Protostars and Planets III*, eds. E. H. Levy & J. I. Lunine (Tucson: University of Arizona Press), 641.
Larson, R. B. 1969, MNRAS, 145, 271.
Larson, R. B. 1992, MNRAS, 256, 641.
Li, Z.-H. 1995, ApJ, 444, 848.
Li, Z.-H., & Shu, F. H. 1996, ApJ, 472, 211.
Lizano, S., & Shu, F. H. 1989, ApJ, 342, 834.
Lovelace, R. V. E., Wang, J. C. L., & Sulkanen, M. E. 1987, ApJ, 315, 504.
Maloney, P. 1988, ApJ, 334, 761.
McKee, C. F., & Zweibel, E. G. 1995, ApJ, 440, 686.
McLaughlin, D. E., & Pudritz, R. E. 1996, ApJ, 469, 194.
McLaughlin, D. E., & Pudritz, R. E. 1997, ApJ, 476, 750.
Myers, P. C., & Fuller, G. A. 1992, ApJ, 396, 631.
Myers, P. C., Fuller, G. A., Goodman, A. A., & Benson, P. J. 1991, ApJ, 376, 561.
Myers, P. C., & Goodman, A. A. 1988, ApJ, 329, 392.
Ouyed, R., & Pudritz, R. E. 1997a, ApJ, in press.
Ouyed, R., & Pudritz, R. E. 1997b, ApJ, in press.
Ouyed, R., Pudritz, R. E., & Stone, J. M. 1997, Nature, 385, 409.
Parker, N. D., Padman, R., & Scott, P. F. 1991, MNRAS, 252, 442.
Patel, K., & Pudritz, R. E. 1994, ApJ, 424, 688.
Pelletier, G., & Pudritz, R. E. 1992, ApJ, 394, 117.
Penston, M. V. 1969, MNRAS, 144, 425.
Pudritz, R. E. 1990, ApJ, 350, 195.
Pudritz, R. E., Wilson, C. D., Carlstrom, J. E., Lay, O. P., Hills, R. E., & Ward-Thompson, D. 1996, ApJ, 470, L123.
Salpeter, E. E. 1955, ApJ, 121, 161.
Shu, F. H. 1977, ApJ, 214, 488.
Shu, F. H., Adams, F. C., & Lizano, S. 1987, ARA&A, 25, 23.
Shu, F. H., Najita, J., Wilkin, F., Ruden, S. P., & Lizano, S. 1994, ApJ, 429, 781.
Silk, J. 1995, ApJ, 438, L41.

Time-Dependent Protostellar Accretion and Ejection

R. N. Henriksen

Department of Physics, Queen's University at Kingston, Kingston, ON K7L 3N6, Canada

Abstract. Numerical simulations of protostellar collapse combined with observations of molecular outflows of different apparent ages indicate that the rate at which a stellar mass is accumulated is variable, at least in the earliest stage. This contradicts the standard opinion stemming from the Shu model for isothermal self-similar accretion in which this rate is constant. In fact none of the usual "hyperbolic" self-similar solutions contain an observationally satisfactory accretion history, although some aspects of the Larson/Penston solution are a better fit to the simulations. And yet observation and simulation together suggest that a Shu accretion phase does develop in the later protostellar stages. I give here a mathematical formulation of the simplest time-dependent accretion problem which shows how a novel self-similar accretion can exist that is compatible both with the observations and simulations of the early state, and with the asymptotic development of the Shu accretion mode.

1. Introduction

The major missing link in our understanding of the life cycle of stars is the evolution from interstellar cloud to protostellar "birthline" (*e.g.*, Stahler, 1994). Observations in the IR and sub-millimetre bands combined with molecular spectroscopy suggest that this phase is associated with simultaneous infall and outflow, but the actual evolutionary sequence leading from pre-stellar "cloud cores" to T Tauri or Herbig Ae/Be stars remains unclear.

André *et al.* (1993) have developed an interesting method of locating the ensemble of embedded *protostellar* objects in an age-accretion-rate plane. In essence, the theoretical idea that ejection is proportional to accretion (see Henriksen *et al.*, 1997, hereafter HAB, for a summary and appropriate references) allows the momentum efflux in molecular outflow to yield a measure of the accretion rate onto the object. Moreover sub-millimetre measurement of the mass remaining in the envelope surrounding the core provides an inverse measure of the age of the source. We suppose that all of the mass was initially in "pre-stellar" form (*i.e.*, in the envelope or cloud) so that "time" for the source may be measured by an increasing ratio of stellar core mass to envelope mass.

This approach led Bontemps *et al.* (1996) to confirm that the classification of protostars into class 0, I, II is in fact an age sequence (André & Montmerle, 1994). The most dramatically new result of their method however is to distinguish the youngest protostars, that is those of class 0, as being the most efficient at producing outflow per unit bolometric luminosity. Since the bolometric luminosity at a given accretion rate is proportional to the ratio of core mass to core radius (which is smallest for the class 0 sources), this result suggests that in fact the accretion rate peaks for these young objects.

The Shu (1977) self-similar evolution from the singular isothermal sphere (SIS) and the Larson/Penston type (L/P) self-similar solutions (see Whitworth & Summers, 1985, hereafter WS, for an extended discussion of all isothermal globally self-similar solutions) predict constant accretion rates from the moment that the stellar core first forms (taken to be $t = 0$ in most studies including this one). This conflicts with the deduction from observations cited above and with numerical simulations made by Foster & Chevalier (1993) and by Hunter (1977). These simulations of collapse from a Bonnor/Ebert isothermal sphere give an initially peaked accretion rate (*i.e.*, at $t = 0$) that declines only asymptotically (provided the initial sphere is extensive enough) to the Shu rate. The L/P accretion rate is closer to the simulated rate initially, but it has neither the correct time dependence nor the correct peak amplitude.

The HAB paper proposes that the resolution of this difficulty lies in the existence of a type of self-similar evolution different from the global "hyperbolic" class discussed above. It arises when the flow has become supersonic, and its form is determined by the gravitational influence of the mass distribution existing at this epoch rather than by the pressure profile. As such it is elliptic in character.

HAB also point out that recent observations of the density profiles in prestellar cores at sub-millimetre wavelengths (Ward-Thompson *et al.*, 1994; André *et al.*, 1996) reveal flattened cores and r^{-2} halos, rather than the SIS profile. This is confirmed quite independently by the ISO observations of the core B2 in Ophiucus by Abergel *et al.* (1996). They infer that at least in some cases, initial core formation may proceed by the homologous collapse of a homogeneous central cloud region rather than by the launch of an outward-going expansion wave at $t = 0$, as is the case in the "inside-out" collapse models that start from the SIS. This suggestion is supported by recent numerical calculations by Tomisaka (1996) that generalize the Bonnor/Ebert instability to include the presence of a magnetic field.

In the next section, I discuss the now classical problem of spherically symmetric isothermal collapse of self-gravitating spherical shells, but in a Lagrangian formulation. I will show that if a central region of the cloud develops a flat density profile as it becomes supersonic, then the self-similarity discussed by Henriksen (1989, hereafter H89) for pressureless collapse should arise there. This is the new mode alluded to above.

2. Theory

In the Lagrangian treatment of spherical isothermal collapse, we introduce the current radius of a shell $R(t, r)$ as a function of time and of a fiducial shell radius r taken as a label at some epoch. Then energy conservation for each shell yields the equation (*e.g.*, H89)

$$\frac{(\partial_t R)^2}{2} - \frac{Gm(r)}{R} + (a_s)^2 \mathcal{H}(t, r) = E(r), \qquad (1)$$

where G is Newton's constant, $m(r)$ is the mass inside the shell labelled r, a_s is the sound speed throughout the material, $\mathcal{H} \equiv \ln \rho/\rho_0(r)$ and E is the specific energy of a given shell. The mass profile at the fiducial epoch is related to the density at this epoch by $m(r) = 4\pi \int_0^r \rho_0(r) r^2 dr$, and the current density $\rho(t, r)$ is given therefore by

$$\rho(t,r) = \frac{r^2 \rho_0(r)}{R^2 \partial_r R}. \tag{2}$$

We shall assume first that the system begins from rest at $t = t_0 < 0$ at the fiducial epoch. Then $E = -Gm(r)/r$ but there is presumably a small deviation in $\rho_0(r)$ from that of the Bonnor/Ebert limiting isothermal sphere in order to perturb the system from equilibrium. The discussion proceeds by introducing the dimensionless radius $\mathcal{S}(t, r) \equiv R/r$ and then by considering various cases.

The introduction of the variable $\mathcal{S}$ to the energy equation yields

$$(\partial_t \mathcal{S})^2 - \frac{\xi^2}{(\Delta t)^2} \frac{1}{\mathcal{S}} + \frac{2\eta^2}{(\Delta t)^2} \mathcal{H} = -\frac{\xi^2}{(\Delta t)^2}, \tag{3}$$

where

$$\xi(t,r) \equiv \sqrt{\frac{2Gm(r)}{r^3}} \Delta t,$$

$$\eta(t,r) \equiv \frac{a_s}{r} \Delta t. \tag{4}$$

Here we have set $\Delta t \equiv t - t_0$.

In general, instead of Δt and r, it is convenient to use the scaled variables ξ and η. However, in the special case where the initial condition is that of the SIS, these two variables are not independent since $m(r) \propto r$. In this case, we use $\xi = \sqrt{2/\alpha_s}\,\eta$ and r as independent variables (which can in fact be done in general) to obtain (HAB) the energy equation (3) in the form

$$(\partial_\xi \mathcal{S})^2 - 1/\mathcal{S} + \alpha_s^2 \mathcal{H}(\xi, r) = -1, \tag{5}$$

since $2(\eta/\xi)^2 \equiv a_s^2 r/Gm(r) \equiv \alpha_s^2$ and, in this case, $\partial_t \equiv (\xi/\Delta t)\partial_\xi$. We note that α_s is rigorously constant when $m(r) \propto r$.

The density follows from equation (2) as

$$e^{\mathcal{H}} = \frac{1}{\mathcal{S}^3 \left(1 + \frac{\partial \ln \mathcal{S}}{\partial \ln r} - \frac{1}{D(r)} \frac{\partial \ln \mathcal{S}}{\partial \ln \xi}\right)}, \tag{6}$$

where

$$\frac{-1}{D(r)} \equiv \frac{\partial \ln \xi}{\partial \ln r} = \frac{3}{2}\left(\frac{1}{3}\frac{d \ln m(r)}{d \ln r} - 1\right),$$

so that $D = 1$ for the SIS.

It is now clear from equations (5) and (6) that the SIS can develop entirely in the self-similar form (Shu, 1977) $\mathcal{S} = \mathcal{S}(\eta)$, $\mathcal{H} = H(\eta)$, provided only that it is perturbed so that $(d^2S/d\xi^2)_{t=t_0}$ is independent of r. This is of course a rather special case that depends heavily on the scale-free SIS initial condition.

There is in fact a parametrically stable family of trans-sonic self-similar solutions that contains the Shu and L/P solutions (WS) as limiting cases. In general, these solutions satisfy initial conditions that are rather different from the SIS. It is easy to see, however, from the requirement that α_s be constant that there must nevertheless be a fiducial epoch in the development of all solutions with this wave-like or hyperbolic symmetry when $m(r) \propto r$, that is $\rho \propto r^{-2}$. Here, r is redefined to be the current radius of a shell at this instant. Moreover, this can only be at $t = 0$ (*cf.*, Hunter, 1977; WS), when the ingoing pressure wave has reached the centre and the central singularity has just formed, since at this moment the memory of initial conditions has been erased everywhere in the supersonic region exterior to the wave front.

Thus, the epoch $t = 0$ is the natural fiducial epoch for the non-Shu type isothermal self-similar solutions. However, one might object that since the gas is in motion, $-Gm/E$ will no longer be the shell radius at the fiducial epoch as we require to maintain the -1 on the right of equation (5). However, both Gm/r and E are constant for all shells at this epoch [from equation (1) since the velocity is constant—WS] and so the actual Lagrangian label a (which would replace r everywhere in our equations) that we must use to maintain the -1 is simply the shell radius rescaled by a constant factor (H89). This constant factor may be absorbed by the arbitrary constants present in the variables ξ and η, so that we may go on treating the Lagrangian label as the shell radius in hyperbolic self-similarity. Moreover, this complication can largely be ignored even if m/r is not constant at the fiducial epoch (but see H89) for shells that are about to enter the central singularity because, just as for the Friedmann equation in cosmology, their motion is adequately described by the energy equation when $E = 0$. Then for such shells, equation (3) has zero on the right.

Now the numerical simulations (*ibid*) that begin from the perturbed Bonnor/Ebert sphere do not fit precisely *any* of this family of solutions, at least for the accretion rate in the central shells just before $t = 0$. Nevertheless, it also seems from the latter simulations that some kind of localized self-similarity does appear in the central regions just before $t = 0$ *in which the velocity does not deviate greatly from the L/P solution in a region of supersonic flow*. This self-similarity is playing the role of "intermediate asymptote" as in Barenblatt & Zel'dovich (1972) rather than being global in character.

We turn then to seek a more general type of intermediate asymptote in which there is *no* finite epoch at which $m(r) \propto r$ everywhere. In this case, the variables ξ and η are independent and we may write equations (3) and (2) respectively quite generally as,

$$(\eta\partial_\eta S + \xi\partial_\xi S)^2 - \frac{\xi^2}{S} + 2\eta^2 \mathcal{H} = -\xi^2, \tag{7}$$

$$e^{\mathcal{H}} = \frac{1}{S^3(1 - \frac{1}{D(r)}\frac{\partial \ln S}{\partial \ln \xi} - \frac{\partial \ln S}{\partial \ln \eta})}.$$

Moreover, for the purposes of analyzing these equations, it is worth noting that D may be written as,

$$\frac{1}{D(r)} = 1 + \frac{1}{2}\frac{d\ln\alpha_s^2}{d\ln r}. \tag{8}$$

An inspection of equations (7) shows immediately that there is no exact solution that depends solely on η unless ξ/η or α_s is constant, which was the case already identified above. If, however, α_s is very large so that the terms involving ξ explicitly might be dropped, an approximate solution is possible provided $\mathcal{H} < 0$. Thus, if the scaling or fiducial density ρ_0 is the density profile at $t = 0$, then we must expect this approximate solution to break down sometime *before* this core formation epoch. Notice, however, from the second of equations (7) that whenever $S = S(\eta)$ is approximately true, then to the same approximation the scaled density profile is also simply a function of η. This distinguishes it from the solution to follow (Lynden-Bell & Lemos, 1988; H89) unless D or α_s is strictly constant.

We turn therefore to study the possible solutions in which $S = S(\xi)$. The first of equations (7) shows us that we require α_s to be small *and* the logarithmic derivative of S relative to η at constant ξ to be small relative to the corresponding derivative with respect to ξ at constant η. To begin with the second requirement, one calculates easily that

$$[d\ln\xi]_{\eta=\text{const}} \equiv -\frac{1}{2}d\ln\Delta t\left[1 - \frac{\partial\ln m}{\partial\ln\Delta t}\bigg|_{\eta=\text{const}}\right], \tag{9}$$

$$[d\ln\eta]_{\xi=\text{const}} \equiv \frac{1}{3}d\ln\Delta t\left[1 - \frac{\partial\ln m}{\partial\ln\Delta t}\bigg|_{\xi=\text{const}}\right]. \tag{10}$$

The order of magnitude of the corresponding partial derivative in equations (7) is inversely proportional to these differentials for a given change in $\ln S$.

We now invoke the freedom to redefine r for the central shells as being the radius of the shells at some pre-core formation epoch $-t_{\text{ff}}(N) > t_0$ (HAB). Consider first a region wherein $m(r) \propto r^{(1+\epsilon)}$, which is a region where the exact hyperbolic self-similarity applies as $\epsilon \to 0$. When $\eta = \text{const}$ then $r \propto \Delta t$ and one calculates from equation (9) that the right-hand side (RHS) is of $O[(\epsilon/2)d\ln\Delta t]$. When $\xi = \text{const}$ then $r \propto (\Delta t)^{[2/(2-\epsilon)]}$ and one calculates from equation (10) that the RHS is of $O[(\epsilon/6)d\ln\Delta t]$. That is, the partial derivatives in equations (7) are of comparable magnitude, and no self-similar reduction of the equations appears, except when ξ and η become proportional as $\epsilon \to 0$.

We are more interested here in a region wherein $m(r) - m_N \propto r^{(3-\epsilon)}$, which corresponds to a constant central mass m_N plus a nearly uniform "halo". Then when $\eta = \text{const}$, $r \propto \Delta t$ once more, and equation (9) shows that the RHS is finite as $\epsilon \to 0$. However when $\xi = \text{const}$ then $r \propto (\Delta t)^{2/\epsilon}$ and we find the RHS of equation (10) to be of $O(6/\epsilon)$ as $\epsilon \to 0$. Hence this differential diverges and the corresponding partial derivative in equations (7) becomes vanishingly small compared to $\partial S/\partial\ln\xi$. Thus for such shells the principle dependence of S is on ξ.

To complete this proof that elliptic or "ξ self-similarity" can arise dynamically in such regions, we must be able to drop the term in $\mathcal{H}$ from the first of equations (7). This requires $\alpha_s^2\mathcal{H}$ to be small compared to $1/S(\xi)$. From the definition, we can arrange that α_s^2 be small by insisting that m_N be such that $Gm_N/r \gg \alpha_s^2$ for the set of shells in question. As the collapse continues, $\mathcal{H}$ increases only logarithmically, α_s^2 remains small for this domain in r, and $1/S$ increases rapidly. Thus, once established, this type of self-similarity should continue to hold.

Hence, the equations to be solved for shells passing through this "window" of dynamic self-similarity near the singular core (*e.g.*, HAB) are the "zero-pressure" equations [although the derivation shows that we really require locally supersonic motion since $(\partial_\xi S)^2 \gg \alpha_s^2 \mathcal{H}$]:

$$[\partial_\xi S(\xi)]^2 = \frac{1}{S(\xi)}, \tag{11}$$

$$e^{\mathcal{H}} = \frac{1}{S(\xi)^3 [1 - \frac{1}{D(r)} \frac{\partial \ln S(\xi)}{\partial \ln \xi}]}. \tag{12}$$

The first of these equations is the familiar "Friedmann equation" for de Sitter cosmology (the energy of the shells is taken to be 0 since we are only interested in them as they accrete onto the core), except that the independent variable is ξ (H89). Since, moreover, we are interested here in collapse, the appropriate boundary condition is a "forward-looking" $S = 0$ at say $\xi = \xi_s$. The second equation is *not* in general self-similar in the sense of giving a scaled density that depends only on ξ (Lynden-Bell & Lemos, 1988; H89). This turns out to be the essential reason for the varying accretion rate (HAB). The accretion rate only tends toward a constant that is equal to the Shu value, when sufficient halo mass to dominate the constant m_N having an inverse-square density profile has been accreted. At this point, $D(r)$ is also constant for the shells not yet accreted and so the global isothermal self-similarity is re-established for $t > 0$ until finally the shells at the edge of the distribution are accreted.

I feel that this analysis justifies the dynamical emergence of the "ξ" or elliptic type self-similarity in the shells that are close to being accreted *provided that the appropriate mass distribution can arise dynamically*. For in the isothermal sphere (*e.g.*, Chandrasekhar, 1939), α_s^2 is never small. In fact, v is exactly what Chandrasekhar calls $1/v$, and v never exceeds 3 (*ibid* p. 169). It tends to zero at the centre of the isothermal equilibrium. Consequently, between the initial equilibrium and the dynamic onset of the elliptic self-similarity, we require a substantial accumulation of mass in the central regions accompanied by supersonic motion. This might be achieved by the formation of a central core, but we also require a flattish "shoulder" region to exist between the core and inverse-square halo as discussed above. Under these conditions the elliptic self-similarity arises in the shells accreting onto the core, but of course we have not proved that these occur. Fortunately, the observations of flattened density profiles in pre-stellar cloud centres (Ward-Thompson *et al.*, 1994), the quasi-static simulations (*e.g.*, Basu & Mouschovias, 1995; Tomisaka, 1996) with magnetic fields and the numerical simulations of Foster & Chevalier (1993) and of Hunter (1977) all suggest that these conditions are possible under the right conditions. Moreover, it can be shown that the fastest growing perturbation mode about an already centrally flattened isothermal cloud is such as to increase the mass in the flattened region. I take these arguments together as being a reasonable justification for the use of equations (11) and (12) in HAB to explain the class 0 behaviour.

3. Discussion

In this paper I have given a demonstration of the mathematical existence of a gravitationally driven, elliptic, dynamic self-similarity in shells that are just accreting. These solutions permit an explanation of the class 0 bipolar outflows in terms of a transitory peak accretion rate, and the later protostellar classes are understood in terms of a gradual decline to the Shu accretion rate (see HAB). However, certain physical constraints on the mass distribution are required for these solutions to appear. We have not shown these to hold with the same rigour as employed in understanding the localized self-similarity, since they depend on a complex physical history. However, observations, numerical simulations and linear analysis all suggest that in fact these conditions do arise in the transition from isothermal equilibrium cloud to protostellar core-halo object.

Acknowledgments. I thank David Clarke and his organizing committee for inviting me to this meeting and to NSERC for their financial support. Philippe André should be thanked especially for inspiring this work.

References

Abergel, A., et al. 1996, A&A, 315, L329.
André, P., Montmerle, Th. 1994, ApJ, 420, 837.
André, P., Ward-Thompson, D., & Barsony, M. 1993, ApJ, 406, 122.
André, P., Ward-Thompson, D., & Motte, F. 1996, A&A, 314, 625.
Barenblatt, G. E., & Zel'dovich, Ya. B. 1972, Ann. Rev. Fluid Mech., 4, 285.
Basu, S., & Mouschovias, T. Ch. 1995, ApJ, 453, 271.
Bontemps, S., André, P., Terebey, S., & Cabrit, S. 1996, A&A, 311, 858.
Chandrasekhar, S. 1939, An Introduction to the Study of Stellar Structure, (Chicago: University of Chicago Press).
Foster, P. N., & Chevalier, R. A. 1993, ApJ, 416, 303.
Henriksen, R. N. 1989, MNRAS, 240, 917 (H89).
Henriksen, R. N., André, P., & Bontemps, S. 1997, A&A, in press (HAB).
Hunter, C. 1977, ApJ, 218, 834.
Lynden-Bell, D., & Lemos, J. P. S. 1988, MNRAS, 233, 197.
Shu, F. 1977, ApJ, 214, 488.
Stahler, S. W. 1994, in The Cold Universe, eds. Th. Montmerle et al., (Paris: Editions Frontières), 201.
Tomisaka, K., 1996, PASJ, 48, L97.
Ward-Thompson, D., Scott, P. F., Hills, R. E., & André, P. 1994, MNRAS, 268, 276.
Whitworth A., & Summers, D. 1985, MNRAS, 214, 1 (WS).

Gas-Dynamical Simulations of the Circumstellar and Interstellar Media

James M. Stone

Department of Astronomy, University of Maryland, College Park, MD 20742-2421, U.S.A.

Abstract. Numerical simulations of the dynamics of the circumstellar and interstellar media have played a fundamental role in our understanding of a very wide variety of phenomena. Numerical algorithms for hydrodynamics, and even magnetohydrodynamics, are mature enough that robust and flexible "community codes" are available. Applications of these codes to astrophysical problems generally can be classified into three types: (1) studies of the non-linear stages of dynamical instabilities, (2) numerical experiments designed to test a hypothesis, and (3) detailed modeling of astronomical observations. In each case, the motivation and goals of numerical simulations are different. Examples are used to illustrate each mode of investigation, including (1) studies of the non-linear evolution of the magnetorotational instability in weakly magnetized accretion disks, (2) the angular momentum transport associated with vertical convection in unmagnetized disks, and (3) the formation of "bullets" in stellar outflows.

1. Introduction

The evolution of matter in the majority of circumstellar and interstellar environments is best described by a macroscopic, continuum approach based on the equations of gas dynamics. Examples drawn from studies of the circumstellar medium include the structure and evolution of stellar winds, accretion disks, and the early stages of supernova remnants, while examples based on studies of the interstellar medium proper include the investigation of the internal dynamics of molecular clouds, star-forming regions, and the galactic disk as a whole. However, because of the mathematical complexity of the equations which govern the evolution of these systems [the equations of hydrodynamics in the simplest case, the equations of magnetohydrodynamics (MHD) if the medium is magnetized, or the equations of radiation hydrodynamics if the radiation field constitutes a significant fraction of the energy or momentum density in the system], it is clear that time-dependent, multidimensional and non-linear phenomena require numerical methods for detailed investigation. It must be understood that numerical methods are simply another tool for generating solutions to the governing equations, a tool which is no less important, and in many cases much more powerful, than analytic methods. For this reason, the contributions that numerical solutions have made to our understand of the dynamics of these systems is so vast, it is well beyond the capacity of one reviewer to cover (this would be akin to reviewing the role that, say, the Fourier transform has played in astrophysics). Instead, I will try to give a subjective overview of the different kinds of gas dynamical problems that can be best attacked with numerical methods, illustrated with examples drawn largely from my own experience. I hope this will help clarify the ways in which users of numerical methods think about their own simulations, and help guide future workers in the selection and use of numerical methods to solve their own problems.

The vast range of astrophysical environments in which numerical techniques can be applied (as illustrated by the examples given above) implies that a vast range of physical processes must be included, in general. For example, in many circumstances magnetic fields are dynamically important to the system, thus in practice one must solve the equations of MHD. The additional degrees of freedom in MHD (characterized, *e.g.*, by the additional wave families in magnetized fluids) produce much richer and complex dynamical evolution. Similarly, radiation fields add additional degrees of freedom and therefore result in richer solutions. Microphysical effects such as heat conduction, resistivity, optically-thin cooling, ionization and recombination can all affect the dynamics and therefore may be important to follow. It is clear that an essential task for the investigator is to incorporate all of the important physical processes as is possible into each study. Of course, this is offset by the practical consideration that numerical algorithms that can solve the increasingly complex systems of equations encountered by the addition of more physics are often immature, complex, and difficult to implement. As described below, well-tested algorithms for the fundamental system of equations at least are available.

2. Numerical Methods

Fortunately, in recent years, a wide variety of numerical algorithms for gas dynamics have become available, either through descriptions in the literature, or even through release of full codes to the "public-domain". Both grid-based (finite-difference) and particle-based (SPH) algorithms are available. Codes which use explicit time integration (appropriate for following the evolution for a few dynamical times), and time-implicit techniques (appropriate for following evolution for very many dynamical times) are available. Some examples include the ZEUS code [for multidimensional hydrodynamics (Stone & Norman, 1992a), MHD (Stone & Norman, 1992b), and radiation hydrodynamics (Stone *et al.*, 1992)], the TITAN code (for one-dimensional radiation hydrodynamics using an adaptive grid), the VH-1 code (for multidimensional hydrodynamics using the PPM algorithm), and various forms of SPH code (*e.g.*, TreeSPH, Hernquist & Katz, 1989). In addition to these relatively mature and complete codes, a wide variety of papers have appeared describing new algorithms which provide improvements over existing techniques. In fact, many of these papers appear in this conference proceeding.

Of course, the primary advantage of using existing codes is avoiding the often considerable investment in time in implementing algorithms. In fact, with the emergence of "community codes" like those mentioned above, we may finally be past the stage where each new graduate student interested in solving hydrodynamical problems must reinvent the wheel to complete his or her thesis. On the other hand, using existing code does not negate the need for careful testing of the algorithms on the application at hand. Often, when using an existing code, it is used in a new region of parameter space in which the algorithms have not been tested, or perhaps the code will be used on a new machine architecture that will affect its performance, or new physics may be added to an already existing code. In each case, the code must be rigorously tested to ensure any changes that are made are working as expected.

Since any given numerical algorithm (say, a particular finite-difference technique) is simply a mechanism for constructing a solution to the partial differential equations (PDEs) which describe the system, the purpose of testing can be thought of as simply measuring how close the solution to the finite-difference equations generated by the computer is to the actual solution of the underlying PDEs. Mathematically, we would like to put a bound on the error (or difference) between these two solutions. There are two ways in which such a bound can be measured. The first is to compare a numerical solution at some fixed resolution to a known analytic solution to a given test problem, the second is to compare the two numerical solutions to the same problem computed at two different resolutions.

In the first case, the bound can be formulated as an error norm between the analytic and numerical solutions. Studying how the error norm changes with numerical resolution can also be used to measure the convergence rate of the numerical scheme. Since an analytic solution is available, convergence of the numerical solution means the algorithm must be converging to that solution. In the second case, an error norm between the solutions at two different resolutions can again be used to measure both the magnitude and (using at least three resolutions) the convergence rate. In this case, there is no guarantee the algorithm is converging to the *correct* solution to the PDEs. Examples of the use of these methods to test one particular code, the ZEUS code, are given in Stone & Norman (1992a; 1992b) and Stone *et al.* (1992).

3. Examples

Following Norman *et al.* (1991), numerical simulations in astrophysics can generally be classified into three categories:

1. Studying the non-linear regime of known phenomena (for example, the non-linear saturation of a dynamical instability).

2. Testing theoretical hypotheses through numerical experiments. The hypothesis may or may not be well defined.

3. Modeling observational data in as much detail and as realistically as possible.

Thinking about which category to which a given study belongs is often useful for identifying the importance and/or limitations of numerical simulations in obtaining quantitative results. For example, in the first case, numerical techniques are not just an option, they are essential. On the other hand, in the second case, there is often a great deal of freedom in specifying the initial state and parameter space is often large, so that one must try to constrain the experiments as carefully as possible to obtain quantitative, meaningful results. Finally, in the third case one must first decide what aspects of the observations are to be modeled: reproducing the observations in every detail is generally fruitless.

To make these general comments more concrete, examples are used to illustrate each category below.

3.1. Mode 1: Studying non-linear phenomena

Often, numerical methods are the only way to generate non-linear solutions to the equations of gas dynamics. Thus, the utility and importance of numerical techniques to this class of problems are obvious.

Perhaps the best illustration of the need for numerical techniques comes from studies of the non-linear evolution and saturation of dynamical instabilities. Since astrophysical fluids are not amenable to laboratory experiments (which represents the only other means of studying the non-linear stage), many examples of instabilities studied via numerical methods exist in the literature. Here, I will focus on studies of the magnetorotational instability in accretion disks (Balbus & Hawley, 1991; 1992, collectively hereafter BH).

It has long been known that classical viscosity alone could not account for the angular momentum transport and mass accretion rates in accretion disks implied by observations: some sort of anomalous viscosity, whose magnitude is characterized by the dimensionless "α"-parameter, is required. Values of α between 0.1 and 1.0 are inferred from observations. Turbulence in the disk is usually suspected as being the culprit responsible for producing angular momentum transport. However, since hydrodynamic disks are linearly stable, the puzzle becomes: what mechanism produces the turbulence?

By performing a linear stability analysis of a weakly-magnetized Keplerian accretion disk, BH showed there exists a powerful local instability which might ultimately be the source of angular momentum transport in disks. However, definitive answers to such questions require an understanding of the properties of the fully saturated instability in disks. Thus, the instability has been the subject of a number of two- and three-dimensional numerical MHD simulations carried out by Hawley & Balbus (1991; 1992), Hawley et al. (1995; 1996), Brandenburg et al. (1995), and Stone et al. (1996).

The results of these studies demonstrate that in three-dimensions, the instability results in MHD turbulence characterized by a power spectrum which is similar to Kolmogorov. There is strong evidence that the instability acts like a dynamo, amplifying and maintaining weak seed fields at values near equipartition over time scales which greatly exceed the dissipation time. In stratified accretion disks, buoyancy does not appear to be an important saturation mechanism; rather, local dissipation processes dominate the non-linear saturation. However, the buoyant rise of an amplified magnetic field does give rise to a vertical structure in stratified disks consisting of a weakly-magnetized and strongly-turbulent central core embedded in a strongly-magnetized corona. Moreover, the MHD turbulence generated by the instability is associated with vigorous angular momentum transport, producing α values between 0.01 and 0.1.

All of these important results could only have been discovered through numerical investigations. The study of turbulence in fluids, especially magnetized fluids, is simply too complex a subject to make much headway with simplified analytic theories, a fact which is evidenced by the very large community using direct numerical simulations to study turbulence in terrestrial fluids. In fact, now that MHD turbulence has been clearly implicated as the transport mechanism, it is likely that future studies of transport processes in accretion disks will now be based primarily on numerical methods for MHD. Important questions that remain to be tackled include the global structure of the magnetic field generated by the instability, the emission properties of MHD unstable disks, and the relationship between the internal dynamics of the disk and the production of winds and outflows from the disk via MHD mechanisms.

3.2. Mode 2: Testing hypotheses

The second *modus operandi* of numerical investigations is to carry out campaigns of simulations to test a physical hypothesis, or to improve one's intuition about how a physical system behaves in certain circumstances. Stated another way, the numerical experimenter asks: "What happens to the system if it is initialized or perturbed in the following way...?" Often, however, this line of investigation can be very time-consuming. In order to produce quantitative conclusions, one must take great care to ensure the questions being asked are well posed.

As an example of the method, one can again turn to the question of angular momentum transport processes in accretion disks. As stated above, unmagnetized Keplerian disks are linearly stable, thus the only viable means to produce hydrodynamic turbulence in the disk is through non-linear instabilities. It is known primarily through wind tunnel experiments that linearly stable Cartesian shear layers are subject to non-linear shear instabilities. Thus, although no analytic calculation or laboratory experiment exists to support the hypothesis, it has long been asserted that Keplerian disks, by analogy to shear layers, must also possess non-linear instabilities. A mechanism is required to produce non-linear perturbations within the disk, and one attractive idea is that vertical convective motions could be important in optically-thick disks. Fortunately, these ideas can now be tested from first principles using three-dimensional hydrodynamical simulations of the dynamics of convectively-unstable accretion disks.

The angular momentum transport associated with vertical convection in disks has been studied by means of three-dimensional numerical hydrodynamical simulations by Cabot & Pollack (1992), Stone & Balbus (1996), and Cabot (1996). Typically, the simulations begin with a convectively-unstable vertical profile which spans several scale heights in the disk. Convection motions grow and saturate, and the angular momentum transport rate in the disk can be measured directly from the volume-averaged Reynolds stress $\langle \rho V_R \delta V_\phi \rangle$, where V_R and V_ϕ are the velocity fluctuations in the radial and azimuthal directions respectively. Generally, it is found that convective motions are not self-sustaining, but instead heating at the midplane of the disk is required to keep the convective cells going. Shear stretches the cells in the azimuthal direction, producing structures which resemble long (in the azimuthal direction) and narrow (in the radial direction) sheets. Figure 1 shows the angular momentum transport rate expressed as the dimensionless "α"-parameter over 40 orbits of evolution of vertical convective motions in a Keplerian disk. Several points are immediately obvious: first, the angular momentum transport rate is very small and undergoes large fluctuations, and second (and somewhat surprisingly), the time-averaged

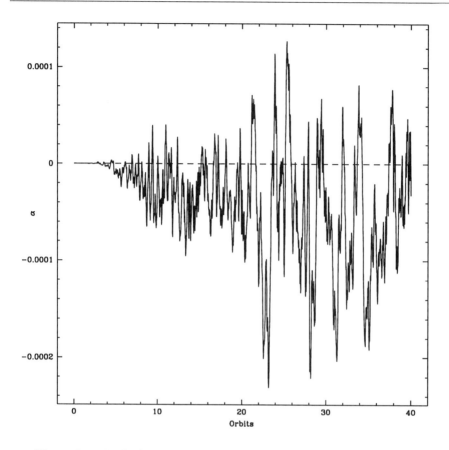

Figure 1. Evolution of the Reynolds stress normalized by the pressure at the disk midplane (equivalent to α) over 40 orbits in a disk in which vertical convection is driven. Note α is both very small and, on average, negative.

value of α is negative. Thus, direct numerical simulations of non-linear convective motions show that not only is there very little angular momentum transport, but in addition it is directed inward!

The surprising result that vertical convection produces inward transport of angular momentum, obtained solely through numerical simulations, motivated further study of the association of hydrodynamic turbulence and transport in disks by Balbus *et al.* (1996). By performing a Reynolds decomposition of the equations of hydrodynamics in a disk, these authors were able to develop a simple analytic theory that shows in a Keplerian disk, the kinetic energy associated with the fluctuating part of the azimuthal velocity would only *increase* (and therefore be associated with instability) if angular momentum transport was *inward*, in agreement with the simulations. Moreover, they were able to predict that for disks with rotation curves that follow a power-law profile, $\Omega(R) = \Omega_0 R^{-q}$ with $q \geq 2$, the reverse would be true, *i.e.*, instability would be associated with *outward* transport. Further numerical experiments also confirmed this prediction. Finally, numerical experiments of Cartesian shear layers demonstrated that if, and only if, non-linear perturbations are supplied, instability and vigorous spanwise momentum transport results, in agreement with the wind tunnel experiments.

Thus, direct numerical simulations have played an essential role in clarifying the non-linear dynamics of hydrodynamic disks. In fact, it could be argued that the simulations "discovered" that hydrodynamical turbulence will be associated with inward transport in Keplerian disks, which was then subsequently confirmed by analytic studies [although see Ryu & Goodman (1992) for a physical argument that convection should result in inward transport which predates the simulations]. Usually, numerical and analytic methods complement each other in the reverse order.

3.3. Mode 3: Modeling observations

A final category of numerical simulations is the detailed modeling of astronomical observations. Often, simulations of this kind are the most difficult, since astrophysical systems are complex, and modeling them requires accounting for *all* the physical processes that effect the radiative emission (including dynamical, microphysical, and radiative transfer effects). Of course, it is always better to be predictive than postdictive. For example, observations of the rings around SN1987A indicate that the progenitor was surrounded by a dense belt of circumstellar gas. The blast wave from the supernova is expected to impact the belt sometime between 2002 and 2005. A number of studies are underway to predict what the emission properties of a strong shock impacting the dense gas will be, all of which can eventually be compared to the actual observations (Blondin *et al.*, 1996; Borkowski *et al.*, 1997). No doubt surprises are in store.

Let me be selfish, however, and illustrate simulations motivated by astronomical observations with another problem from my own experience: explaining the "bullets" discovered in the massive star forming region OMC-1. Deep infrared images of the region collected by Allen & Burton (1993) show emission knots which extend radially away from the embedded infrared source IRc-2 (thought to be a massive star in the late stages of formation). The brightest emission features are located at the tips of the knots. Moreover, the velocity of the knots inferred from matching radiative shock models indicate they were all produced roughly 1,000 years ago within 2 arcseconds of IRc-2. The simplest interpretation would then be some sort of explosive event accelerated the knots into the circumstellar medium, much like shrapnel from a bomb. In fact, such "bullets" seem to be associated with outflows from a wide variety of sources, including planetary nebulae (Harrington & Borkowski, 1994), and supernovae (Strom *et al.*, 1996).

However, the idea that dense fragments of gas can be accelerated to highly supersonic velocities and remain intact has been challenged by numerical simulations of the interaction of strong shocks with dense clouds (Klein *et al.*, 1994; Xu & Stone, 1995; Jones *et al.*, 1996). Without exception, these studies show that dense clouds are fragmented by the acceleration process on a time scale which is comparable to the shock crossing time in the interior of the cloud. A more likely explanation, therefore, is that rather than being accelerated from rest, the "bullets" form *in situ* in an already supersonic flow. Stone *et al.* (1995)

GAS DYNAMICAL SIMULATIONS OF THE ISM

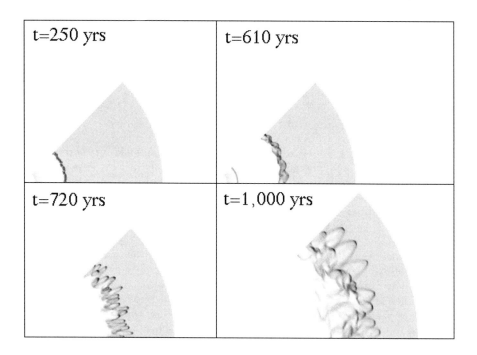

Figure 2. Snapshots of the evolution of the density in a powerful stellar wind as it sweeps-up ambient gas and, at the same time, undergoes a sudden increase in terminal velocity.

provided one demonstration of how such a process could occur. A dense shell of circumstellar material swept up by a stellar outflow will fragment if it is accelerated either by (1) an increase in the ram pressure of the stellar wind, or (2) a decrease in the ambient density which is steeper than R^{-2} (Weaver et al., 1977). Figure 2 shows the results of hydrodynamical simulations that demonstrate the process. A dense shell is swept up by a slow stellar wind in the first frame at a time of 250 years. After 500 years, the wind velocity is increased by a factor of ten, and the higher-velocity wind begins to sweep up a second shell interior to the first. The high-velocity shell impacts the first, accelerating and fragmenting it, so that by 1,000 years, discrete, well-formed "bullets" are evident that can be compared directly with observations.

The simulations, therefore, demonstrate the physical mechanism that can produce bullets in stellar outflows. Moreover, they are used to model the morphology of the observed infrared emission in OMC-1. Much more effort would be required to model the observations in more detail, for example by modeling the detailed microphysics of the gas to compute line-emission profiles and intensity ratios.

4. Summary

Numerical methods have become a fundamental tool for investigating gas dynamical processes in astrophysics. The computer is every bit as important to solving these problems as pencil and paper. As such, reviewing all the contributions of numerical techniques to our understanding of the dynamics of the interstellar and circumstellar medium is intractable. Instead, I have focused on general comments on the use of numerical simulations illustrated with specific examples drawn largely from my own experience.

I have argued that numerical investigations can be classified into three kinds of approaches: (1) studying non-linear processes, (2) numerical experiments, and (3) modeling observations. Because numerical simulations require considerable investments of time to initialize, run, and analyze, it is very useful to think critically about how each simulation will improve one's understanding of a given problem, and whether there is another problem that can be solved more efficiently. Stated another way, always go after the "low-hanging fruit". My experience has been that studies of the first category are the most productive.

Some of the biggest advances have come from adding more physics to the problem. For example, our understanding of the internal dynamics of accretion disks has taken an enormous step forward once the importance of magnetic fields was appreciated. Of course, the trick is to add only the next most important piece to the puzzle; trying to solve the full problem on the first try (the "everything-including-the-kitchen-sink" approach) generally leads to confusion rather than enlightenment.

Finally, never forget the importance of testing your methods, especially if someone else wrote the code you are using. Try to think of ways to use your application as a test problem (for example, when studying the non-linear regime of an instability, check that the growth rates in the linear regime agree with analytic predictions). Given the continued improvements in algorithms and hardware, there appear to be few problems in astrophysics whose solution will not depend on the results provided by carefully formulated, accurate, and realistic numerical simulations.

Acknowledgments. I thank my colleagues Steve Balbus, Charles Gammie, and John Hawley for their invaluable contributions to the work described here.

References

Allen, D. A., & Burton, M. G. 1993, Nature, 363, 54.
Balbus, S. A., & Hawley, J. F. 1991, ApJ, 376, 214.
Balbus, S. A., & Hawley, J. F. 1992, ApJ, 400, 610.
Blondin, J. M., Lundquist, P., & Chevalier, R. 1996, ApJ, 472, 257.
Borkowski, K. J., Blondin, J. M., & McCray, R. 1997, ApJ., 477, 281.
Brandenburg, A., Nordlund, A., Stein, R., & Torkelsson, U. 1995, ApJ., 446, 741.
Cabot, W., & Pollack, J. 1992, Geophys. Astrophys. Fluid Dyn., 64, 97.
Cabot, W. 1996, ApJ, 465, 874.
Harrington, J. P., & Borkowski, K. J. 1994, BAAS, 26, 1469.
Hawley, J. F., & Balbus, S. A. 1991, ApJ, 376, 223.
Hawley, J. F., & Balbus, S. A. 1992, ApJ, 400, 595.
Hernquist, L., & Katz, N. 1989, ApJS, 70, 419.

Jones, T., Ryu, D., & Tregillis, I. 1996, ApJ, 473, 365.
Klein, R., McKee, C., & Colella, P. 1994, ApJ, 420, 213.
Norman, M. L., Clarke, D. A., & Stone, J. M. 1991, Computers in Physics, 5, 138.
Ryu, D., & Goodman, J. 1992, ApJ, 388, 438
Stone, J. M., & Norman, M. L. 1992a, ApJS, 80, 753.
Stone, J. M., & Norman, M. L. 1992b, ApJS, 80, 791.
Stone, J. M., Mihalas, D., & Norman, M. L. 1992, ApJS, 80, 815.
Stone, J. M., Hawley, J. F., Evans, C. E., & Norman, M. L. 1992, ApJS, 388, 415.
Stone, J. M., & Balbus, S. A. 1996, ApJ, 464, 364.
Stone, J. M., Hawley, J. F., Gammie, C. F., & Balbus, S. A. 1996, ApJ, 463, 656.
Stone, J. M., Xu, J., & Mundy, L. 1995, Nature, 377, 315.
Strom, R., Johnston, H. M., Verbundt, F., & Aschenback, B. 1995, Nature, 373, 590.
Weaver, R., McCray, R., Castor, J., Shapiro, P., & Moore, R. 1977, ApJ, 218, 377.
Xu, J., & Stone, J. M. 1995, ApJ, 454, 172.

The Role of Weak Magnetic Fields in Kelvin-Helmholtz Unstable Boundary Layers

T. W. Jones, Joseph B. Gaalaas

School of Physics and Astronomy, University of Minnesota, Minneapolis, MN 55455, U.S.A.

Dongsu Ryu[1]

Department of Astronomy and Space Science, Chungnam National University, Daejun 305-764, Korea

Adam Frank

Department of Physics and Astronomy, University of Rochester, Rochester, NY 14627-0171, U.S.A.

Abstract. We report on high-resolution numerical simulations of Kelvin-Helmholtz unstable shear layers threaded by weak magnetic fields. The simulations demonstrate that even very weak fields can have a major impact on the non-linear evolution of the flows. Fields caught in vortices that form from the instability will be amplified and then relaxed through tearing-mode reconnection. That process can stabilize the boundary layer and/or add substantially to energy dissipation within it.

1. Introduction

Magnetic fields thread almost everything in the universe, while many environments are also fully ionized so that they are highly conducting. Consequently, it is very important to understand how conducting, magnetized fluids behave; *i.e.*, to understand astrophysical magnetohydrodynamics. Rapid improvements in computers, together, over the past decade, with development of robust and accurate algorithms for numerical magnetohydrodynamics, have recently opened up this topic to study in ways that were simply not realistic before (*e.g.*, Stone & Norman, 1992; Dai & Woodward, 1994; Ryu & Jones, 1995; Balsara, 1997). While it is obvious that we want to study many MHD problems that are immediately astrophysical in content; *e.g.*, radio galaxies (Clarke *et al.*, 1989), accretion disks (Hawley *et al.*, 1996), star formation (Basu & Mouschovias, 1995) and supernova remnants (Jun & Norman, 1996), it is perhaps more important at this stage that we focus some attention on understanding the relevant basic MHD processes. While there is a broad awareness of the likely importance of "strong" magnetic fields (generally, those whose pressure is comparable to or exceeds the

---

[1] also, Department of Astronomy, University of Washington, Seattle, WA 98195-1580, U.S.A.

gas pressure), relatively little effort has gone into trying to determine what significance one may attach to the presence of a substantially weaker field. Probably the best known hint that such roles exist comes from the fact that a vanishingly small magnetic field threading an accretion disk may make the disk unstable to form a turbulent structure (Balbus & Hawley, 1991).

Our task in the work described here is to explore one relatively simple, but highly relevant MHD problem that touches on both of the above themes; that is, it highlights some basic MHD physics, and reveals the essential role that can be played by a magnetic field that is energetically so weak that it would ordinarily be considered negligible. This is the problem of the Kelvin-Helmholtz instability in a shear layer threaded by a weak magnetic field. We consider for now a periodic section of a flow and limit ourselves to cases with an initially uniform magnetic field. We begin with flows that exhibit a planar symmetry; that is, we calculate in "2 1/2" dimensions, although we have begun a set of calculations that are fully "3-D". Even with these restrictions, we are in position to explore some aspects of such fundamental issues as magnetic reconnection and flux expulsion from vortices. Results so far from this work are discussed in detail through two published papers (Frank *et al.*, 1996; Jones *et al.*, 1997). Ours is not the first attempt to study the Kelvin-Helmholtz instability in MHD (*e.g.*, Miura, 1990; Malagoli *et al.*, 1996), but we believe these papers constitute the most thorough analysis to date of many of the important issues.

2. The Calculations

Cosmic plasma flows should mostly be described by large kinematic and magnetic Reynolds numbers, so we apply numerical solutions to the ideal equations of MHD. Thus, we depend on numerical dissipation for magnetic reconnection, for example. Our code uses an explicit, "TVD" scheme that is second-order accurate in both space and time (Ryu & Jones, 1995; Ryu *et al.*, 1995). It is fully conservative of mass, momentum and energy and maintains conservation of magnetic flux to machine accuracy as well. Thus, while we cannot define a quantitative model for the numerical viscosity and resistivity, the exchange of energy and momentum among the various components is internally consistent. There is also good evidence in the literature that conservative, monotonic schemes such as this one do a good job of approximately representing physical dissipative processes that are expected to take place on scales smaller than those of the grid (*e.g.*, Porter & Woodward, 1994). We find reasonable convergence in our results towards such important global flow properties as total energy dissipation when we conduct calculations on finer grids (Frank *et al.*, 1996). In addition, we find close resemblance of vortex behaviors to resistive MHD calculations of high magnetic Reynolds numbers. Thus, we are confident that the general behaviors we see are reliable indicators of weakly dissipative MHD.

As mentioned in the introduction, we compute the evolution of a periodic section of a shear layer that is uniform in all properties except for the velocity field. The latter is represented by a hyperbolic tangent distribution, so that the flow faces left in the top of our grid and faces right below. The total velocity change is equal to the sound speed of the plasma. The wave phase speed is zero in this situation, so the patterns that form are at rest with respect to the grid.

The width of the initial transition, $2a$, is kept small compared to the size of the computational box, L, so that reflections from the top and bottom boundaries are minimized. In practice $a/L = 1/25$. The magnetic field lies projected at some fixed angle to the computational plane. We have considered the full range of projection angles. A linear, normal mode perturbation of wavelength equal to the width of the box is applied to this flow (Miura & Pritchett, 1982). Higher perturbation mode numbers can be applied, but the results are identical.

3. Results

In the absence of a magnetic field component in computational plane, the flow evolves into a single, large "Cat's Eye" vortex that spans the computational box. If the numerical dissipation is small, this vortex is stable and will spin for a very long time, until viscous dissipation eventually converts all the kinetic energy into thermal energy. (The box is a closed system, so that outcome is inevitable, even if it is a long time in coming.) Since the cat's eye is stable and long lived, we refer to it and other similar states in the MHD flows as "quasi-steady relaxed states".

The initial growth of the instability can be explained simply by Bernoulli's principle; that is, if there is a corrugation in the flow boundary, flow lines are squeezed along ridges and expanded over valleys. Thus, ridges are "lifted" up by reduced pressure and valleys (which are ridges in the other fluid) are "pushed" down. As demonstrated a long time ago, if the magnetic field has sufficient tension, it can stabilize the boundary (Chandrasekhar, 1960). In particular, for the configuration we have used, a field strength that leads to an Alfvén speed in excess of twice the velocity change across the shear layer will produce a stable flow. For a field only slightly weaker than this critical value, the boundary will be linearly unstable. But, as the field lines are stretched (because a corrugated boundary is longer than a flat one), their tension increases and stabilizes the boundary before it forms into the cat's eye vortex mentioned above. Our interest lies in situations where the field is too weak to prevent formation of the cat's eye. Thus, we are dealing with cases that appear at first glance to behave as in hydrodynamics, so that the magnetic field is not important. This is a misconception, however.

We find two distinct patterns of behavior that apply to this regime. In Jones et al. (1997), we characterized these as "weak field, disruptive" behavior and "very weak field, dissipative" behavior. An example of the "quasi-steady, relaxed" state for each behavior is illustrated in Figure 1. In each case magnetic field lines are wrapped into the cat's eye vortex structure that forms as the Kelvin-Helmholtz instability becomes non-linear. This stretches field lines around the perimeter of the vortex. Eventually, those field lines are brought back onto themselves, so that an "X" point forms. This is unstable to tearing-mode magnetic reconnection. When this happens, field lines develop a new topology that isolates magnetic flux within the vortex and reestablishes the field around the exterior into something similar to its initial configuration. The moment of the first major "reconnection event" is, in our simulations, generally the moment when the total magnetic energy reaches its highest value. We note, however, that

Weak Magnetic Fields in Boundary Layers 149

(a)

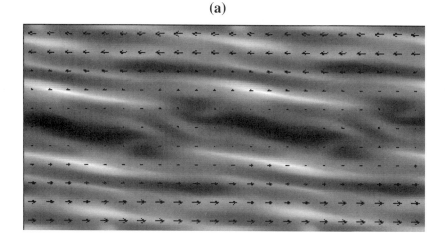

(b)

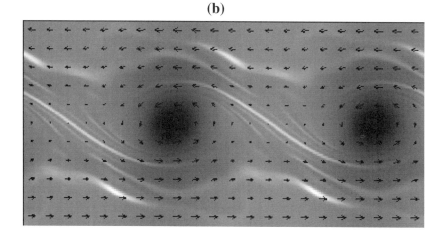

Figure 1. Two images (doubled) of magnetic pressure with velocity vectors superposed for flows that have reached a "quasi-steady relaxed state". High tones are strong fields. Field lines follow fairly well the pressure patterns. (a) "Weak" field outcome, in which magnetic reconnection leads to disruption of the flow vortex. (b) "Very weak" field outcome, in which magnetic reconnection leads primarily to expulsion of magnetic field from inside the vortex and to enhanced dissipation of kinetic energy.

in all these cases the magnetic energy is always less than a few percent of the total.

The two distinct behaviors we mentioned above depend on the magnitude of the magnetic tension where reconnection tears the field. If the tension there is comparable to the Reynolds stress in the vortex, $\sim \rho v^2$, then the formation of the new magnetic topology leads to disruption of the vortex. This may involve a single event if the field is strong enough, or may come through a series of partial disruptions, if the field is not quite strong enough to destroy the vortex at once. Once the vortex is destroyed, the flow relaxes into a broadened, laminar and marginally-stable shear layer. In effect, the field has stabilized the flow through a dynamical realignment of the velocity and magnetic vector fields that converts kinetic into thermal energy. The shear layer is noticeably heated, so that it contains most of the entropy generated through the relaxation process. The relaxed magnetic field is almost uniform, with a total energy very close to its initial value. There is a hint in some of the simulations that flux tubes are forming that bound the new shear layer. Remaining, weak magnetic structures seem to be linearly-polarized Alfvén waves.

For still weaker fields, the cat's eye vortex remains, but a succession of reconnection events on a time scale of the vortex rotation period enhances dissipation of the vortex over what would develop without the field. In that case, field is continuously drawn into the perimeter of the vortex and strengthened. Most of that magnetic energy is then quickly converted into redirected motions and heat through reconnection. The enhanced dissipation rate is consistent with the rate at which magnetic energy is being generated by stretching around the vortex. Thus, it seems clear that the field is the responsible agent for the dissipative excess. We note, as well, that as we increase the numerical resolution in a given simulation, the rate of energy dissipation *increases* while the effective numerical viscosity and resistivity are diminished. This increased dissipation happens because at higher resolution it is easier for the flows to form the magnetic topologies that are tearing-mode unstable. The reconnection requires that the so-called "Lundquist number" be large, and this is easier to achieve at higher resolution.

Finally, we note in those cases where the vortex is maintained, that magnetic field lines within the vortex are almost completely destroyed. This is the well-known "flux expulsion" phenomenon that was studied long ago by Weiss (1966) for example. It is clear from our simulations that this expulsion results not from a steady magnetic diffusion from the interior of the vortex, but from driven, unsteady reconnection, through the tearing mode.

Acknowledgments. This work has been supported at the University of Minnesota by the NSF through grants AST-9318959 and INT-9511654, by NASA grants NAGW-2548 and NAG5-5055 and by the University of Minnesota Supercomputer Institute. Work by DR at CNU is supported in part by the Seom Scholarship Foundation. We are grateful to B.-I. Jun for stimulating discussions.

References

Balbus, S., & Hawley, J. 1991, ApJ, 376, 214.

Balsara, D. 1997, preprint.
Basu, S., & Mouschovias, T. Ch. 1995, ApJ, 452, 386.
Chandrasekhar, S. 1960, *Hydrodynamic and Hydromagnetic Stability*, (New York: Oxford).
Clarke, D. A., Norman, M. L., & Burns, J. O. 1989, ApJ, 342, 700.
Dai, W., & Woodward, P. R. 1994, J. Comput. Phys., 111, 354.
Frank, A., Jones, T. W., Ryu, D., & Gaalaas, J. B. 1996, ApJ, 460, 777.
Hawley, J. F., Gammie, C. F., & Balbus, S. A. 1996, ApJ, 464, 690.
Jones, T. W., Gaalaas, J. B., Ryu, D., & Frank, A. 1997, ApJ, in press (June 10).
Jun, B.-I., & Norman, M. L. 1996, ApJ, 465, 800.
Malagoli, A., Bodo, G., & Rosner, R. 1996, ApJ, 456, 708.
Miura, A. 1990, Geophys. Res. Lett., 17, 749.
Miura, A., & Pritchett, P. L. 1982, JGR, 87, 7431.
Porter, D. H., & Woodward, P. R. 1994, ApJS, 93, 309.
Ryu, D., & Jones, T. W. 1995, ApJ, 442, 228.
Ryu, D., Jones, T. W., & Frank, A. 1995, ApJ, 452, 785.
Stone, J. M., & Norman, M. L. 1992, ApJS, 80, 791.
Weiss, N. O. 1966, Proc. Royal Soc. London, A, 293, 310.

The Jeans Condition: A New Constraint on Spatial Resolution in Simulations of Self-Gravitational Hydrodynamics

Richard I. Klein[1], J. Kelly Truelove[2,3] & Christopher F. McKee[2]

Department of Astronomy, University of California, Berkeley, CA 94720-3411, U.S.A.

Abstract. A key step in the star-formation process is fragmentation. We demonstrate with a new 3-D adaptive mesh refinement hydrodynamics code that perturbations arising from discretization of the equations of self-gravitational hydrodynamics can grow into fragments, a process we term *artificial fragmentation*. This solution-contaminating effect can be significantly reduced and possibly eliminated by ensuring the ratio of cell size to Jeans length, which we call the *Jeans number*, $J \equiv \Delta x / \lambda_J$, is kept below 0.25. We refer to this constraint as the *Jeans condition*. In particular, we show its necessity in star-formation calculations. We demonstrate that the fragmentation in several prominent published calculations is artificial and that collapse to filaments results without further fragmentation. The Jeans condition has important implications for numerical studies of self-gravitational hydrodynamics.

1. Introduction

Jeans' analysis of the linearized equations of one-dimensional, inviscid, isothermal, self-gravitational hydrodynamics for a medium of infinite extent (Jeans, 1928) revealed that perturbations on scales larger than the Jeans length, λ_J, are unstable. Thermal pressure cannot resist self-gravity on such scales, and runaway collapse results.

All methods for numerical solution of the equations of self-gravitational hydrodynamics (hereafter gravitohydrodynamics, or GHD) produce results containing finite errors. These errors act as perturbations to the exact solution. Jeans' analysis therefore suggests that errors correlated on scales exceeding λ_J grow unstably, leading to artificial condensations within the gas. Note that while the source of the perturbations is numerical, the driver of their growth is a physical instability.

---

[1] also, Lawrence Livermore National Laboratory, University of California, Livermore CA 94550, U.S.A.

[2] also, Institute of Geophysics and Planetary Physics, Lawrence Livermore National Laboratory, University of California, Livermore CA 94550, U.S.A.

[3] also, Department of Physics, University of California, Berkeley, CA 94720-7300, U.S.A.

The Jeans Condition 153

Star formation simulations that evolve isothermal collapse across many orders of magnitude in density are especially vulnerable to the growth of artificial perturbations. To simulate collapse from a near-equilibrium initial state, such simulations often start with a spherical cloud in which the temperature and density yield $\lambda_J \sim R$, where R represents the initial cloud radius. During isothermal collapse, the minimum λ_J in the cloud steadily decreases as the density rises. In many published calculations, the minimum λ_J decreases to the scale of the spatial resolution, Δx, and shorter, creating artificial fragmentation. Here we present evidence to support this assertion and assess its impact on calculations of prominent problems in the literature. The fragmentation process may be key to establishing the stellar initial mass function. It is imperative that growth of errors introduced by discretization of the underlying equations be effectively removed from such studies so that the final outcomes of simulations can be clearly related to the initial conditions.

2. Artificial Fragmentation in Multiple-Grid Calculations

GHD collapse problems are notorious for their resolution requirements. Consider a 3-D simulation of uniform cloud collapse across 9 orders of magnitude in density that has a mere 10 zones spanning the cloud radius in the end state. Use of a single Cartesian discretization of the computational volume would require a total of more than 10^{12} cells. This is an implausible number of cells for supercomputer speed and memory to accommodate.

The powerful technology of adaptive mesh refinement (AMR) bridges this gap. Employing AMR, our code discretizes the computational volume onto multiple grids at multiple levels of resolution instead of a single grid at a single level of resolution. It automatically and dynamically inserts and removes grids at different levels during the course of a calculation placing fine resolution only where it is needed. An enormous savings relative to single-grid codes results. Truelove *et al.* (1997; in preparation) describe the development and application of the first 3-D GHD code utilizing the AMR methodology introduced by Berger & Colella (1989).

Using AMR, we have discovered the effect of *artificial fragmentation*, in which perturbations originating in the discretization of the underlying physical equations grow to produce fragments. Consider the Gaussian cloud problem studied by Boss (1991), the isothermal collapse of a 1 $M_\odot$ spherical cloud. The $R = 5 \times 10^{16}$ cm cloud is initially set in solid-body rotation at $\Omega = 10^{-12}$ rad s^{-1} and is given a Gaussian radial density dependence, $\rho(r) = \rho_c e^{-(r/R_1)^2}$, with $\rho_c = 1.7 \times 10^{-17}$ g cm^{-3} and $R_1 = 0.58R$. The sound-speed is set to $c_s = 0.19$ km s^{-1}. To stimulate or "seed" fragmentation, Boss applied a sinusoidal perturbation to the cloud: $\rho \to \rho \times 1.1 \cos(2\phi)$, where ϕ is the azimuthal angle. In the absence of this perturbation, there should be no fragmentation; any fragments that appear in its numerical solution are entirely caused by the discretization of the underlying equations.

In testing our code, we reproduced the fragmentation results of Burkert & Bodenheimer (1996, hereafter BB96) on the Gaussian cloud problem by mimicking the static grid structure of their calculation. *When we repeated the run without the $m = 2$ perturbation, we still found fragmentation. Despite the lack*

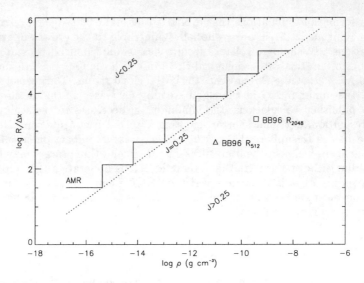

Figure 1.

of a seed perturbation in the input initial conditions, perturbations appeared and grew into fragments of similar characteristics to those in the original run.

Our tests revealed the effect to be resolution-dependent. In GHD a natural length scale exists: the Jeans length. This crucial physical scale must be resolved by the grid that is used to solve the GHD equations. We define the *Jeans number*

$$J \equiv \frac{\Delta x}{\lambda_J} \qquad (1)$$

as a dimensionless measure of the resolution. It is straightforward to use AMR to insert dynamically ever-finer resolution into a calculation to ensure that the ratio of cell size to Jeans length remains below a given maximum value at all times and at all locations within the computational volume.

We made four calculations of the perturbation-free variation of the Gaussian cloud problem in which we used AMR to ensure that J at all spatial locations in the simulation remained below a fixed maximum value $J_{\max} = 0.5, 0.375, 0.25,$ and 0.125, respectively. We denote a resolution $\Delta x = R/N$ by R_N. In each run, we began with a resolution of R_{32}, yielding $J = 0.05$ as the initial maximum value in the cloud. At each time step, the code monitored the subsequent evolution of J and inserted refined grids of 4 times smaller Δx in any region where J reached the fixed value of $J_{\max}$ for the run. As J at that point increased again, the process repeated indefinitely. Figure 1 illustrates the dependence of resolution $R/\Delta x$ upon density utilized in these runs and indicates the manner in which $J < J_{\max}$ was maintained. The points labeled BB96 indicate the resolution of two published calculations at their peak densities.

Figure 2 shows equatorial slices of the log of the density in two different runs at the times at which they each displayed $\log \rho_{\max} \sim -9.5$. The artificial fragmentation in the $J_{\max} = 0.5$ case (*left*) is striking, while traces of perturbation growth in the $J_{\max} = 0.25$ case (*right*) are much more subtle.

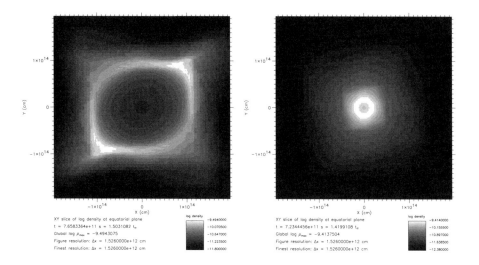

Figure 2.

When the spatial extent of a fluctuation exceeds the Jeans length, the perturbation grows. Our results indicate that to follow collapse across the half dozen or more orders of magnitude separating mean molecular cloud core densities from the breakdown of isothermality requires $J_{\max} \leq 0.25$. We term this new constraint the *Jeans condition*. For this problem, enforcement of $J_{\max} = 0.25$ above $\log \rho = -9.4$ requires use of an *effective* resolution of $R_{131,072}$ or 10^{15} cells.

The Jeans condition arises directly from the GHD PDEs. The growth of numerical perturbations on scales greater than λ_J is thus a *physical* effect that will manifest itself in any GHD code. The magnitude of the perturbations can be reduced through a decrease of the Jeans number. The use of $J_{\max} \leq 0.25$ suppresses discretization perturbation growth well into the density regime—$\rho > 10^{-12}$ g cm^{-3}—at which astrophysical clouds become adiabatic and runaway density growth subsides.

3. Implications for Gaussian and Uniform Cloud Problems

Poorly understood fragmentation appears in several published solutions to two prominent problems: (1) the Gaussian cloud studied by Boss (1991), and (2) the closely related uniform cloud first treated by Burkert & Bodenheimer (1993, hereafter BB93). The discovery of artificial fragmentation prompts us to reexamine these problems to assess the impact of this effect on published solutions.

The Gaussian cloud problem examined by Boss (1991) has been revisited many times (Boss, 1993; Klapp *et al.*, 1993; BB96; Truelove *et al.*, 1997). BB96 treated it with their fixed multiple-grid code. They made two runs in which the finest resolution was fixed at R_{512} and $R_{2,048}$, respectively, with the latter using higher resolution only in a limited region about the origin. Below, we

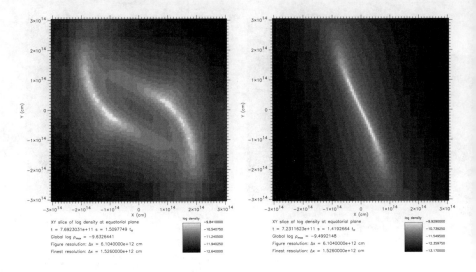

Figure 3.

describe three approaches we took to study this problem and assess the presence of artificial fragmentation in BB96.

(1) We carried out a series of calculations using increasingly higher resolution. We performed runs that began with R_{32} and utilized AMR to maintain $J \leq J_{\max}$. Figure 3 illustrates the results of a run at $J_{\max} = 0.5$ (*left*) which produced a binary and $J_{\max} = 0.25$ (at roughly $\log \rho_{\max} \sim -9.5$, *right*) which resulted in a single bar. We performed an additional run in which we universally doubled the resolution by starting with R_{64} and utilizing $J_{\max} = 0.125$. We carried this highly resolved run to $\log \rho_{\max} = -11.2$ and again found only the single-bar structure seen in the right panel of Figure 3. The R_{32} & $J_{\max} = 0.5$ run had already displayed binary fragmentation by this peak density. We conclude that the binary formation found using R_{32} & $J_{\max} = 0.5$ is artificial, and collapse to a single filamentary singularity represents the correct converged solution. In general, as shown by Inutsuka & Miyama (1992; 1997), perturbed, unstable, inviscid, isothermal clouds form filaments that increase in density without bound.

A study in which the resolution is globally increased by proceeding to both higher initial resolution and lower $J_{\max}$ must be made to determine adequate convergence of quantities of interest. We stress that the Jeans condition is necessary but not sufficient for convergence, since hydrodynamics alone can yield sub-λ_J structure that is poorly resolved at $J = 0.25$.

(2) We repeated the R_{512} calculation of BB96 both with and without the initial $m = 2$ perturbation. In the perturbation-free case, any resulting fragmentation that appears must arise from numerical fluctuations generated by discretization of the underlying equations. The left panel of Figure 4 shows our reproduction of BB96's R_{512} results for the logarithm of the density in the equatorial plane. We began with R_{32} and used AMR to enforce $J_{\max} = 0.25$. J eventually exceeded 0.25 in high-density regions on the R_{512} grid. The AMR grid structure closely

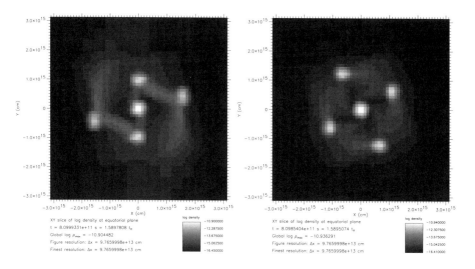

Figure 4.

approximated the fixed grids used by BB96. We present a fuller comparison of results in Truelove et al. (1997; in preparation). The evolution we found agrees well with the data of BB96. The right panel of Figure 4 shows the same calculation except for removal of the $m = 2$ perturbation from the initial conditions. The similarity in gross morphology—e.g., the number of fragments and size scale—is manifest. The signature of the $m = 2$ perturbation appears in the relative displacement of fragments between the figures, but clearly it did not control the essentials of the fragmentation to this point in the evolution. These data indicate that, using a grid structure similar to BB96, fragmentation occurs in the *absence* of any intentionally applied perturbations, and this fragmentation bears a strong resemblance to that found when the perturbation is included. This strongly suggests that the fragmentation found by BB96 is artificial.

(3) We calculated the maximum J reached in the $R_{2,048}$ calculation of BB96. J reached 0.25 when $\log \rho$ rose to -11.8. As the density increased, $J_{\max} = 3.65$. This estimate of BB96's $J_{\max}$ is plotted in Figure 1. Thus they violated the Jeans condition by over an order of magnitude. Again, this indicates that the fragmentation found by BB96 is artificial.

Similarly, we carried out extensive calculations of the uniform cloud problem satisfying the Jeans condition. We found a well resolved binary connected by a thin bar without any evidence of the multiple fragments BB93 found along the bar. We find that BB93 violated the Jeans condition resulting in artificial fragmentation.

4. Conclusions

In this paper we demonstrated that:

1. Perturbations arising from discretization can grow to fragments in a GHD simulation, a process we term *artificial fragmentation*.

2. Artificial fragmentation can be eliminated across the isothermal density regime by ensuring the ratio of cell size to Jeans length, which we term the *Jeans number*, $J \equiv \Delta x / \lambda_J$, is kept at $J \leq 0.25$ throughout a simulation. We refer to this constraint as the *Jeans condition*.

3. When we solved the Gaussian cloud problem keeping $J \leq 0.25$, collapse to a filamentary singularity resulted in agreement with the predictions of Inutsuka & Miyama (1992; 1997) and in marked contrast to previous work.

4. Two methods indicate that artificial fragmentation occurred within prominent published calculations leading authors to false conclusions about the relation of initial to final states. We found (1) fragmentation in the absence of input perturbations and (2) the occurrence of $J > 0.25$ in these calculations.

It is only by carefully ensuring that the Jeans condition is satisfied throughout the evolution of a collapse calculation that one can remove the contaminating effect of artificial fragmentation and begin to address the relationship of initial conditions to end states. We believe these results will have important implications for numerical studies of star formation and cosmology.

Acknowledgments. Research on star formation by RIK and CFM is supported in part by a grant from NASA's Astrophysics Theory Program to the Center for Star Formation Studies. RIK was additionally supported under the auspices of the US Department of Energy at the Lawrence Livermore National Laboratory under contract W-7405-Eng-48. The authors wish to thank the Lawrence Livermore Institute of Geophysics and Planetary Physics and the Pittsburgh Supercomputing Center for their support.

References

Berger, M. J., & Colella, P. 1989, J. Comput. Phys., 82, 64.
Boss, A. P. 1991, Nature, 351, 298.
Boss, A. P. 1993, ApJ, 410, 157.
Burkert, A., & Bodenheimer, P. 1993, MNRAS, 264, 798 (BB93)
Burkert, A., & Bodenheimer, P. 1996, MNRAS, 280, 1190 (BB96)
Inutsuka, S., & Miyama, S. M. 1992, ApJ, 388, 392.
Inutsuka, S., & Miyama, S. M. 1997, ApJ, submitted.
Jeans, J. H. 1928, *Astronomy & Cosmogony*, (London: Cambridge University Press).
Klapp, J., Sigalotti, L. Di G., & de Felice, F. 1993, A&A, 273, 175.
Myhill, E. A., & Boss, A. P. 1993, ApJS, 89, 345.
Truelove, J. K., Klein, R. I., McKee, C. F., Holliman, J. H., Howell, L. H., & Greenough, J. A. 1997, ApJ, submitted.

Warm Interstellar Molecular Hydrogen

P. G. Martin

CITA, University of Toronto, 60 St. George St., Toronto, ON M5S 3H8, Canada

Abstract. A significant fraction of the gas in the interstellar medium in our Galaxy is in the form of molecular hydrogen (H_2), rather than atomic hydrogen, particularly in the dense regions harboring nascent stars. Historically, H_2 has been elusive because cold (10 K) H_2 has no detectable emission. However, when warmed up to about 1,000 K, H_2 emits characteristic spectral lines in the infrared; there exists now a rich variety of observations of warm regions "lit up" in this fashion, including supernova shocks, bipolar outflows, and photodissociation regions near hot young stars. The detailed pattern of emission-line strengths depends on the gas density as well as the temperature, offering a unique and direct probe into the basic physical environment of this predominant species. Numerical simulation of these regions requires as basic input state-to-state rate coefficients for excitation and dissociation of H_2 by collision partners H, He, and H_2. Even for these relatively simple systems involving the most abundant element in the universe, this problem is not easily solved. The interstellar medium is so far from equilibrium that appropriate rate coefficients cannot be deduced from laboratory measurements—the same situation as for much of astrochemistry. Thus, first-principles calculations must be made for "cosmic" applications. An extensive computational program exploiting cluster (parallel) computing to remedy this situation is described. Illustrative results of collisions of H_2 with H_2 showing both non-dissociative energy transfer and dissociation are presented.

1. Introduction

1.1. H_2 in the interstellar medium of the Galaxy

A major feature of our Milky Way Galaxy is the flattened disk of stars, which includes the Sun. Within the disk there is an important interstellar medium. An excellent multiwavelength introduction to the various components can be seen at http://adc.gsfc.nasa.gov/mw/milkyway.html. Stars are prominent in the optical representation from the Digitized Sky Survey and in the near infrared (1.25, 2.2, and 3.5 μm) from the Diffuse Infrared Background Experiment instrument on the Cosmic Background Explorer. Further in the infrared [12, 60, and 100 μm imaging from the Infrared Astronomical Satellite (IRAS)], most of the emission is from interstellar dust re-emitting absorbed starlight. Atomic hydrogen is mapped throughout the Galaxy using the 21-cm line, which gives velocity information as well, as in the recent Leiden-Dwingeloo Survey of H I.

Cold clouds of interstellar gas are seen as shadows against background thermal X-ray emission in the soft X-ray imaging (0.25, 0.75, and 1.5 keV) from the Röntgen Satellite. Diffuse gamma ray emission is also a tracer of material in the interstellar medium. The gamma rays are produced by two processes: i) a cosmic ray colliding with an interstellar proton (whether as an ion, atom, or in a molecule) creates a π^0 meson, which subsequently decays to two gamma rays with a broad spectrum peaked near 70 MeV, half the rest mass energy, and extending as a power law to energies as high as 100 GeV; ii) bremsstrahlung of cosmic ray electrons on nucleons, giving a power law dominating at low energies which fades to about equal to case i) at 70 MeV.

Neither of these indirect methods identifies H_2 regions uniquely. The traditional tracer of H_2 is the CO molecule, whose $1 \to 0$ rotational transition has been mapped throughout the Milky Way at Columbia University and Cerro Tololo, with resolution comparable to the original H I surveys.

Multiwavelength data lead to a deeper understanding of the interrelationships of the various components of the Milky Way and so are being acquired at increasingly higher resolution and sensitivity. A quadrant of the Galactic plane is being imaged with 1' resolution in the 21-cm line and adjacent continuum (with polarization) by the Synthesis Telescope at the Dominion Radio Astrophysical Observatory (DRAO; http://www.drao.nrc.ca/web/survey.html). The resulting panoramic mosaics are to be used in conjunction with CO data for the same region from Five Colleges Radio Astronomy Observatory, lower frequency continuum imaging (DRAO, Mullard Radio Astronomy Observatory, Beijing Astronomical Observatory), and HiRes reanalysis of the IRAS data.

1.2. Formation of H_2

A significant fraction of the gas in the interstellar medium in our Galaxy (about half near the Sun, increasing with decreasing Galactocentric radius) is in the form of molecular hydrogen (H_2) rather than atomic hydrogen, particularly in the dense regions harboring nascent stars. The formation of H_2 via gas phase processes is too slow and so we look at grain surfaces as catalysts. The formation rate (cm^3 s^{-1}) is $F \approx sv\Sigma/2$, where s is the sticking coefficient (~ 1), v is the thermal velocity ($\sim 10^5$ cm s^{-1} at 80 K), and Σ is the projected surface area of dust ($\sim 10^{-21}$ cm^2 per H nucleon, found from the grain size distribution derived from the interstellar extinction curve; *small* grains are important—also for photoelectric heating). Evaluating for these numbers and typical hydrogen density $n_H \sim 100$ cm^{-3} in "clouds", the formation time scale $\tau = 1/(n_H F) \approx 10^7$ yr—very short by Galactic standards. Clearly, destructive processes (§ 2.1) must be important in establishing a steady-state molecular fraction.

1.3. Outline

Historically, H_2 has been elusive because cold (10 K) H_2 has no detectable emission. In § 2, I discuss various spectroscopic aspects of H_2 as an introduction to how H_2 is actually detected, most widely by infrared rotation-vibration transitions in a warm gas. When the H_2 lines are detectable, it is possible to probe the physical conditions in the gas directly: radial velocity (via Doppler shift); density n and temperature T (via relative line strengths; § 2.3); column density N_{H_2} (via absolute line strengths); global cooling (and heating); dissociation.

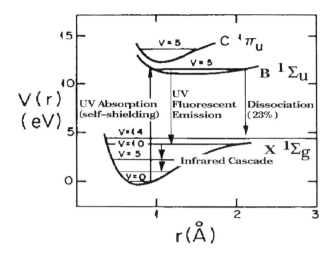

Figure 1. Interaction potential energy as a function of internuclear separation r in H_2 for the ground and two excited states. Electronic transitions, such as the $(5,0)$ Lyman band shown on left, have energies less than the H ionization limit (13.6 eV). Absorption leads either to fluorescent ultraviolet emission plus an infrared cascade or to transitions to the vibrational continuum, whence dissociation (23% overall probability for Lyman bands; much less for the Werner bands).

For detailed interpretation, it is necessary to carry out astrophysical simulations of the excitation conditions in the line-emitting regions. Key ingredients are collisional rate coefficients, which need to be computed from first principles as described in § 3. This requires enormous computational resources (§ 4). In § 5, I discuss how these can be mustered on a network of "workstations", using specifically Parallel Virtual Machine (PVM) software. Some preliminary results for collisions of H_2 with H_2 (denoted $H_2 + H_2$) are presented in § 6.

2. Spectroscopy: Direct Observations of H_2

2.1. Electronic transitions in the ultraviolet

Electronic transitions are highlighted in Figure 1. Ultraviolet absorption measurements of diffuse or translucent interstellar clouds projected against hot stellar continua have established N_{H_2} and physical conditions ($T \sim 80$ K). In gas near hot stars, such as in H II regions or bright reflection nebulae and even in the diffuse background, ultraviolet fluorescent emission can be detected. More accessible, in the near infrared, is the cascade through the vibration-rotation $[(v, j)]$ levels of the ground electronic state. A significant fraction of the time, ultraviolet absorption leads to dissociation; this is what regulates the steady-state abundance of H_2, with self-absorption being an important factor.

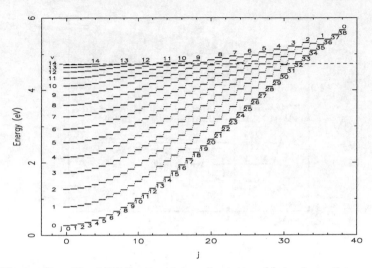

Figure 2. The 348 vibration (v) and rotation (j) levels in the ground electronic state of H_2. The energy spacing in j is fairly close to that in v, unlike for heavier molecules like CO. The dissociation energy is indicated by the dashed line.

2.2. Warm H_2: infrared emission in shocks and photodissociation regions (PDR)

The ground electronic state has 348 vibration (v) and rotation (j) levels (see Figure 2). In a warm gas in excess of 1,000 K, the (v, j) levels are collisionally excited. H_2 then emits via quadrupole transitions in the infrared. Excitation by ultraviolet pumping can be important too (see § 2.1), with collisional redistribution among the (v, j) levels in regions above the critical density (see § 2.3). There now exists a rich variety of observations of warm regions "lit up" in this fashion, including supernova shocks, bipolar outflows, and photodissociation regions near hot young stars (Orion provides good examples: OMC1 outflow shocks; Orion A PDR). A prominent line in the K passband is the 1–0 S(1) line at 2.12 μm, shown in Figure 3. Common upper-level triplets like this are useful for differential extinction measurements. The lowest energy rotational transition 0–0 S(0) at 28.2 μm has now been detected by the Infrared Space Observatory.

2.3. Critical density

The populations of the (v, j) levels are influenced by a competition between radiative and collisional de-excitation. For example, consider the population of the (1, 3) level of H_2. The sum of the Einstein A spontaneous emission coefficients gives $\Sigma A = 8.3 \times 10^{-7}$ s^{-1}; the lifetime is ~ 14 d. Collisional depopulation (at 2,000 K via collisions with H), using the sum of the collisional rate coefficients, yields $\Sigma \gamma = 1.3 \times 10^{-10}$ cm^3 s^{-1}. [This detailed result can be checked roughly as follows: collision cross section $\sigma \sim 10^{-16}$ cm^2 ($r \sim 0.6$ Å) and speed $v \sim 10^6$ cm s^{-1}, so that $\gamma \sim \sigma v \sim 10^{-10}$ cm^3 s^{-1}.]

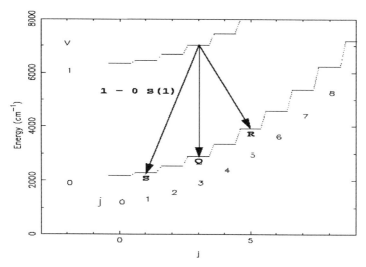

Figure 3. Detail showing three of the many infrared transitions. P and R branches are forbidden by *quadrupole* selection rules.

The critical density is defined to be $\Sigma A/\Sigma\gamma$ which evaluates as 6×10^3 cm^{-3} for the above example. Populations are most sensitive to n and T near the critical density. Note the important coincidence that for a warm gas the critical densities for many levels are in a range that is *interesting astronomically*!

3. Collisional Thermal Rate Coefficients

I am engaged in an ambitious effort relating to collisional thermal rate coefficients for H_2, key aspects of which are described in Figure 4. In the interstellar medium, the potential collision partners are H, another H_2, and He. The target H_2 molecule has 348 possible initial states (Figure 2) and so there are *many* state-to-state coefficients (see, *e.g.*, § 4).

3.1. Interaction potential or potential energy surface (PES)

Interaction energies on an m-dimensional grid are found via *ab initio* quantum calculations. For H_2, $m = 1$ (see, *e.g.*, Figure 1). Collisions H + H_2 require the H_3 PES for which $m = 3$. Most challenging is the new focus, $H_2 + H_2$; the H_4 PES has $m = 6$ and required some 20,000 *ab initio* calculations. A suitable six-dimensional analytic function has been fit (hundreds of parameters) to these data (the required analytic derivatives of this function have been derived too).

3.2. Collision cross sections

Collision cross sections are computed via quasi-classical trajectories (QCT). Given an initial (v, j) state and energy E, the other initial conditions (orientation, rotational and vibrational phase, impact parameter) are *Monte Carlo* selected. The trajectory (N-body dynamics with $N = 3$ or 4) is computed using forces derived from the quantum PES. The trajectory outcome is then analyzed,

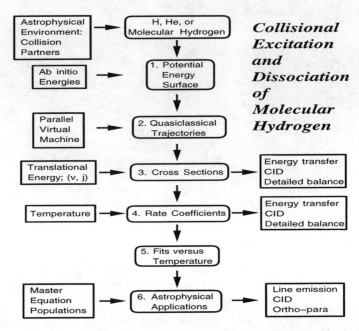

Figure 4. Systematic approach to a comprehensive solution of basic H_2 collisional interactions.

assigning quantum numbers v'' and j''. Outcomes are binned into appropriate quanta (v', j') to evaluate state-to-state cross sections $\sigma(E)$ for $(v, j) \to (v', j')$. See for example Figure 5.

3.3. Thermal rate coefficients

QCT calculations of $\sigma(E)$ are repeated for many E, then integrated over E using a Maxwellian distribution for the appropriate temperature T; this gives $\gamma(T)$ for $(v, j) \to (v', j')$. This integration is carried out for many values of T. An analytic fit to γ as a function of T is found for ease of use in simulations.

At the meeting itself, an $H + H_2$ example was presented (omitted here for lack of space) showing these steps for the $(0, 0) \to (0, 2)$ transition, and emphasizing the importance of using an accurate PES (Boothroyd et al., 1996).

4. Production Requirements of QCT

For $H + H_2$ (Martin & Mandy, 1995), this may be summarized as follows: 348 initial (v, j) states $\times$ 100 values of $E \times$ 4,000 Monte Carlo selections required about 10^8 trajectories. At about 1 s per trajectory on a Sparc 2 (a relevant benchmark at the time), this took the equivalent of about 2.5 cpu yr. The product is about 180,000 state-to-state thermal rate coefficients (obeying detailed balance) at each of 22 temperatures, with analytic functions fit over the range 600–20,000 K (good for shocks and PDR's) for each $\gamma(T)$.

My current project, $H_2 + H_2$ collisions, is *much more demanding computationally*. For one thing, the H_4 PES and its derivatives are harder to evaluate and so the four-body trajectories are individually 50–100 times more costly. Furthermore, there are 348 initial (v, j) states not only for the target H_2 but now also for the collision partner (the projectile H_2). For even a single state of the collision partner, a product comparable to that for $H + H_2$ would take ~ 125 Sparc 2 cpu yr. A faster processor (*e.g.*, DEC Alpha or SUN Ultra 1) would still take 5 cpu yr, again for just a single state of the collision partner. This highlights the *need for "parallel" computations*.

5. Parallel Production using PVM

Parallel computations for the trajectories have been implemented on a network of workstations using PVM (Parallel Virtual Machine), public domain software from Oak Ridge National Laboratory. Among the features of this implementation are: *i*) the Monte Carlo nature of the problem lends itself to a master-slave model; *ii*) it is coarse-grained parallel at the trajectory level; *iii*) the master determines a list of which trajectories to run, farms them out to the slaves one by one, keeps track of which have been completed, and archives the results; *iv*) there is low network traffic per trajectory: a few initial conditions and a few results; *v*) the master-slave interaction is completely asynchronous; *vi*) slaves are cpu-bound (using more than 95% of cycles on unloaded nodes) which leads to effective utilization of "idle" cycles; *vii*) slaves have a low impact on other users of workstations: slaves run at low priority ("nice" 19) and the executable and PVM demons are small (typically 3.4 Mb and 1.4 Mb, respectively); *viii*) load-balancing software actively monitoring the load on each node can effectively idle a slave using various criteria; *ix*) the master is robust against slave slow-down or outright failure: when the list to be farmed out is exhausted and the master detects that a given trajectory has not completed, another slave is given the same task; and *x*) slaves can run on different architectures.

Even in the development stages, sustained speeds of 100's of Mflop s^{-1} were possible on two separate networks of workstations (statistics were shown for one with 60 slow nodes and another in CITA with a heterogeneous collection of faster machines). This has already produced some interesting results (§ 6). The particular parallel implementation scales well and can take advantage of new architectures and equipment brought on the network for other purposes. For final production runs I am installing a dedicated cluster of stripped-down Pentium-based PC's (running LINUX); benchmarks indicate that this is the most cost-effective way to achieve the required Gflop s^{-1} performance.

6. A Glimpse at New Results for $H_2 + H_2$ Collisional Interactions

6.1. Energy transfer

For given initial conditions (target and energy), the distribution of products is sensitive to the choice of collision partner (Figure 5). The upper portion shows the case in which the collision partner is an H atom. Two panels are needed: the left is for non-reactive outcomes, and the right is for the situation in which

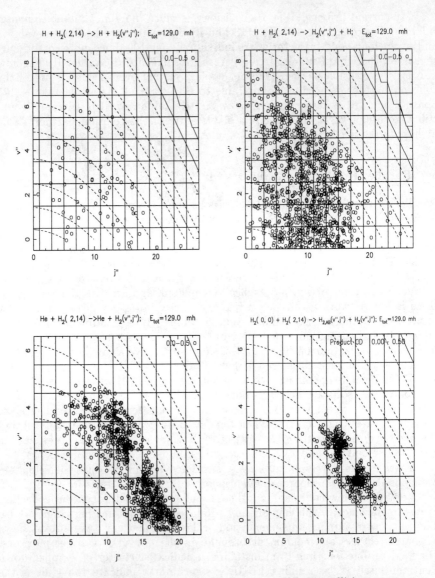

Figure 5. Sensitivity of trajectory outcomes to different collision partners under similar conditions: target $H_2(2,14)$, same total energy [internal energy relative to $(0,0)$ plus translational] of 129 mhartree (compared to 164.6 mhartree for the dissociation limit of one H_2 molecule), and impact parameter range for trajectories restricted to $0 - 0.5$ Å (full range considered, but not presented here). Dashed contours for equally-spaced energies (solid for dissociation limit); extra contour at total energy [through $(5,16)$]. Upper: H collision partner. Non-reactive on the left, with elastic outcomes in the $(2,14)$ bin not plotted. Reactive (exchange: para H_2 bins shown) on the right. Lower: He (left panel) and $H_2(0,0)$ collision partners.

exchange occurred. For central trajectories like these, exchange is the more probable process (921 out of 1,000 trajectories, compared to only 74) but the relative prominence decreases with impact parameter. For both exchange and non-reactive outcomes, the products are widely distributed in (v'', j'') compared to the original $(2, 14)$, occupying states up to the maximum energy allowed [the *extra* energy contour, crossing bin $(5, 16)$]. Because of this type of behavior, no restriction should be placed on the types of collisional transition. The cumulative result of many different transitions with substantial Δj and/or Δv can have an important quantitative effect on the overall excitation of the molecule. Thus, we have recommended including all collisional transitions among the (v, j) states in simulations (Martin *et al.*, 1996), rather than trying to guess which ones might be the most important in different environments.

In the lower portion of Figure 5, results for He as collision partner are shown on the left side (there is of course no exchange channel). Of 1,000 trajectories, 814 were inelastic [not in the $(2, 14)$ bin]. Compared to the above case with H, the products are not so spread out in (v'', j''), tending to lie along a contour of equal energy (diagonal transitions). High energy states approaching the maximum energy available are simply not accessed. Nevertheless, the spread is substantial so that the interaction cannot be described by only a few state-to-state transitions. Final judgement on He should await results with a more accurate HeH_2 PES, under development.

When the collision partner is $H_2(0, 0)$, I find results like those shown in Figure 5 on the lower right (there is no exchange channel until the energy rises to close to the dissociation energy). The outcomes for 457 of 1,000 trajectories were inelastic with respect to the target (CD) molecule, moving it outside the $(2, 14)$ bin. The products from the target are even less spread out than for the case with He. The collision partner (AB) becomes slightly excited, but typically only rotationally to $(0, 2)$. Of the three cases studied, the outcomes are closest to being elastic. The new results are tantalizing. It appears that it might be possible to restrict the number of transitions studied for $H_2 + H_2$.

In the three-way comparison, $H_2(0, 0)$ appears closer in behavior to He than to H (the zero-order expectation). But there is much more to be said than space permits here: impact parameter dependence; effects of "structure" and "size" of the collision partner; dependence on mass, velocity, momentum, *etc*. And of course, many different initial states and energies have to be explored.

6.2. Collisionally-induced dissociation (CID)

I have examined the results of trajectories that had sufficient energy to dissociate at least one of the molecules. Most H_2 initial states in these exploratory calculations have been the ground states, $(0, 0)$ or $(0, 1)$. Figure 6 shows the cross section for CID, comparing three collision partners. For collision partner H_2, the dynamical threshold energy appears to be close to energetic (~ 0.17 hartree), like the case with collision partner H. However, there is a dramatic difference in scale. With H, the cross section increases more steeply and is some two orders of magnitude higher in the 0.2–0.3 hartree range; the corresponding γ's at temperatures of several thousand K are similarly higher. Thus, even a tiny fraction of H relative to H_2 in the gas could have a dramatic effect. Dissociation in astrophysical conditions often occurs preferentially from highly excited states (following

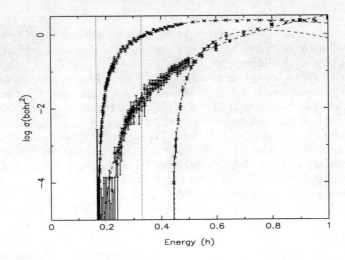

Figure 6. CID cross section for target $H_2(0,0)$ with collision partners H (curve on left), He (on right), and $H_2(0,0)$ as a function of translational energy. Vertical lines correspond to energetic thresholds for dissociating one or two H_2.

ladder climbing) even with sub-thermal populations (*e.g.*, Martin *et al.*, 1996), and so determining the relative behavior in this situation will be important.

Most of the $H_2 + H_2$ dissociations observed in the lower part of the energy range shown occurred by exchange. Many more exchanges occurred, with a somewhat lower apparent dynamical threshold, but the molecules remained bound. CID for $He + H_2$, with no exchange channel, has a significantly elevated dynamical threshold; it appears that the existence of an exchange channel is important in keeping the dynamical threshold for dissociation close to energetic.

Whether or not an H_2 molecule behaves the same as He as a collision partner depends on the energy and type of transition; this is rooted in fundamental differences in the shapes of the relevant potential energy surfaces, which will be explored. For CID at least, as opposed to energy transfer, the collisional interaction $H_2 + H_2$ is closer in nature to $H + H_2$ than to $He + H_2$.

Acknowledgments. I am grateful for the collaborative efforts of my colleagues and for grant support from NSERC of Canada.

References

Boothroyd, A. I., Keogh, W. J., Martin, P. G., & Peterson, M. R. 1996, J. Chem. Phys., 104, 7139.

Martin, P. G., & Mandy, M. E. 1995, ApJ, 455, L63.

Martin, P. G., Schwarz, D. H., & Mandy, M. E. 1996, ApJ, 461, 265.

GRAPESPH with Fully Periodic Boundaries: Fragmentation of Molecular Clouds

Ralf Klessen

Max-Planck-Institut für Astronomie, Königstuhl 17, 69117 Heidelberg, Germany

Abstract. Smoothed particle hydrodynamics (SPH) is a Lagrangian method using a particle approach to solve the hydrodynamical equations. This makes it suitable for combination with the special hardware device GRAPE. Because of its specific design, GRAPE cannot treat periodic particle distributions directly. However, many astrophysical problems require periodic boundary conditions. This limitation of GRAPESPH can be overcome by computing a periodic correction force for each particle (Ewald correction) on the host computer using a PM scheme. We apply this method for studying the dynamical evolution and the fragmentation process in the interior of giant molecular clouds.

1. The Special Purpose Hardware Device GRAPE

GRAPE is a special purpose hardware device, which calculates the forces and the potential in the gravitational N-body problem by direct summation on a specifically designed chip with high efficiency (Sugimoto et al., 1990; Hut, these proceedings). The force law is hard-wired to be a Plummer law. GRAPE furthermore returns a list of nearest neighbors for each particle. This feature makes it suitable for use in SPH (Bate et al., 1995; Steinmetz, 1996).

2. Implementing Periodic Boundaries—the Ewald Method

Since forces are computed by direct summation, GRAPE can only handle isolated N-particle systems, which limits its applicability to strongly self-gravitating systems. Many astrophysical topics, like cosmological simulations or studying the fragmentation process in the interior of molecular clouds, demand more complex boundary conditions.

In 1921, P. Ewald suggested a method to compute the forces in an infinite, periodic particle distribution. Solving Poisson's equation, he realized that convergence can be improved considerably by splitting the Green's function into a short-range and a long-range part and treating the first in real space and the latter one in Fourier space. (For a more recent article, see Hernquist et al., 1991.)

We calculate the periodic forces pairwise on a grid. Since we are interested in a force correction, we subtract the contribution of the isolated pair. Using an FFT, we transform this table into Fourier space and obtain the correct

Green's function for the correction term. (This is done once at the begin of the simulation.)

At each time step, we get the isolated solution via GRAPE. We then assign the particle distribution to a density field defined on a grid, transform this field into Fourier space, convolve it with the corrective Green's function and transform back into real space. Assigned back onto the particles, these values are added to the isolated solution to obtain the correct periodic forces.

This approach has several advantages compared to pure PM methods:

- As a Poisson solver and for computing the neighbor list needed for SPH, GRAPE is very efficient since these operations are hard-wired.

- SPH (and GRAPE as well) is Lagrangian and can resolve strongly-structured fields with high density contrast, since it is not bound to a mesh. The influence of the boundaries on strongly self-gravitating clumps is minute. Thus it is sufficient to compute the periodic correction on a relatively coarse grid (typically 32^2 mesh points) whereas GRAPE takes care of the high density.

- To reduce computational cost spent on the Fourier transform, we compute the FFT only when a minimum number of particles has to be moved (in an individual time step scheme) and use the old values in between.

As always, there are some drawbacks. The method of force computation on the GRAPE board and in the PM scheme are completely different. This can lead to misalignment when combining the forces returned by GRAPE and by the PM routine. This misalignment is mostly caused by the discretization in the particle assignment in the PM scheme. It is negligible for the inner parts of the mesh, but can be as large as a few percent at the border of the grid.

However, this misalignment effect can be minimized by randomly shifting the simulation box through the periodic particle distribution at each call of the FFT. As long as one keeps the periodicity, this does not alter the physics of the simulation and prevents the errors in the border cells from adding coherently. On average, the errors cancel and the particles stay close to their theoretical trajectories.

A more detailed discussion of the method described here and its numerical stability is given in Klessen (1997).

3. Fragmentation of Molecular Clouds

The dynamical evolution of a molecular cloud is very complicated and far from being understood in detail. Density-size or linewidth-size relations indicate that the cloud is to a large extent supported by "turbulent motion". Observation of Zeeman splitting indicate the presence of magnetic fields. Feedback mechanisms from newly-formed stars, outflows, stellar winds, ionization fronts and finally supernovae, produce shells and bubbles and deposit huge amounts of energy and momentum into the interstellar medium. Observations of molecular clouds thus reveal extremely complex structure; a hierarchy of clumps, subclumps, filaments, and shells on all scales have been resolved.

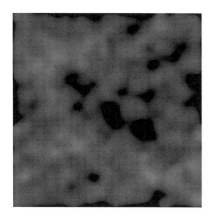 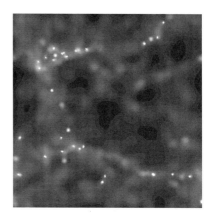

Figure 1. Evolution of a Gaussian density field with $P(k) \propto 1/k^2$: Initial condition (left) and after one free-fall time (right). The density scale is logarithmic and equal in both images.

As a first step to assess the problem of fragmentation and clump formation, we neglect many of the above effects and start in the most basic way: we follow the dynamical evolution of isothermal gas in a region in the interior of a molecular cloud, starting from an initial density distribution to the formation of self-gravitating clumps. To prevent global collapse, we assume periodic boundary conditions. This interplay between gravity and gas pressure by itself will already produce hierarchical filamentary structure (see de Vega et al., 1996).

As an illustration, Figure 1 shows two snapshots of a simulation with 100,000 particles. The initial state (left image) is a Gaussian density field with power spectrum $P(k) \propto 1/k^2$ and evolves into a network of filaments and knots. The image on the right depicts the system after one free-fall time[1]. The gas temperature was chosen to be very low so gravity dominates strongly over gas pressure and a large number of collapsing cores form.

A more detailed investigation will be described elsewhere (Klessen & Burkert, in preparation) and future work will attempt to include a more elaborate physical description into the numerical model.

References

Bate, M. R., Bonnell, I. A., & Price, N. M. 1995, MNRAS, 277, 362.
Ewald, P. P. 1921, Ann. Physik, 64, 253.
de Vega, H. J., Sanchez, N., & Combes, F. 1996, Nature, 383, 139.
Hernquist L., Bouchet, F. R., & Suto, Y. 1991, ApJS, 75, 231.
Klessen, R. 1997, submitted to MNRAS.
Steinmetz, M. 1996, MNRAS, 278, 1005.
Sugimoto, D., et al. 1990, Nature, 345, 33.

---

[1] as defined for the isolated system: $\tau_{\rm ff} = \sqrt{\frac{3\pi}{32 G \langle \rho \rangle}}$.

Origin of Episodic Outflows in YSO and AGN

Rachid Ouyed

Department of Physics and Astronomy, McMaster University, Hamilton, ON L8S 4M1, Canada

Abstract. Energetic, highly-collimated, bipolar jets are observed in active galactic nuclei, in regions of star formation, and in binary stellar systems that harbour black holes or white dwarfs. Jets are also observed to be intrinsically time-dependent phenomena whose episodic eruptions lead to moving knots in the flow. We demonstrate, using time-dependent magnetohydrodynamical simulations, that magnetic fields anchored in a gaseous accretion disk orbiting a star or black hole can accelerate and collimate a bipolar jet which originates from the surface of the disk. Moreover, for certain magnetic field geometries, we find that episodic eruptions originate within the jet close to the central object.

1. Introduction and Set Up

One of the fundamental questions raised by observations of astrophysical jets concerns the mechanism for initiating and driving the outflows. Among the scenarios currently being actively investigated are magnetohydrodynamic (MHD) models in which magnetized accretion disks, rotating around central objects (proto-stars, black holes or white dwarfs), are involved. All of the analytical models to date are based on solutions to the *time-independent* equations (*e.g.*, Blandford & Payne, 1982, hereafter BP; Heyvaerts & Norman, 1989, hereafter HN; Pelletier & Pudritz, 1992, hereafter PP). Therefore, important features observed in nature such as episodic outflows (Eislöffel & Mundt, 1992; Davis *et al.*, 1995 to cite only a few) or the dynamics during the initiation of these winds, cannot be investigated. A *time-dependent* approach to this problem is necessary, but because of the complexity of the MHD equations, direct numerical simulation is currently the only effective tool.

We present 2.5-dimensional time-dependent simulations of the evolution of non-relativistic outflows from Keplerian accretion disks orbiting low-mass protostars or black holes accreting at sub-Eddington rates. We use the ZEUS-2D code as presented in Ouyed (1996). We adopt cylindrical coordinates (r, ϕ, z) with the central object at the origin, and take the z-axis to be perpendicular to the disk, such that the *surface* of the disk lies in the $z = 0$ plane of our coordinate system. The gas is injected at a very small speed ($v_{\rm inj} = 10^{-3}\, v_{\rm K}$) from the surface of the disk (a fixed boundary in our simulations) into a cold corona. The corona is in stable equilibrium and is supported by Alfvénic turbulent pressure (Dewar, 1974).

In order to allow a simple, self-consistent, initial coronal state to be established, we consider initial configurations whose Lorentz force $\mathbf{J} \times \mathbf{B} = 0$. The two magnetic configurations we have chosen both have $\mathbf{J} = 0$ which is the simplest way of satisfying this condition. The next step consists of choosing an appropriate magnetic potential $\mathbf{A} = A_\phi \, \mathbf{e}_\phi$, because then the divergence-free condition for axisymmetric flows is automatically satisfied. In this scheme, the components of the magnetic field for axisymmetric MHD are: $B_r = -\partial A_\phi / \partial z$ and $B_z = \partial r A_\phi / r \partial r$. Two particularly straightforward initial magnetic configurations are presented here. The magnetic configuration is extended smoothly into the disk where the toroidal magnetic field is taken to scale inversely with the disk radius (PP).

The detailed analytical and numerical approaches to our problem, as well as many results, are given in Ouyed (1996), Ouyed et al. (1997), Ouyed & Pudritz (1997a, hereafter OPI; 1997b, hereafter OPII). A dynamic view of the simulations presented here can be found at:

http://www.physics.mcmaster.ca/Grad_Ouyed/ROuyed.html.

2. Stationary Outflows

We start with a potential configuration given by (OPI)

$$A_\phi = \frac{\sqrt{r^2 + (z_d + z)^2} - (z_d + z)}{r}, \quad (1)$$

where z_d is the dimensionless disk thickness.

The simulation shows that the gas is centrifugally accelerated through the Alfvén and the fast magnetosonic (FM) surfaces and collimated into cylinders parallel to the disk axis. The collimation of the outflow is caused by the pinch force exerted by the dominant toroidal magnetic field generated by the outflow itself. Beyond the FM surface, we found that a "Hubble" flow is present; $v_z \propto z$. The velocities achieved in our simulations are of the order 180 km/s for our standard young stellar object (a 0.5 $M_\odot$ proto-star) and of order 10^5 km/s for our standard active galactic nuclei (a $10^8 \, M_\odot$ black hole). Our jet solutions, dominated mainly by the poloidal kinetic energy ρv_p^2, are very efficient in magnetically extracting angular momentum and energy from the disk. We estimate the ratio of the disk accretion rate to the wind mass flux rate to be of the order of $\dot{m}_a / \dot{m}_w \simeq 6.0$. Our stationary outflows have many similarities to steady state models of MHD disk winds (BP, HN, PP).

3. Episodic Outflows

Here, the initial magnetic field lines are taken to be uniform and parallel to the disk axis (z-axis). That is (OPII),

$$A_\phi = r. \quad (2)$$

Because of the gradient force in the non-linear torsional Alfvén waves generated by the rotor at the footpoints of the field lines, the initial magnetic configuration opens up in a narrow region located at $r_i < r < 8r_i$ with r_i being the innermost radius of the disk. Within this narrow region, a wind is ejected from the field lines that have opened to less than the critical angle ($\simeq 60^o$), as expected from the centrifugally driven wind theory (BP, PP). Our simulations show that the strong toroidal magnetic field generated recollimates the flow towards the disk's axis and, through magnetohydrodynamic (MHD) shocks, produces knots (Ouyed & Pudritz, 1993; Ouyed & Pudritz, 1994). The knot generation mechanism occurs at a distance of about $z \simeq 8r_i$ from the surface of the disk. Knots propagate down the length of the jet at speeds less than the diffuse component of the outflow. The knot generator is episodic, and is inherent to the jet.

4. Conclusion

Our two simulations, while typical of several dozen that we have carried out, still only probe a small part of parameter space. The outflow velocities achieved in the episodic model are still too small to match the fastest structures seen in observed YSO jets. These simulations have much smaller scales than can be directly imaged by HST. The major constraint in our study is that the accretion disk is not allowed to respond to the changing torques exerted by the jet. Simulations which treat both the production and collimation of an MHD outflow (as presented here), and which follow the internal dynamics of the disk self-consistently, are highly desirable but require computational resources beyond those available for this study. We believe, however, that these suggested improvements will not qualitatively change the rather simple physics of episodic jet formation that emerges from our present simulations.

Acknowledgments. We thank Jim Stone, David Clarke and Patricia Monger for all their help and support during the course of this project. R.O. acknowledges the financial support of McMaster University.

References

Blandford, R. D., & Payne, D. R. 1982, MNRAS, 199, 883 (BP).
Davis C. J., Mundt, R., Eislöffel, J., & Ray, T. P. 1995, AJ, 110, 766.
Dewar, R. L. 1970, Phys. Fluids, 13, 2710.
Eislöffel, J., & Mundt, R. 1992, A&A, 263, 292.
Heyvaerts, J., & Norman, C. 1989, ApJ, 347, 1055 (HN).
Ouyed, R., & Pudritz, R. E. 1993, ApJ, 419, 255.
Ouyed, R., & Pudritz, R. E. 1994, ApJ, 423, 753.
Ouyed, R. 1996, Ph.D. Thesis, (McMaster University).
Ouyed, R., Pudritz, R. E., & Stone, J. M. 1997, Nature, 385, 409.
Ouyed, R., & Pudritz, R. E. 1997a, ApJ, in press (OPI).
Ouyed, R., & Pudritz, R. E. 1997b, ApJ, in press (OPII).
Pelletier, G., & Pudritz, R. E. 1992, ApJ, 394, 117 (PP).

Part IV
GALACTIC DYNAMICS

The GRAPE-4, a Teraflops Stellar Dynamics Computer

Piet Hut
Institute for Advanced Study, Princeton, NJ 08540, U.S.A.

Abstract. Recently, special-purpose computers have surpassed general-purpose computers in the speed with which large-scale stellar dynamics simulations can be performed. Speeds up to a Teraflops are now available for simulations in a variety of fields, such as planetary formation, star cluster dynamics, galactic nuclei, galaxy interactions, galaxy formation, large-scale structure and gravitational lensing. Future speed increases for special-purpose computers will be even more dramatic: a Petaflops version, tentatively named the GRAPE-6, could be built within a few years, whereas general-purpose computers are expected to reach this speed somewhere in the 2010–2015 time frame. Boards with a handful of chips from such a machine could be made available to individual astronomers. Such a board, attached to a fast workstation, will then deliver Teraflops speeds on a desktop, around the year 2000.

1. Introduction

Computational physics has emerged as a third branch of physics, grafted onto the traditional pair of theoretical and experimental physics. At first, computer use seemed to be a straightforward off-shoot of theoretical physics, providing solutions to sets of differential equations too complicated to solve by hand. But soon the enormous quantitative improvement in speed yielded a qualitative shift in the nature of these computations. Rather than asking particular detailed questions about a model system, we now use computers more often to model the whole system directly. Answers to relevant questions are then extracted only after a full simulation has been completed. The data analysis following such a virtual lab experiment is carried out by the computational physicist in much the same way as it would be done by an experimenter or observer analyzing data from a real experiment or observation.

With this shift from theory to experimentation, computers have become important laboratory tools in all branches of science. There is one striking difference, though, between the use of a computer and that of other types of lab equipment. Whereas laboratory tools are typically designed for a particular purpose, computers are usually bought off the shelf, and used as is, without any attempt to customize them to the particular usage at hand. In contrast, it would be unthinkable for a astronomy consortium to build a new observatory around a huge pair of binoculars, as a simple scaled-up version of commercial bird-watching equipment.

The reason for this difference in buying pattern has nothing to do with an inherent difference between the activities of computing, experimenting, or observing. Building a special-purpose computer is not more difficult than building a telescope, or any other major type of customized laboratory equipment. Rather, the difference in attitude has everything to do with the fact that our computational ability has gone through an extraordinary period of sustained rapid exponential growth in speed.

Imagine that binoculars would grow twice as powerful every one or two years. If that were the case, astronomers might as well simply buy the latest model binoculars, and use those for their observations. Planning to build a big telescope would be self-defeating: in the ten or so years it would take to design and build the thing, technology would have progressed so much that commercial binoculars would out-perform the special-purpose telescope.

Over the last forty years, computer speed has exponentially increased. As a result, there has never been a particularly great need for physicists to design and build their own computer. As with all cases of exponential growth, this tendency will necessarily flatten off. How and when this flattening will occur is difficult to predict. This will depend on technological and economic factors that are as yet uncertain. But it is already the case that increase in computer speed is significantly more modest than what could be expected purely from the ongoing miniaturization of computer chips. This trend, in the case of general-purpose computers, will be discussed briefly in § 2. Various alternatives, in the form of special-purpose computing equipment, are mentioned in § 3. One such alternative, the GRAPE family of special-purpose computer hardware, is reviewed in § 4. Some astrophysical applications of these GRAPE machines are discussed in § 5. A preview of coming GRAPE attractions is presented in § 6.

2. General-Purpose Computers

After mainframes and minicomputers turned out to be no longer cost-effective, some time around the early-to-mid eighties, the only general-purpose computers used in physics were workstations and supercomputers. At first, there was an enormous gap in performance between the two types of machines, but over the last fifteen years this gap has narrowed steadily.

For example, during the eighties, supercomputers increased in speed by about a factor of 10^2, while microprocessors saw an increase of a factor of 10^3. The main reason was that workstations at first were rather inefficient, requiring many machine cycles for a single floating point operation. With increased chip size, this situation improved rapidly. In contrast, the first supercomputers, built in the mid seventies, were designed specifically to deliver at least one new floating point result for each clock cycle, through the use of pipelines.

Although the speed of the floating point components for supercomputers has continued to increase over the years, most of the increase in their peak speed has been realized through increasing the number of processing units. This increase in parallelism has made the sharing of memory by different processors increasingly cumbersome, involving significant hardware overhead: a full interconnect between N processors and a central memory bank requires an amount of additional hardware that scales as N^2.

In contrast, the much faster speed-up of microprocessor-based workstations has been possible exactly because there was (as yet) no need for parallelism. Throughout the eighties, chips did not contain enough transistors to allow floating point operations to be performed on a single chip in one cycle. Therefore, personal computers used to have a special floating point accelerator chip, in addition to the central processor chip, and even this accelerator typically needed several cycles even for the simplest operations of addition and multiplication. As a result, increase in the number of transistors per chip translated linearly into an increase in speed.

However, this situation changed as soon as it became possible to put a complete computer on a single chip, including a floating point unit with the capability of producing a new output every cycle. While in itself a great achievement, this capability also creates new trouble. From this point on, the scaling of general-purpose computers, based on microprocessors, will become less favorable, for the following reasons.

With further miniaturization, a single chip will soon contain several floating point units, with an extremely fast on-chip interconnect. These interconnections, however, require a significant "real estate" overhead on the chip: many extra components have to be added to the chip in order to implement the administrative side of this fast communication efficiently. In addition, the off-chip communication with the main memory is far slower, forming a bottleneck.

As a result of both factors, a shrinking in feature size by a factor two no longer guarantees a speed-up of a factor ~ 8, but rather $\sim 2\text{-}4$. In the eighties, when the feature width would become a factor two smaller, four times as many transistors would fit on one chip, and in addition the shrinking of the size of the transistors by a factor two would allow a clock speed nearly twice as high as before. However, this gain of a factor of eight from now on will be offset by a communication penalty of a factor $\sim 2\text{-}4$. The conclusion is that microprocessors are now facing the same problem of increasing "internal administrative bureaucracy" that supercomputer processors have had to deal with for the last twenty years.

3. Special-Purpose Computing Equipment

3.1. Special-purpose computers

Until the late seventies, almost all scientific calculations were carried out on general-purpose computers. Around that time, microprocessors began to offer a better price-performance ratio than supercomputers. By itself, this was not very helpful to a physicist, given the fact that a single microprocessor could only offer a speed of 10 kflops or so, peanuts compared to the supercomputers of those days, with peak speeds above 100 Mflops. The key to success was to find a way to combine the speed of a large number of those cheap microprocessors.

This was exactly what several groups of physicists did in the eighties. They took large numbers of off-the-shelf microprocessors, and hooked them up together. Building these machines was not too hard, and indeed raw speeds at low prices were reached relatively easily. The main problem was that of software development. To get a special-purpose machine to do a relevant physics calculation and to report the results in understandable form, provided formidable challenges. For example, writing a reasonably efficient compiler for such a machine was a tedious and error-prone job. In addition, developing application programs was no simple task either.

An interesting and somewhat unexpected development has been the commercialization of these machines, originally built by and for physicists. The design of most of the current highly-parallel general-purpose computers has been directly or indirectly influenced by the early special-purpose computers. This blurring of the distinction between special-purpose and general-purpose computers may continue in the future, when demand for higher peak speeds will force increasing parallelization to occur.

This development reflects the fact that the so-called special-purpose machines in physics actually attacked a general type of problem: how to let many individual processors cooperate on a single computational task. The fact that the applications have been rather specialized in many cases (to particle physics, astrophysics, or hydrodynamics) is less important than the fact that each application required a carefully balanced strategy at dynamic inter-processor communication. As a result, the experience gained from the development of both hardware and software for special-purpose computers has turned out to be very helpful for the development of their general-purpose counterparts as well.

3.2. Special-purpose accelerators

In the late eighties, an alternative model was developed. Following the example of some special-purpose components designed as back-end processors in radio telescopes, the idea was advanced to design special hardware components to speed up critical stages within large-scale simulations, most of which would still be delegated to general-purpose workstations.

A similar idea had already been employed for general-purpose computers as well. In the early eighties, personal computers would come with a central processor that could handle floating-point calculations only in software, at rather low efficiency. Significant speed-up, of an order of magnitude, could be obtained by including a so-called floating-point accelerator, at only a fraction of the cost of the original computer. Another example is the use of graphics accelerators in most modern personal computers.

Building a special hardware accelerator for a critical segment of a physics simulation is another example of this general approach. In this way, the good cost-performance ratio of special-purpose hardware can be combined with the flexibility of existing workstations, without much of a need for special software development. This approach can be compared to using hand-coded assembly-language or machine-code for an inner loop in an algorithm that otherwise is programmed in a higher-level language—the difference being that this inner loop is now realized directly in silicon.

4. The GRAPE Project

4.1. Prehistory

In 1984, a group of astrophysicists and computer scientists built the digital Orrery, a 10 Mflops special-purpose computer designed to follow the long-term evolution of the orbits of the planets (Applegate *et al.*, 1985). For that purpose, ten processors were connected in a ring, one for each planet (or test particle). The processors were designed around an experimental 64-bit floating-point chip set developed by HP. Each chip could perform one floating point operation in 1.25 μs. A central controller sent instructions to all processors at each machine cycle.

A few years later, results from the Orrery lead to the important discovery of the existence of a positive Lyapunov coefficient for the evolution of the orbit of Pluto, which was interpreted as a sign of chaos (Sussman & Wisdom, 1988).

Besides the question of the long-term evolution of planetary orbits, there were many other problems in gravitational dynamics that required far more than the typical speed available to astrophysicists in the mid-eighties. While significant speed-up was obtained with the introduction of more efficient algorithms (*e.g.*, Barnes & Hut, 1986; 1989), many problems in stellar dynamics could not be effectively tackled with the hardware available at that time.

Among those problems, the most compute-intensive was the long-term simulation of star clusters past core collapse. The record in that area in the late eighties was held by Makino & Sugimoto (1987) and Makino (1989), for N-body calculations with $N = 1,000$ and $N = 3,000$, respectively. Unfortunately, the computational costs for these types of calculations scales roughly with N^3, which meant that realistic simulations of globular clusters, with N in the range $10^5 \sim 10^6$, were still a long way off.

The only hope to make significant progress in this area was to make use of the fastest supercomputers available, in the most efficient way possible. Therefore, the next step we took was a detailed analysis of the algorithms available for the study of dense stellar systems (Makino & Hut, 1988; 1990), following the earlier analysis given by Makino (1986).

Our analysis showed that the best integration schemes available, in the form of Aarseth's individual time step predictor-corrector codes (Aarseth, 1985), were close to the theoretical performance limit. Based on these results, we predicted that a speed of order 1 Teraflops would be required to model globular star clusters, and to verify the occurrence of gravothermal oscillations in such models (Hut *et al.*, 1988).

Unfortunately, such speeds were not commercially available in those days, and it was clear that they would not be available for another ten years or so. The fastest machine that we could lay our hands on was the Connection Machine CM-2, which was first being shipped by Thinking Machines in 1987. In the Fall of that year, Jun Makino and I spent a few months at Thinking Machines, to perform an in-depth analysis of the efficiency of various algorithms for stellar dynamics simulations on the CM-2.

The results were somewhat disappointing (Makino & Hut, 1989), in that most large-scale simulations could utilize only $\sim 1\%$ of the peak-speed of the CM-2. As a result, even with a formidable peak speed of tens of Gigaflops, most of our simulations only obtained a speed of a few hundred Megaflops, when scaled up to a full CM-2 configuration. The main reason for its poor performance was the slowness of the communication speed compared to the speed of the floating point calculations.

Since we needed a Teraflops in order to study gravothermal oscillations and other phenomena in dense stellar systems, it was rather disheartening that we could not even reach an effective Gigaflops. And given the typical increase in speed of supercomputers, by a factor of ~ 10 every five years, it seemed clear that we would have to wait till well after the year 2000, before being able to compute at an effective Teraflops speed.

In reaction to our experiences, Sugimoto took up the challenge and formed a small team at Tokyo University to explore the feasibility of building special-purpose hardware for stellar dynamics simulations. This group started their project in the Spring of 1989, resulting in the completion of their first machine in the Fall of that same year (Ito *et al.*, 1990).

4.2. The GRAPE family

The name GRAPE stands for GRAvity PipE, and indicates a family of pipeline processors that contain chips specially designed to calculate the Newtonian gravitational force between particles. A GRAPE processor operates in cooperation with a general-purpose host computer, typically a normal workstation. The force integration and particle pushing are all done on the host computer, and only the inter-particle force calculations are done on the GRAPE. Since the latter require a computer processing power that scales with N^2, while the former only require $\propto N$ computer power, load balance can always be achieved by choosing N values large enough.

The development history of the GRAPE series of special-purpose architectures shows a record of rapid performance improvements (see Table 1). The limited-precision GRAPE-1 achieved 240 Mflops in 1989; its successor, the GRAPE-3, reached 15 Gflops in 1991. Over 30 GRAPE-3 systems are currently in use worldwide in applications (such as tree codes and SPH applications) where high numerical precision is not a critical factor.

A prototype board of the full-precision GRAPE-2 achieved 40 Mflops in 1990. The full GRAPE-4 system reached 1.1 Teraflops (peak) in 1995. Individual GRAPE-4 boards, delivering from 3 to 30 Gflops depending on configuration, are currently in use at five institutions around the world.

A third development track is represented by the GRAPE-2A and MD-GRAPE machines, which include a user-loadable force look-up table that can be used for arbitrary central force laws (targeted at molecular dynamics applications). Overall, the pace of development has been impressive: 10 special-purpose machines with a broadening range of applications and a factor of 4,000 speed increase in just over six years.

The GRAPE-4 developers have won the Gordon Bell prize for high-performance computing in each of the past two years. In 1995, the prize was awarded to Junichiro Makino and Makoto Taiji for a sustained speed of 112 Gflops, achieved using one-sixth of the full machine on a 128k particle simulation of the evolution of a double black-hole system in the core of a galaxy. The 1996 prize was awarded to Toshiyuki Fukushige and Junichiro Makino for a 332 Gflops simulation of the formation of a cold dark matter halo around a galaxy, modeled using 768k particles on three-quarters of the full machine.

Table 1
Summary of GRAPE Hardware

Limited-Precision Data Path

Machine	Year	Peak Speed	Notes
GRAPE-1	1989	240 Mflops	Concept system, GPIB interface
GRAPE-1A	1990	240 Mflops	VME interface
GRAPE-3	1991	15 Gflops	48 Custom LSIs, 10 MHz clock
GRAPE-3A	1993	5 Gflops/board	8 chip version for distribution, 20 MHz, PCB implementation

Full-Precision Data Path

Machine	Year	Peak Speed	Notes
GRAPE-2	1990	40 Mflops	IEEE precision, commercial chips
HARP-1	1993	180 Mflops	"Hermite" pipeline
HARP-2	1993	2 Gflops	Evaluation system of the custom chips to be used in GRAPE-4
GRAPE-4	1995	1.1 Tflops	The Teraflops GRAPE, 1692 pipelines

Arbitrary Force Law

Machine	Year	Peak Speed	Notes
GRAPE-2A	1992	180 Mflops	Force look-up table
MD-GRAPE	1995	4 Gflops	Custom chip with force look-up table

4.3. Using the GRAPE

Modifying an existing program to use the GRAPE hardware is straightforward, and entails minimal changes. Subroutine and function calls (written in C or FORTRAN) to the GRAPE hardware replace the force-evaluation functions already found in existing N-body codes.

Communication between host and GRAPE is accomplished through a collection of about a dozen interface routines. The force evaluation code which is replaced typically consists of only a few dozen lines at the lowest level of an algorithm. Thus, using the GRAPE calls only for small, localized changes which in no way inhibit future large-scale algorithm development.

The GRAPE interface has been successfully incorporated into the Barnes-Hut tree algorithm (Barnes & Hut, 1986; Makino, 1991) and the P^3M scheme (Hockney & Eastwood, 1988; Brieu et al., 1995).

Here is a typical code fragment for the Newtonian force calculation on a workstation:

```
       subroutine accel_workstation
       do 10 k =1,ndim
         do 20 i=1,nbody
           accnew(i,k)=0.0
 20      continue
 10    continue
       do 30 i=1,nbody-1
         do 40 j=i+1,nbody
           do 50 k = 1,3
             dx(k)=pos(k,j)-pos(k,i)
 50        continue
           r2inv=1.0/(dx(1)**2+dx(2)**2+dx(3)**2+eps2)
           r3inv=r2inv*sqrt(r2inv)
           do 60 k=1,3
             accnew(k,i)=accnew(k,i)+r3inv*mass(j)*dx(k)
             accnew(k,j)=accnew(k,j)-r3inv*mass(i)*dx(k)
 60        continue
 40      continue
 30    continue
       end
```

To use the GRAPE, all that has to be done is to replace the inner loop of the force calculations by a few special function calls in order to offload the bulk of the computation onto the GRAPE hardware:

```
       subroutine accel_grape
       call g3init()
       xscale = 1.0d0/1024
       call g3setscales(xscale, mass(1))
       call g3seteps2(eps2)
       call g3setn(nbody)
         do 20 i=1,nbody
           call g3setxj(i-1,pos(1,i))
           call g3setmj(i-1,mass(i))
 20      continue
       nchips=g3nchips()
       do 30 i=1,nbody,nchips
         ii = min(nchips, nbody - i + 1)
         call g3frc(pos(1,i),accnew(1,i),pot(i),ii)
 30    continue
       call g3free
       end
```

5. Some Astrophysical Applications

In this brief review, there is no room for an exhaustive review of the scientific results that have been obtained with the few dozen GRAPE machines that have been installed in a number of different research institutes around the world. In addition to the four fields listed below, the GRAPEs have been used in a variety of other areas, for example to study the role of exponential divergence of neighboring light trajectories on gravitational lensing, the formation of large-scale structure in the Universe, the role of violent relaxation in galaxy formation, and the effectiveness of hierarchical merging in galaxy clusters.

5.1. Planet formation.

Ida & Makino (1992a; 1992b) used the GRAPE-2 to investigate the evolution of the velocity distribution of a swarm of planetesimals, with an embedded protoplanet. They confirmed that equipartition is achieved and that therefore runaway growth should take place, along the lines suggested by Stewart & Wetherill (1988).

Kokubo & Ida (1995) used the HARP-2 (a smaller prototype of the GRAPE-4) to simulate a system of two protoplanets and many planetesimals. They found that the separation between two planets tends to grow to roughly 5 r_H (the Hill radius). They coined the term "orbital repulsion" for this phenomenon, and provided a qualitative explanation for its occurrence.

Kokubo & Ida (1996a) used the GRAPE-4 to simulate planetary growth assuming perfect accretion, where any physical collision leads to coalescence. They started with 3,000 equal-mass planetesimals. After 20,000 orbits, they found that the most massive particle had become 300 times heavier, while the average mass of the particles increased by only a factor of two. Kokubo & Ida (1996b) extended these calculations. They showed that several protoplanets are formed and grow while keeping their mutual separations within the range 5–10 r_H. Their results strongly suggests that orbital repulsion has determined the present separation between the outer planets.

5.2. Star cluster evolution.

The first scientific result obtained with the GRAPE-4 was the demonstration of the existence of gravothermal oscillations in N-body simulations. Predicted more than ten years earlier by Sugimoto & Bettwieser (1983), they were found by Makino (1996a), and presented by him at the I.A.U. Symposium 174 in Tokyo, in August 1995 (Makino, 1996b). Using more than 32,000 particles, he was also able to confirm the semi-analytical predictions made by Goodman (1987). The calculation took about two months, using only one quarter of a full GRAPE-4, running at a speed of 50 Gflops.

We are currently exploring ways to couple stellar dynamics and stellar evolution in one code, in order to perform more realistic simulations of star cluster evolution. Based on steller evolution recipes implemented by Portegies Zwart & Verbunt (1996), we have carried out a series of increasingly realistic approximations (Portegies Zwart et al., 1997a; 1997b; in preparation); see our web site with a movie that shows a star cluster, as an evolving N-body system side-to-side with its correspondingly evolving H-R diagram, at http://casc.physics.drexel.edu.

5.3. Density profiles of galactic nuclei.

Ebisuzaki et al. (1991) used the GRAPE-2 to simulate the merging of two galaxies, each with a central black hole, using up to 4,096 + 2 particles. They found an increase in core radius as a result of the heating of the central regions caused by the spiral-in of the two black holes.

Makino & Ebisuzaki (1996) used the GRAPE-4 to study hierarchical merging, in which the merger product of one pair of galaxies was used as a template for constructing progenitors for the next simulation of merging galaxies. They used more than 32,000 particles. They found the ratio between the core radius and the effective radius to converge to a value depending on the mass of the black holes.

However, it turned out that 32k particles were not enough. Makino (1997) performed a similar type of calculation with 256k particles, and found a core structure which was rather different from that obtained in the previous 32k runs. In particular, he found the volume density of stars to decrease in the vicinity of the black hole binary in the 256k runs, and ascribed this to the "loss cone" effect predicted by Begelman et al. (1980).

Fukushige & Makino (1997) used the GRAPE-4 to simulate hierarchical clustering, using an order of magnitude more particles than in previous studies. They found that the central density profiles are always steeper than $\rho \sim r^{-1}$. They interpreted the observed shallower cusps as the result of the spiral-in of the central black holes from the progenitor galaxies, involved in the merging process.

5.4. Interactions between galaxies.

Okumura et al. (1991) used the GRAPE-1 to investigate the structure of merger remnants formed from encounters between two Plummer models on parabolic orbits, using 16,000 particles. They determined the non-dimensional rotation velocity V_{max}/σ_0, where V_{max} denotes the maximum rotation velocity and σ_0 is the velocity dispersion at the center. They found typical values of ~ 0.6 for merging at large initial periastron separations. Their result is in good agreement with the observation of large ellipticals, which show a rather sharp cutoff in the distribution of V_{max}/σ_0 around 0.6.

Makino & Hut (1997) used the GRAPE-3A to simulate more than 500 galaxy encounters, in order to determine their merger rate as a function of incoming velocity, for a variety of galaxy models. They characterized the overall merger rate in a galaxy cluster by a single number, derived from their cross sections by an integration over encounter velocities in the limit of a constant density in velocity space. In addition, they provided detailed information concerning the reduction of the overall encounter rate through tidal effects from the cluster potential as well as from neighboring galaxies.

6. Coming Attractions: the GRAPE-6

In the GRAPE-4, once all pipelines are filled, each chip produces one new interparticle interaction (corresponding to ~ 60 floating-point operations) every three clock cycles. For a clock speed of 30 MHz, a peak chip speed of ~ 0.6 Gflops is achieved. The GRAPE-4 chips represent 1992 technology (1 μm fabrication line width). Even if no changes were made in the basic design, advances in fabrication technology would permit more transistors per chip and increased clock speed, enabling a 50–100 MHz, 10–30 Gflops chip with 1996 (0.35 μm line width) technology, and a 100–200 MHz, 50–200 Gflops chip with 1998 (0.25 μm) technology. Based on these projected performance improvements, a total of $\sim 10^4$ GRAPE-6 chips of 100 Gflops each could be combined to achieve Petaflops speeds by the year 2000, for a total budget of 10 million dollars. We have recently completed an initial "point design study" of the feasibility of constructing such a system (McMillan et al., 1996). This study was funded by the NSF, in conjunction with NASA and DARPA, as part of a program aimed at paving the way towards Petaflops computing.

While planning to build a hardwired Petaflops-class computational engine, we are also investigating complementary avenues, based on the use of reconfigurable logic, in the form of Field-Programmable Gate Array (FPGA) chips. The merging of custom LSI and reconfigurable logic will result in a unique capability in performance and generality, combining the extremely high throughput of special-purpose devices with the flexibility of reconfigurable systems. In many applications, gravity requires less than 99% of the computing power. Although the remainder of the CPU time is typically dominated by just one secondary bottleneck, its nature varies greatly from problem to problem. It is not cost-effective to attempt to design custom chips for each new problem that arises. In these circumstances, a FPGA-based system can restore the balance, and guarantee scalability from the Teraflops to the Petaflops domain, while still retaining significant flexibility. Astrophysical applications could include, for example, various forms of Smooth Particle Hydrodynamics (SPH), for applications ranging from colliding stars to the formation of large-scale structure in the Universe.

An additional benefit of the construction of Petaflops-class machines will be the availability of individual chips at reasonable prices, once the main machine has been designed and constructed. A typical GRAPE-6 chip will run at ~ 100 Gflops. A single board with 10 or more chips will already deliver a speed of 1 Teraflops or more, for a total price that is likely to lie in the range of 10,000–20,000 dollars. Hooking such a board up to a workstation will instantly change it into a top-of-the-line supercomputer.

Acknowledgments. I thank Jun Makino and Steve McMillan for their comments on the manuscript. This work was supported in part by the National Science Foundation under grant ASC-9612029.

References

Aarseth, S. J. 1985, in *Multiple Time Scales*, eds. J. U. Brackhill & B. I. Cohen, (New York: Academic), 377.
Applegate, J. H., Douglas, M. R., Gürsel, Y., Hunter, P., Seitz, C. L., & Sussman, G. J. 1985, IEEE Transactions on Computers, C34, 822.

Barnes, J. E., & Hut, P. 1986, Nature, 324, 446.
Barnes, J. E., & Hut, P. 1989, ApJS, 70, 389.
Begelman, M. C., Blandford, R. D., & Rees, M. J. 1980, Nature, 287, 307.
Brieu, P. P., Summers, F. J., & Ostriker, J. P. 1995, ApJ, 453, 566.
Goodman, J. 1987, ApJ, 313, 576.
Ebisuzaki, T., Makino, J., & Okumura, S. K. 1991, Nature, 354, 212.
Fukushige, T., & Makino, J. 1997, ApJ, 477, L9.
Hockney, R. W., & Eastwood, J. W. 1988, *Computer Simulation Using Particles*, (New York: Adam Hilger).
Hut, P., Makino, J., & McMillan, S. 1988, Nature, 336, 31.
Ida, S., & Makino, J. 1992a, Icarus, 96, 107.
Ida, S., & Makino, J. 1992b, Icarus, 98, 28.
Ito, T., Makino, J., Ebisuzaki, T., & Sugimoto, D. 1990, Comp. Phys. Comm., 60, 187.
Kokubo, E., & Ida, S. 1995, Icarus, 114, 247.
Kokubo, E., & Ida, S. 1996a, Lunar Planet. Sci., 27, 683.
Kokubo, E., & Ida, S. 1996b, Icarus, 123, 180.
Makino, J. 1986, in *The Use of Supercomputers in Stellar Dynamics*, eds. P. Hut and S. McMillan (New York: Springer), 151.
Makino, J. 1989, in *Dynamics of Dense Stellar Systems*, ed. D. Merritt (Cambridge: Cambridge University Press), 201.
Makino, J. 1991, PASJ, 43, 621.
Makino, J. 1996a, ApJ, 471, 796.
Makino, J. 1996b, in *Dynamical Evolution of Star Clusters*, I.A.U. Symp. 174, eds. P. Hut & J. Makino (Dordrecht: Kluwer), 151.
Makino, J. 1997, preprint.
Makino, J., & Ebisuzaki, T. 1996, ApJ, 465, 527.
Makino, J., & Hut, P. 1988, ApJS, 68, 833.
Makino, J., & Hut, P. 1989, Comp. Phys. Rep., 9, 199.
Makino, J., & Hut, P. 1990, ApJ, 365, 208.
Makino, J., & Hut, P. 1997, ApJ, in press.
Makino, J., & Sugimoto, D. 1987, PASJ, 39, 589.
McMillan, S. L. W., Hut, P., Makino, J., Norman, M. L., & Summers, F. J. 1996, in *Petaflops Architecture Workshop*, Chap. 6.4.
Okumura, S., Ebisuzaki, T., & Makino, J. 1991, PASJ, 43, 781.
Portegies Zwart, S. F., & Verbunt, F. 1996, A&A, 309, 179.
Portegies Zwart, S. F., Hut, P., & Verbunt, F. 1997a, A&A, in press.
Portegies Zwart, S. F., Hut, P., McMillan, S. L. W., & Verbunt, F. 1997b, submitted to A&A.
Stewart, G. R., & Wetherill, G. W. 1988, Icarus, 74, 542.
Sugimoto, D., & Bettwieser, E. 1983, MNRAS, 204, 19p.
Sussman, G. J., & Wisdom, J. 1988, Science, 241, 433.

On Simulations of Galactic Disks of Stars

Evgeny Griv[1]

Institute of Astronomy and Astrophysics, Academia Sinica, Taiwan

Abstract. This is a preliminary report on an analytical study of some instability modes in a model disk of stars. This disk is studied by employing linear theory to determine its stability against small-amplitude gravity perturbations. The well-elaborated mathematical methods of plasma kinetic theory are utilized, through the studying of dispersion relations.

1. Introduction

Recently, two outstanding discoveries associated with stellar disks of highly flattened galaxies have been made with the help of gravitational N-body simulations. They include the discovery of the long-lived recurrent spiral modes (Sellwood & Lin, 1989; Donner & Thomasson, 1994) and the bending out-of-plane instability of central, practically non-rotating parts of a disk (Combes & Sanders, 1981; Combes et al., 1990; Raha et al., 1991; Griv et al., 1997a). Here, I report on analytical study of these instability modes in a model stellar disk.

1. Understanding how spiral structures develop in galaxies is a fundamental problem in galactic dynamics. There exists, as yet, no satisfactory theory of the origin and conservation of the spiral structure so prominent in the Milky Way and many other giant galaxies. Spiral galaxies are composed of at least three ingredients: stars, interstellar gas, and cosmic rays. The bulk of the luminous mass, probably $\sim 90\%$, is in the stars. Therefore, it seems likely that the spiral structures are intimately connected with the stellar constituent of a galaxy, and that stellar dynamical phenomena play a basic role. If spiral arms in a differentially-rotating, disk-shaped galaxy of stars were material structures, they would wind up in a time scale of a few galactic revolutions ($\sim 10^8$ yr), a time scale which is much less than the age of a typical galaxy ($T \sim 10^{10}$ yr). The large number of spiral galaxies, which must be of age $\sim T$, suggests that spiral structure is a *wave* phenomenon (Lin et al., 1969; Shu, 1970). Accordingly, the spiral pattern is a manifestation of density waves propagating around the galaxy disk with a constant angular velocity. Obviously, a wave theory may explain the long-lived spiral structure only if there is some instability of gravity oscillations which is capable of supplying the growth of initial small-amplitude perturbations up to observable amplitudes (Toomre, 1977).

---

[1]on leave from Physics Department, Ben-Gurion University of the Negev, P.O. Box 653, Beer-sheva 84105, Israel

In conjunction with the problem of spiral structure in a stellar disk, I note that Sellwood & Lin (1989) and Donner & Thomasson (1994) have discovered a new type of small-amplitude, long-lived spiral mode in collisionless N-body models. These modes lead to spiral wave amplitudes far greater than can be explained by particle noise, and the instabilities do not scale with N (the number of particles). Therefore, these instabilities appear to be physical and develop through non-linear changes to the particle distribution function in the neighborhood of a particular speed (Sellwood & Lin, 1989, Figure 7). The modes are driven by gradients in the phase-space density of particles at corotation. Therefore, as suggested by Sellwood & Lin (1989), it strongly resembles Landau excitation in plasmas. Resonant scattering leads to a dynamical "heating" of the N-body system: the disk, which starts with $Q = 1.5$ everywhere, heats to values $Q = 2.5$–3 within ~ 13 orbital periods at the outer edge of the numerical model. Here and below, $Q = c_r/c_T$ is Toomre's stability parameter (Toomre, 1977), c_r is the radial dispersion of random (residual) velocities of particles, $c_T \simeq 3.4 G\sigma_0/\kappa$ is Toomre's critical velocity dispersion (Toomre, 1964), G is the gravitation constant, $\sigma_0(r)$ is the equilibrium surface density, and $\kappa(r)$ is the epicyclic frequency. In my own opinion, the nature of these spiral modes may be explained by the resonant self-excitation of density waves. This resonant wave-particle interaction has already been investigated in detail elsewhere (Griv, 1996; Griv & Peter, 1996; Griv & Yuan, 1996; Griv et al., 1997b); in § 2, I summarize these results. It will be shown that as a result of wave-particle interaction at corotation, a Jeans-stable disk is still unstable to oscillatory growing perturbations. It is similar to the instability (overstability) of the Cherenkov-type (the inverse Landau damping effect) in magnetized plasmas, and it is a result of the nature of the differential rotation of the galactic system. As actual flat galaxies of stars rotate differentially, the oscillating instability can be considered to be a long-term generating mechanism for propagating spiral density waves, thereby leading to spiral-like patterns in normal galaxies.

2. The bending (or "fire-hose") instability of a thin stellar disk which leads to an aperiodic wave growth along the vertical direction (perpendicular to the galactic plane) has been predicted by Toomre (1966) using moment equations. Toomre considered the collisionless analog of the Kelvin-Helmholtz instability in an infinite, non-rotating sheet of stars. It was demonstrated that the bending instability is driven by the stellar "pressure" anisotropy: all instabilities in the non-rotating slab of the buckling kind may be avoided provided the ratio of the velocity dispersion ("temperature") along the minor axis c_z to the velocity dispersion in the equatorial plane c_r (i.e., the disk thickness) is large enough. The bending instability was also discovered independently by Kulsrud et al. (1971) and Mark (1971) using kinetic theory (see also Polyachenko & Shukhman, 1979). Fridman & Polyachenko (1984) have discussed the role of this instability in explaining the existence of maximum oblateness in practically non-rotating elliptical galaxies and the formation of the bulges of disk-shaped galaxies of stars. Merritt & Hernquist (1991) observed the bending instability in N-body simulations of non-rotating stellar systems. They confirmed Fridman & Polyachenko's hypothesis that all pressure-supported models of elliptical galaxies are unstable to bending modes when their elongations exceed $\sim 1 : 3$ or, correspondingly, their velocity dispersion ratio c_r/c_z exceeds ~ 2. It seems

likely that the latter explains why there are no elliptical galaxies with an oblateness exceeding a definite critical value; the largest oblateness is possessed by the galaxies of the Hubble type E7, for which the small axis is about 30% of the large axis. In turn, Combes et al. (1990) and Raha et al. (1991) observed the bending instability in N-body simulations of rotating warm disks which initially developed planar bars. It was stated that this instability may play a part in the formation of triaxial bulges in flat galaxies and an explanation was given why many bulge stars in the Galaxy are less than 10 Gyr old.

In § 3, a slowly-rotating disk representing the central parts of disk-shaped galaxies is studied by employing linear kinetic theory (similar to Kulsrud et al., 1971; Mark, 1971). In contrast to their study, however, a three-dimensional disk is assumed from the beginning. It is clear that because the formation of an out-of-plane bend is predominantly a three-dimensional phenomenon, one needs to analyze the dynamics of three-dimensional perturbations. Note that the firehose instability discussed here is well-known in plasma physics in transferring energy from one degree of freedom to another (perpendicular) degree of freedom.

2. Planar Spiral Modes

Stellar disks of normal galaxies, by definition, are highly flattened, i.e., $R \gg h$, where R is the characteristic radius, $h \approx c_z/\sqrt{\pi G n}$ is the thickness of the system, $n(r,z)$ is the three-dimensional density, and $(\pi G n)^{1/2}$ is approximately the frequency of vertical epicyclic oscillations. In the rotating frame of a disk-like galaxy, the collisionless motion of an ensemble of identical stars of unit mass in the plane of the system can be described by the linearized Boltzmann equation without the integral of interparticle collisions (Lin et al., 1969; Griv, 1996; Griv & Peter, 1996),

$$\frac{df_1}{dt} = \frac{\partial \Phi_1}{\partial r}\frac{\partial f_0}{\partial v_r} + \frac{\partial \Phi_1}{r \partial \varphi}\frac{\partial f_0}{\partial v_\varphi}, \tag{1}$$

where d/dt is taken along the unperturbed orbits of stars in the local rotating frame, f_1 is the small perturbation of the equilibrium distribution function f_0, and v_r and v_φ are the random radial and azimuthal velocities, respectively.

Applying the linear theory of disk stability in the kinetic equation (1), the total gravitational potential of the system was divided into a smooth part, $\Phi_0(r)$, satisfying the equilibrium condition $\partial \Phi_0/\partial r = \Omega^2 r$ [where $\Omega(r)$ is the angular velocity], and a fluctuating part, $\Phi_1(\mathbf{r},t)$, with $|\Phi_1/\Phi_0| \ll 1$ for all $\mathbf{r}$ and t. We consider the special case of waves propagating in the plane of the system under study. This is a valid approximation if one considers perturbations with a radial wavelength greater than h (Shu, 1970; Griv & Peter, 1996).

The perturbed potential, Φ_1, is related to the perturbed surface density of the stellar disk, σ_1, through the asymptotic solution of Poisson's equation as (Lin & Lau, 1979)

$$\sigma_1 = -|k|\Phi_1/2\pi G, \tag{2}$$

where

$$\sigma_1 = \int_{-\infty}^{\infty}\int_{-\infty}^{\infty} f_1 d\mathbf{v}, \quad k = \sqrt{k_r^2 + k_\varphi^2}$$

and where k is the wave number in the plane, k_r and k_φ are the radial and azimuthal wave numbers: $|\tan\psi| = |k_\varphi/k_r| \lesssim 1$. We can choose, as the steady-state distribution function, f_0, the Schwarzschild distribution (Shu, 1970; Griv & Peter, 1996).

Equation (1) describing the perturbed motion of stars under gravitational perturbations can be solved formally by the method of characteristics or "integrating over unperturbed particle orbits" (Shu, 1970; Griv, 1996; Griv & Peter, 1996; Griv & Yuan, 1996). The problem of particle motion in the disk of a flat galaxy is equivalent to the problem of the motion of an electrically charged particle in a given electromagnetic field, in which the solution can be decomposed into two parts: the guiding center motion and the epicyclic (Larmor) motion. The postepicyclic equations of motion for an individual disk star have been derived by Griv & Peter (1996).

Carrying out the integrations in equation (1) over t and the velocity space, and equating the result to the asymptotic solution of equation (2), it is straightforward to show that the wave frequency and the wave number are connected through the simplified dispersion relation:

$$\omega_*^2 - \omega_J^2 + i\frac{\omega_* \kappa^2 \gamma}{x e^{-x}} = 0, \tag{3}$$

where $\omega_J^2 \approx \kappa^2 - 2\pi G \sigma_0 (k_*^2/|k|) F(x)$ is the squared Jeans frequency, $F(x) \approx \exp[-(3/4)x]$ is the so-called "reduction factor", $|\omega_*|^2 < \kappa^2$, $x = k_*^2 c_r^2/\kappa^2 \leq 1$, and $k_*^2 = k^2\{1 + [(2\Omega/\kappa)^2 - 1]\sin^2\psi\}$ is the generalized wave number (Griv, 1996; Griv & Peter, 1996). Also in equation (3), the value of $\gamma \propto d\Omega/dr < 0$ has been defined by Griv (1996) and Griv & Peter (1996). The wave frequency was represented in the form $\omega_* \equiv \mathrm{Re}\,\omega_* + i\,\mathrm{Im}\,\omega_*$; the existence of solutions with $\mathrm{Im}\,\omega_* > 0$ implies the kinetic-type instability of hydrodynamically-stable oscillations.

The general dispersion relation (3) describes not only the ordinary gravitational Jeans-type perturbations, but also the Landau growth/damping of the Jeans-stable, natural perturbations $((\mathrm{Re}\,\omega_*)^2 \equiv \omega_J^2 > 0$ and $\gamma \neq 0)$. This dispersion relation generalizes that of Lin and Shu (Lin & Shu, 1966; Lin et al., 1969; Shu, 1970) for non-axisymmetric perturbations, ψ or $m \neq 0$. [In fact, because the latter authors as well as Toomre (1964; 1977) and Mark (1971) were interested only in radial, axisymmetric perturbations, they excluded all non-axisymmetric terms except that in the expression for $\omega_* = \omega - m\Omega$.] In addition, equation (3) describes the influence of the resonant wave-particle interactions on the natural perturbations of the stellar disk (Griv, 1996; Griv & Peter, 1996).

Let us define first the spectrum of Jeans oscillation frequencies, neglecting in equation (3) the term with γ. Then, the solution is $\omega_* \simeq \pm p|\omega_J|$, where $p = 1$ for Jeans-stable ($\omega_J^2 > 0$) perturbations and $p = i$ for Jeans-unstable ($\omega_J^2 < 0$) ones. By using the condition $\omega_J^2 \geq 0$ for all possible k, a local stability criterion can be easily obtained (Griv et al., 1994; Griv & Peter, 1996):

$$c_r \geq c_M \simeq c_T \left\{1 + \left[\left(\frac{2\Omega}{\kappa}\right)^2 - 1\right]\sin^2\psi\right\}^{1/2}. \tag{4}$$

Accordingly, stability of non-axisymmetric (or spiral, $\psi \neq 0$) perturbations in a differentially-rotating disk ($2\Omega/\kappa > 1$) requires a larger velocity dispersion than Toomre's critical value c_T, and the critical radial velocity dispersion is approximately $c_r = (2\Omega/\kappa)c_T \sim 2 \cdot c_T$ (in the limit $\psi \to 90°$ and $2\Omega/\kappa \sim 2$).

Including the term with γ (i.e., the resonant wave-particle interaction), one derives for the real part of wave frequency $\operatorname{Re} \omega_* \approx \pm|\omega_J|$, and for the imaginary part

$$\operatorname{Im} \omega_* \simeq -\frac{\gamma \kappa^2}{2x e^{-x}}. \tag{5}$$

Equation (5) should be used only when the Jeans frequency ω_J is real, that is, $c_r \geq c_M$ [or $\kappa^2 \geq 2\pi G \sigma_0(k_*^2/|k|)F(x)$; equation (4)].

Equation (5) describes the kinetic Landau-type instability of Jeans-stable perturbations in a differentially-rotating disk. The non-axisymmetric, unstable density waves are excited at the corotation resonance $\omega - m\Omega \approx 0$ as a result of the wave-particle interaction, and the spiral pattern will be formed in time $\sim (\operatorname{Im} \omega_*)^{-1}$. It can be shown from equation (5) that the growth rate of this *oscillating* instability increases as $|\omega_J| \to 0$ (Griv, 1996; Griv & Peter, 1996). In other words, as the radial velocity dispersion in the disk approaches approximately the critical value $(2\Omega/\kappa)c_T$ [equation (4)] and, correspondingly, $\omega_J^2 \to 0$, then the growth rate of the oscillating instability becomes sufficiently large, $\operatorname{Im} \omega_* \to \Omega$ (Griv, 1996; Griv & Peter, 1996).

It is natural to explain the phenomenon observed by Sellwood & Lin (1989) and Donner & Thomasson (1994) in N-body experiments as the result of the resonant self-excitation of the spiral density waves discussed above.

3. Off-Plane Bending Instability

As the observations and N-body simulations show, the central very thin ($h \sim 50$ pc) parts of spiral galaxies, say at distances $< 0.6 - 0.8$ kpc from the center, rotate slowly and their local circular velocities of regular rotation become less than the stellar random velocities. In this case, the wave propagation properties in a system correspond to an unmagnetized plasma with unperturbed magnetic field vector and *free-streaming* particle orbits. It may be shown that the Poisson's equation for the perturbed gravitational potential in a three-dimensional disk may be written in the form of Schrödinger's equation

$$\frac{\partial^2 \Phi_1}{\partial z^2} - k^2 \Phi_1 = -\frac{\aleph}{h^2} \operatorname{sech}^2(z/h)\Phi_1, \tag{6}$$

where

$$\aleph = \frac{2\pi G n_0}{\sqrt{2\pi} c_r^3} h^2 \int_{-\infty}^{\infty} \frac{v e^{-v^2/2c_r^2}}{v - \omega/k} dv,$$

$n_0 = \sigma_0/h$, $v = (v_r^2 + v_\varphi^2)^{1/2}$, and Φ_1 is evaluated at $z = 0$ (Griv, 1997).

Two separate cases may be considered in equation (6): the high-frequency limit, $\omega \gg kv$ (or the long-wavelength limit in the sense that $k \ll \omega/v$), and the low-frequency (or the short-wavelength, $k \gg \omega/k$) one, $\omega \ll kv$. In the long-wavelength limit, the integral in equation (6) is easily estimated to give the dispersion relation as the following

$$\omega^2 \simeq \frac{-2\pi G\sigma_0|k| + 3k^2 c_r^2}{1 + |k|h} \quad \text{for} \quad \omega/kc_r \gg 1, \tag{7}$$

where $2\pi G\sigma_0|k| > 3k^2 c_r^2$ and $|k|h \leq 1$. According to equation (7), the system is always gravitationally (Jeans) unstable ($\omega^2 < 0$) against long-wavelength perturbations. The purely imaginary value of ω denotes aperiodic growth, so that wave propagation cannot occur. Clearly, because $\omega^2 < 0$, there are no overstable modes in the limit under consideration.

It is obvious that this Jeans-type instability developing in the equatorial plane leads to an oval bar-like structure. This is because the Jeans instability of a practically non-rotating system is linked with the energetic advantage of reconstruction of the circular shape of a system to produce an elliptical one (in the equatorial plane): the potential energy is only one possible source for the growth of the average perturbation energy. Once such a bar-like structure is formed in the plane, it is subject to the short-wavelength buckling-type instability leading to an off-plane bell-like (or Ω-shaped) bend of the nuclear region of a disk when viewed from the side. Indeed, estimating the integral in equation (6) in this limit, one obtains the following dispersion relation (Griv, 1997):

$$\omega^2 \simeq \frac{k^2 c_r^2}{2\pi G\sigma_0|k|}(2\pi G\sigma_0|k| - k^2 c_r^2) \quad \text{for} \quad \omega/kc_r \ll 1. \tag{8}$$

Accordingly, the system is unstable aperiodically for wavelength

$$\lambda < c_r^2/G\sigma_0 \equiv \lambda_\mathrm{J},$$

where λ_J is the ordinary Jeans wavelength for instability with respect to gravity perturbations in a medium without rotation. This is the fire-hose or bending instability which occurs for short wavelengths, $h < \lambda < \lambda_\mathrm{J}$ (Griv, 1997).

The bending instability is driven by an excess of kinetic energy of stars in the plane of the disk ($\propto c_r^2$) when the ratio of the stellar velocity dispersion in the plane to the vertical velocity dispersion (perpendicular to the disk) is large enough (or the disk thickness is small enough). The instability serves to enhance the vertical velocity dispersion c_z. From equation (8), it is possible to show that the fierce aperiodic instability may be avoided, that is, $\omega^2 > 0$, provided the ratio

$$c_z^2 \gtrsim \frac{c_r^2}{2} \tag{9}$$

(Griv, 1997). Note that the latter condition may be found directly by the analogy with the condition for the fire-hose instability in plasmas.

In closing, it would be natural to suppose that as an almost non-rotating disk of newly formed stars evolves, it should arrive at a state near the limit of stability with respect to the bending instability so that the ratio c_r^2/c_z^2 for a disk of relatively old stars (say, with ages $> 10^8$ yr) comes close to the critical value $\simeq 2$ as predicted by our theory [equation (9)]. It would be interesting to compare our prediction with observations and N-body simulations of central, $r < 1$ kpc regions of spiral galaxies.

Acknowledgments. I am very grateful to Tzi-Hong Chiueh, Alexei M. Fridman, Michael Gedalin, Frank H. Shu, Alar Toomre, and Chi Yuan for useful discussions. Also I am grateful to Jerry Sellwood, who has his own interpretation of N-body simulations, for a question posed in 1994 that improved my understanding of the problems under consideration. The research was supported in part by the Academia Sinica in Taiwan and the Israeli Ministry of Immigrant Absorption.

References

Combes, F., Debbasch, F., Friedly, D., & Pfenniger, D. 1990, A&A, 233, 82.
Combes, F., & Sanders, R. H. 1981, A&A, 96, 164.
Donner, K. J., & Thomasson, M. 1994, A&A, 290, 785.
Fridman, A. M., & Polyachenko, V. L. 1984, *Physics of Gravitating Systems*, Vol. 1, (New York: Springer-Verlag).
Griv, E. 1996, Planet. Space Sci., 44, 579.
Griv, E. 1997, ApJ, submitted.
Griv, E., Chiueh, T., Lin, S.-C., & Peter, W. 1997a, A&A, submitted.
Griv, E., Chiueh, T., & Peter, W. 1994, Physica A, 205, 299.
Criv, E., & Peter, W. 1996, ApJ, 469, 84.
Griv, E., & Yuan, C. 1996, Planet. Space Sci., 44, 1185.
Griv, E., Yuan, C., & Chiueh, T. 1997b, Planet. Space Sci., 45, in press.
Kulsrud, R., Mark, J.-W., & Caruso, A. 1971, Astrophys. Space Sci., 14, 52.
Lin, C. C., & Lau, Y. Y. 1979, Stud. in Appl. Math., 60, 97.
Lin, C. C., & Shu, F. H. 1966, Proc. Natl. Acad. Sci., 55, 229.
Lin, C. C., Yuan, C., & Shu, F. H. 1969, ApJ, 155, 721.
Mark, J.-W. K. 1971, ApJ, 169, 455.
Merritt, D., & Hernquist, L. 1991, ApJ, 376, 439.
Polyachenko, V. L., & Shukhman, I. G. 1979, Soviet Astron., 23, 407.
Raha, N., Sellwood, J. A., James, R. A., & Kahn, F. D. 1991, Nature, 352, 411.
Sellwood, J. A., & Lin, D. N. C. 1989, MNRAS, 240, 991.
Shu, F. H. 1970, ApJ, 160, 99.
Toomre, A. 1964, ApJ, 139, 1217.
Toomre, A. 1966, in *Notes on the Summer Study Program in Geophysical Fluid Dynamics at the Woods Hole Oceanographic Insitution*, p. 111.
Toomre, A. 1977, ARA&A, 15, 437.

A Fokker-Planck Model of Rotating Stellar Clusters

John Girash
Center for Astrophysics, 60 Garden Street, Cambridge, MA 02138, U.S.A.

Abstract. We have developed a two-dimensional orbit-averaged Fokker-Planck model of stellar clusters which expands on spherically-symmetric one-dimensional models to include rotation and ellipticity. Physical effects such as collisions, finite stellar lifetimes and bar formation (*i.e.*, a non-axisymmetric component of the potential) can also be included. The first use of the model is to study the evolution of dense clusters ($\rho_o \simeq 10^7 M_\odot/\mathrm{pc}^3$) that may be expected to have existed at the centres of newly-forming galaxies, with the goal of verifying that angular momentum can be removed from the core of the cluster quickly enough so that rotation no longer prevents the formation of a massive ($\sim 10^2 M_\odot$) object. This could act as the seed black hole for the formation of an AGN.

1. Introduction

Quinlan & Shapiro (1989) developed a Fokker-Planck model of a spherically-symmetric, dense cluster of compact stars, and found "rapid buildup of massive black holes in the cluster core resulting from successive binary mergers and mass segregation." Subsequently, they studied clusters of solar-mass stars and found "it is remarkably easy for massive stars to form through multiple stellar mergers in dense galactic nuclei" (Quinlan & Shapiro, 1990; hereafter QS).

The goal of this project is to generalise the QS approach to two dimensions in order to answer the question: What about rotation? The hoped-for answer is that mergers and mass segregation can still produce a massive object (perhaps $\simeq 10^{2-3} M_\odot$) in the core of the cluster, which could then undergo growth via accretion to reach supermassive size ($\sim 10^{6-8} M_\odot$) within a Hubble time (*e.g.*, David et al., 1987) and be seen as an AGN.

2. The Orbit-Averaged Fokker-Planck Equation

Because of the non-spherical nature of our model, it is not possible to use (E, J_z) as our integrals of the motion as previous two-dimensional work has done (Cohn, 1979). Instead, we use radial (I_1) and tangential (I_2) action variables, which are adiabatic invariants as the potential evolves. Orbit-averaging is then truly a simple average of quantities over the 2π change in (an) angle variable. The orbit-averaged Fokker-Planck equation then takes a particularly simple form:

$$\frac{\partial}{\partial t} f_n = -\frac{\partial}{\partial I_j}[f_n \langle \Delta I_j \rangle] + \frac{1}{2}\frac{\partial^2}{\partial I_i \partial I_j}[f_n \langle \Delta I_i \Delta I_j \rangle] - L_n + G_n - B_n + R_n,$$

where $f_n(I_1, I_2, t)$ are the distribution functions for stars of mass-type n and $\langle . \rangle$ are the drift and diffusion coefficients. L_n and G_n are ad-hoc terms to account for losses and gains caused by stellar mergers, while B_n and R_n are the analogous parameters for stellar evolution. Summation over repeated dummy indices i and j is implied.

The calculation of the coefficients [partially derived by Tremaine & Weinberg (1984) and by Van Vleck (1926) in different contexts] involves expanding the potential in action space and summing over the entire distribution $f = \sum_n f_n$:

$$\langle \Delta I_j \rangle = -2\pi^3 \int dI_1 dI_2 \left(\sum_{k=1}^{2} \ell_k \frac{\partial f}{\partial I_k}\right) \sum_{\ell_1 \ell_2 \ell_3 \pm} \ell_j \, |\Psi_{\ell_1 \ell_2 \ell_3}|^2 \, \delta(\ell_1 \Omega_1 + \ell_2 \Omega_2 \pm \ell_3 \Omega_b),$$

$$\langle \Delta I_i \Delta I_j \rangle = -4\pi^3 \int dI_1 dI_2 \, f \sum_{\ell_1 \ell_2 \ell_3 \pm} \ell_i \ell_j \, |\Psi_{\ell_1 \ell_2 \ell_3}|^2 \, \delta(\ell_1 \Omega_1 + \ell_2 \Omega_2 \pm \ell_3 \Omega_b).$$

In the above, Ω are the orbital frequencies, i.e., $\Omega_1 = \partial E/\partial I_1$, and the ℓ are integers labelling different coefficients $\Psi_{\ell_1 \ell_2 \ell_3}$ in the expansion of the potential in action space. Subscript b signifies "bar", but should be understood to represent any individual element of the potential, e.g., that of one particular "field" star.

The Fokker-Planck approximation remains valid as long as the number $N_n = \int dI_1 dI_2 f_n \gg 1$, and requires that the time step Δt satisfies $t_{\text{dyn}} \ll \Delta t \ll t_r$. The dynamical time $t_{\text{dyn}} <\simeq 10^4$ yr, and the relaxation time $t_r > 10^6$ yr or more.

3. The Potential

We must solve iteratively for the self-consistent potential Φ after each Fokker-Planck time step. To allow for non-sphericity, we first calculate the ellipticity ϵ from the overall net rotational velocity using the virial theorem (Binney & Tremaine, 1987). To make the problem tractable, it is assumed that isodensity surfaces are all of this common ellipticity. The density is thus expressed in terms of the homeoidal radial coordinate $m^2 \equiv R^2 + z^2(1+\epsilon)^{-1}$:

$$\rho(m^2) = \frac{1}{4\pi^2 m} \int \frac{dI_1 dI_2}{I_2} \Omega_1 \sum_n M_n f_n$$

in which M_n denotes the mass of star type n, and from which the new potential run can be calculated. This procedure is iterated until convergence is achieved, typically to $\sim 1\%$, and accelerated by the Aitken "Δ^2" process (Henrici, 1964). A final check of conservation of overall energy is also made. The m^2 grid is variable with each time step, with regions of larger $d\Phi/dm^2$ and $d\rho/dm^2$ being given a greater density of grid points.

The assumption of homeiodal isodensity surfaces, along with f_n, determines the potential Φ. The required integral, however, is too involved to be performed each time the knowledge of potential is needed in the Fokker-Planck coefficient calculation. After testing interpolation schemes and analytic approximations, only the Clutton-Brock self-consistent field method (Hernquist & Ostriker, 1992) proved adequate in both accuracy and speed. The field method is tested at each time step, and if sufficient accuracy cannot be achieved with a reasonable number of expansion terms, the code falls back on direct integration.

4. The Bar

Although some angular momentum is expected to be transported outwards via the effects of shear between the higher-Ω inner regions and the lower-Ω outer regions, we are appealing to a bar-like perturbation in the potential to do most of the angular momentum transfer. So, should the cluster be found to have become unstable to formation of a stellar bar [i.e., be found to satisfy a $T_{\rm rot}/|W|$-style instability criterion such as that of Christodoulou et al. (1995)], a non-axisymmetric component is incorporated into the potential. Combes & Sanders (1981) found that bars form over 1–2 rotational periods, and last for many more (> 10). This has led us to build the non-axisymmetric potential from a fraction ($\sim 10\%$) of the 1 $M_\odot$ population that is forced to orbit in "locked step", sharing a common orbital frequency and phase. We assume there is no transfer of stars from bar to field or vice versa. Combes & Elmegreen (1993) show that the bar frequency is a compromise of the orbital frequencies of its component stars, so our Ω_b is set by conserving either the total energy or total angular momentum of those stars. Allowing the bar distribution to evolve like the field stars avoids any problem of inserting the bar binding energy by hand.

5. Results

It is standard in this game to start with a Plummer sphere distribution. To introduce rotation, we start with the rotation parameter $\lambda \equiv J_{\rm rot}|W_{\rm grav}|^{\frac{1}{2}}/GM_{\rm tot}^{2.5}$, and "shift" the distribution of $f(J)$ so that total J matches desired $J_{\rm rot}$ while conserving number, so that the new $f_\lambda(J) = 0$ for $J <$ some $J_{\rm min}$, and $f_\lambda =$ const $\times f$ for $J > J_{\rm min}$. A typical λ value from cosmological tidal torques is ~ 0.05 (Barnes & Efstathiou, 1987). While this prescription does produce an overall rotation, it does so at the "expense" of the tangential velocity dispersion σ_t, i.e., the fraction of radial orbits is enhanced and the tangential component of the kinetic energy is actually decreased. From a computational "proof-of-concept" point of view, however, this increase in radial orbits is an advantage as it plays into the density dependence of the gravitational relaxation.

Figure 1 details the evolution of the potential run for $N = 10^6$ clusters of 1 $M_\odot$ stars. The initally Plummer-like cluster shows no change over half of a central relaxation time ($\simeq 9 \times 10^6$ yr). Unfortunately, there was not time for a longer run prior to this meeting, but this demonstrates the stability of the staggered distribution/potential updating scheme. The modified cluster was run on a faster machine, so there was time to evolve it for ~ 2 central relaxation times.

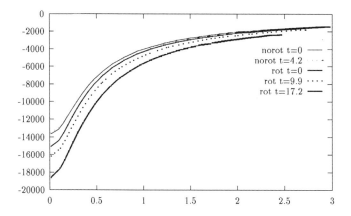

Figure 1. Gravitational potential Φ in units of km/s as a function of equatorial distance r [pc] for an $N = 10^6$ Plummer sphere and a modified cluster of 1 $M_\odot$ stars with core radius 0.3 pc. It is probably wise not to put too much faith in the $t = 17.2$ line, as by that time the energy error $\Delta E/E$ had reached 100%. At $t = 9.9$, $\Delta E/E \simeq 10\%$. The time unit is 0.98×10^6 yr. $t_r(r = 0) \simeq 9 \times 10^6$ yr.

Note how its potential starts out deeper and deepens more rapidly than that of the Plummer cluster, as would be expected when radial orbits are more dominant and stars spend more time in the denser core where two-body relaxation has a greater effect. The corresponding densities are shown in Figure 2; the breaks in $\rho(r)$ are caused by insufficient resolution in the innermost region.[1]

6. Conclusions and "To-do's"

We have demonstrated the stability of this method for modeling the evolution of a dense star cluster using the two-dimensional Fokker-Planck equation, but work needs to be done to balance keeping $\Delta E/E$ small while maintaining a reasonable time step Δt. Ways to accomplish this include allowing a variable Δt when $(\Delta E/E)/\Delta t$ becomes too large (which introduces its own issues of energy error; these may be alleviated using the forced-time-symmetry technique described elsewhere in these proceedings by Piet Hut), or simply by throwing more cycles at the problem. We should also weight the small-r grid more. On the physical side of the problem, the initial conditions need to be chosen for higher $v_{\rm rms}$ (a criticism of QS was that they chose initial conditions favouring massive object formation, but we should at least reproduce their result as a test case), and also not to cool σ_t^2 when rotation is desired. The bar can then be "turned on", as can the effects of stellar mergers and evolution.

---

[1] Further similar results were presented at the meeting, but in light of refinements made since, I will use the remainder of this space to show the equivalent newer results.

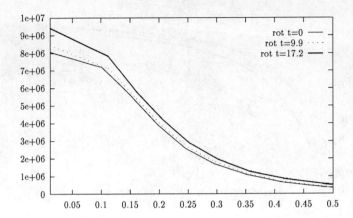

Figure 2. Density $[M_\odot/\mathrm{pc}^3]$ of the inner 0.5 pc of the modified cluster.

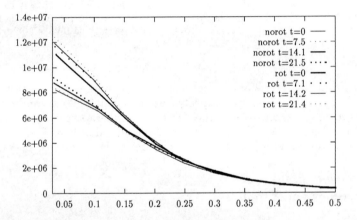

Figure 3. New density runs for similar clusters as in Figure 2.

7. Post-Denouement

In the time between the Kingston meeting and the proceedings deadline, some progress towards these goals has been made. The energy error has been found to be largely dependent upon the differencing scheme used for the distribution functions $f_n(I_1, I_2)$. The Chang-Cooper scheme, which in one dimension guarantees that a Fokker-Planck distribution will remain positive for any Δt, does not generalise to 2-D (essentially because the calculus of extrema becomes more complicated). For the meeting, I used a Chang-Cooper-based method that overcompensated (treating all derivatives as full) but which in later tests proved unsatisfactory, as did undercompensating (using partials). Figure 3 shows the results of a simple fixed differencing weighted 90% towards the forward side, which allowed the 1 $M_\odot$ case to run for more than two central relaxation times (*i.e.*, over four times as long as before) before $\Delta E/E$ reached $\sim 10\%$. One can

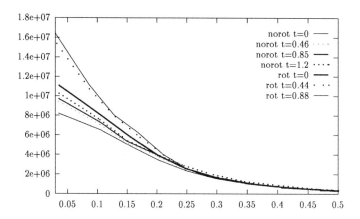

Figure 4. $\rho(r)$ for clusters of 60% $1M_\odot$, 30% $2M_\odot$, 10% $4M_\odot$ by mass.

now see the beginnings of density increase in the core. Also, the break in $\rho(r)$ at $r \simeq 0.1$ pc has been almost eliminated by the use of a better-optimised m^2 grid.

The results for clusters of the same overall mass, but with a mix of 1-, 2- and 4-$M_\odot$ stars is shown in Figure 4. In this case, $\Delta E/E$ reached $\sim 20\%$ after just over one core relaxation time of ~ 0.75 had passed. Here we see that both clusters have larger rates of central density increase than the equivalent all-1 $M_\odot$ clusters, and that the relative increase of the rate for the cluster with enhanced radial orbits (quixotically labelled "rot" as described in § 5) relative to the Plummer-like one ("norot") is much quicker than in the 1 $M_\odot$ case. Both of these differences are as expected when the evolution is dominated by two-body gravitational relaxation.

Acknowledgments. In addition to the insightful support of my advisor George Field, I would like to acknowledge Bill Press, Charles Gammie and Ue-Li Pen for their advice, and NCSA for cycles.

References

Barnes, J., & Efstathiou, G. 1987, ApJ, 319, 575.
Binney, J., & Tremaine, S. 1987, *Galactic Dynamics*, (Princeton: Princeton University Press).
Christodoulou, D. M., Shlosman, I., & Tohline, J. E. 1995, ApJ, 143, 551.
Cohn, H. 1979, ApJ, 234, 1036.
Combes, F., & Elmegreen, B. G. 1993, A&A, 271, 391.
Combes, F., & Sanders, R. H. 1981, A&A, 96, 164.
David, L. P., Durisen, R. H., & Cohn, H. 1987, ApJ, 316, 505.
Henrici, P. 1964, *Elements of Numerical Analysis*, (New York: Wiley).
Hernquist, L. & Ostriker, J. P. 1992, ApJ, 386, 375.
Quinlan, G. D., & Shapiro, S. L. 1989, ApJ, 343, 725.
Quinlan, G. D., & Shapiro, S. L. 1990, ApJ, 356, 483.
Tremaine, S., & Weinberg, M. D. 1984, MNRAS, 209, 729.
Van Vleck, J. H. 1926, *Quantum Principles and Line Spectra*, (Washington, D.C.: National Research Council).

Stability Properties of Spherical Models with Central Black Holes

Andres Meza & Nelson Zamorano

Universidad de Chile, Facultad de Ciencias Físicas y Matemáticas, Departamento de Física, Casilla 487-3, Santiago, Chile

Abstract. We show preliminary results for the stability properties of a set of isotropic, spherical, cuspy models with a central black hole. These models have a luminous mass density that varies as $r^{-\gamma}$ at small radii and as r^{-4} at large radii. We have investigated the stability to radial and nonradial modes using an N-body code. The black hole is represented as a fixed external potential. Our simulations show that in this case, there are no signs of instabilities.

1. Introduction

Recent Hubble Space Telescope photometry has revealed that almost all early-type galaxies have central surface brightness cusps (Crane *et al.*, 1993; Ferrarese *et al.*, 1994; Forbes *et al.*, 1995), consistent with the existence of a black hole (BH), or at least a massive dark object at its center. These findings have increased considerably the interest of theorists in the study of cuspy models. One important family of spherical models with a central divergent density has been explored independently by Dehnen (1993) and Tremaine *et al.* (1994). In particular, the latter authors have studied the dynamical properties of isotropic models with a BH, in which the phase-space distribution function (DF) depends only on the energy per unit mass, $f = f(E)$.

It is known that isotropic spherical models are stable to radial and nonradial perturbations if $df/dE < 0$ (Antonov, 1962; Sygnet *et al.*, 1984). Unfortunately, the isotropic spherical models with a central BH investigated by Tremaine *et al.* (1994) fail to satisfy this condition. We have studied the stability properties of this family of models using an N-body code based on the self-consistent field method by Hernquist & Ostriker (1992).

2. Models and Methods

The γ-models are discussed in detail by Dehnen (1993) and Tremaine *et al.* (1994). Here we give a brief summary of its properties. These models have a mass density of the form

$$\rho(r) = \frac{3-\gamma}{4\pi} \frac{1}{r^\gamma (r+1)^{4-\gamma}}, \quad 0 \leq \gamma < 3. \tag{1}$$

Spherical Models with Central Black Holes

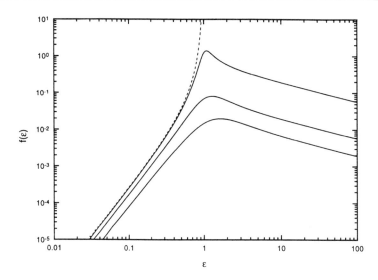

Figure 1. Distribution function f as a function of the relative energy ε for the model with $\gamma = 1$, containing a central black hole of mass $\mu = 0$ (dashed line), 0.01, 0.1, and 0.3. The total mass in stars is 1.

Models are normalized so that the total mass is unity. The models of Jaffe (1983) and Hernquist (1990) are recovered for $\gamma = 2$ and $\gamma = 1$, respectively.

The gravitational potential associated with the density given by equation (1) is obtained from the Poisson equation, and has the simple form

$$\Phi(r) = \begin{cases} \ln\left(\dfrac{r}{r+1}\right), & \text{for } \gamma = 2, \\ -\dfrac{1}{2-\gamma}\left[1 - \left(\dfrac{r}{r+1}\right)^{2-\gamma}\right], & \text{for } \gamma \neq 2. \end{cases} \quad (2)$$

Tremaine et al. (1994) have obtained a family of isotropic models with a central BH of mass μ. In their approximation, the sole effect of the BH is to modify the gravitational potential $\Phi(r)$ to

$$\Phi^*(r) \equiv \Phi(r) - \frac{\mu}{r}. \quad (3)$$

In general, the DF must be obtained numerically. Figure 1 shows the DF for the $\gamma = 1$ model with several values of the BH mass μ. The dashed line corresponds to the isotropic model without BH (Hernquist, 1990). This model satisfies the Antonov criterion $df/dE < 0$, and thus is stable to radial and nonradial modes. However, for the other models, df/dE changes sign at $E \sim -1$. Therefore, the Antonov criterion is not applicable to them.

We have studied the stability properties of these models using an N-body code based on the self-consistent field method developed by Hernquist & Ostriker

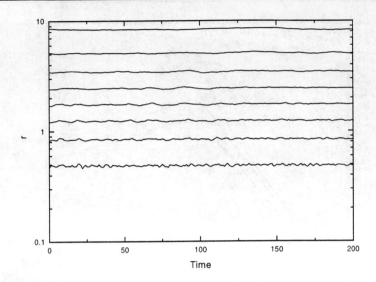

Figure 2. Evolution of the radii of the shells containing 10%, 20%, ... of the mass for the model with $\gamma = 1$ and a black hole of mass $\mu = 0.01$.

(1992; see also Clutton-Brock, 1973). We represent the BH as a fixed softened potential

$$\Phi_{BH}(r) = -\frac{\mu}{\sqrt{r^2 + \epsilon^2}}, \qquad (4)$$

where we adopt $\epsilon = 10^{-2}$. Simulations were done employing $N = 50,000$ equal-mass particles and expansions up to $n_{max} = 4$ and $l_{max} = 2$ for the radial and angular functions, respectively.

3. Results

3.1. Radial stability

The radial stability of the model with $\gamma = 1$ was studied retaining only the $l = 0$ terms in the expansion of the angular variables. Figure 2 shows the evolution of the radii of the shells containing 10%, 20%, ... of the mass for the model with $\gamma = 1$ and a BH of mass $\mu = 0.01$. We observe no change in the radial distribution of matter. Similar results were obtained for the same model with a BH of mass $\mu = 0.1$.

3.2. Nonradial stability

To study the stability to nonradial modes, we have used an iterative algorithm to estimate the axial ratio of the particle distribution from the modified inertia tensor

$$I_{ij} = \sum \frac{x_i x_j}{a^2}, \quad a^2 = x^2 + \frac{y^2}{q_1^2} + \frac{z^2}{q_2^2}. \qquad (5)$$

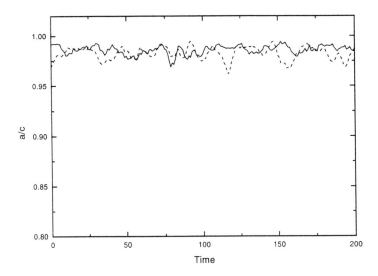

Figure 3. Evolution of the minor to major axis ratio for the model with $\gamma = 1$ and a central black hole of mass $\mu = 0.01$ (dashed line) and 0.1 (solid line).

Figure 3 shows the evolution of the minor to major axis ratio for the model with $\gamma = 1$ containing a central BH of mass $\mu = 0.01$ and 0.1. These values were obtained fitting an ellipsoid at radius $r_{\text{med}} = 5$, the radius which encloses $\sim 70\%$ of the initial total mass. In both cases, we conclude that the system is stable.

4. Conclusions

We have found that a fixed softened potential used to represent a central BH apparently does not affect the stability properties of isotropic models, independent of the BH mass. This result shows that spherical systems can be stable even if $df/dE > 0$ (see Henon, 1973).

However, this conclusion may change if the BH is taken as a massive particle. In this case, the movement of the BH would induce multipolar perturbations in the density of the system which, in turn, could trigger nonradial instabilities. We are planning to incorporate this scenario in a new series of N-body simulations.

Acknowledgments This work was partially supported by DTI E-3646-9312 and FONDECYT 1950271 (NZ) and by a grant from CONICYT and FONDECYT 2950005 to A. M.

References

Antonov, V. A. 1962, Vestnik Lenningradskogo University, 1996; English translation in *IAU Symposium 127, Structure and Dynamics of Elliptical Galaxies*, ed. T. de Zeeuw (Dordrecht: Reidel), 531.

Clutton-Brock, M. 1973, Ap&SS, 23, 55.
Crane, P., *et al.*, 1993, AJ, 106, 1371.
Dehnen, W. 1993, MNRAS, 265, 250.
Ferrarese, L., van den Bosch, F. C., Ford, H. C., Jaffe, W., & O'Connell, R. W. 1994, AJ, 108, 1598.
Forbes, D. A., Franx, M., & Illingworth, G. D. 1995, AJ, 109, 1988.
Henon, M. 1973, A&A, 24, 229.
Hernquist, L. 1990, ApJ, 356, 359.
Hernquist, L., & Ostriker, J. P. 1992, ApJ, 386, 375.
Jaffe, W. 1983, MNRAS, 202, 995.
Sygnet, J. F., Des Forets, G., Lachieze-Rey, M., & Pellat, R. 1984, ApJ, 276, 737.
Tremaine, S., Richstone, D. O., Yong-Ik, B., Dressler, A., Faber, S. M., Grillmair, C., Kormendy, J., & Lauer, T. R. 1994, AJ, 107, 634.

Numerical Simulation of the Formation of Compact Groups of Galaxies

Nicholas J. White & Alistair H. Nelson
Department of Physics and Astronomy, University of Wales, College of Cardiff, Cardiff CF2 3YB, Wales, U.K.

Abstract. The results of numerical simulations of compact groups of galaxies are summarized. Starting from a clumpy primordial gas cloud, the collapse to form a group similar to the observed Hickson galaxy groups is followed for a series of different initial states. The galaxy-like objects ("globs") in each group are spinning discs of gas supported by rotation against their own gravity. During the simulations, tidal disruptions between the globs are seen to be common, leading to polar rings and glob mergers. About 30% of the groups also produced gravitationally-slung escapees.

1. Introduction

Hickson (1982) compiled a catalogue of compact groups of galaxies after searching the red plates of the Palomar sky survey. The selection criteria employed for this catalogue were that there should be more than four galaxies with magnitude less than the brightest magnitude plus three, the distance of the nearest non-member galaxy satisfying the magnitude criterion should be greater than three times the group radius, and that the mean surface brightness inside the group radius should be less than 26. One puzzle concerning these groups is their apparent longevity in that it might be expected that they should have, within the lifetime of the member galaxies, merged into a single galaxy. Here we report on a series of numerical experiments aimed at elucidating whether "primordial" compact groups can form and survive for sufficiently long times, thus allowing them to be observed today.

2. The Simulations

The code employed to perform the simulations was (the now standard) SPH Tree-code. A version of this was constructed by Davies (1991), and Davies & Nelson (1991) have shown this code to be adequate for dealing with large dynamical ranges and asymmetries in self-gravitating fluids. Thirty-seven simulations were performed which contained only baryonic gas. In these, 6,000 gas particles were evenly distributed among 20 spherical clouds, with each cloud having a statistically uniform particle density distribution. Of these 20 clouds, 19 had radii of 0.25 dimensionless units, and were randomly embedded within the remaining cloud which had a radius of unity. Each of these clouds was given a

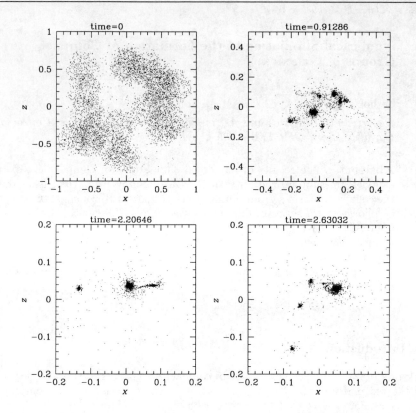

Figure 1. The evolution of the gas content of one of the runs containing dark matter. The time of the final frame scales to just over 6.5 Gyr.

bulk random velocity and the system as a whole was set in solid-body rotation with both quantities ranging from 0.05 to 0.4 in dimensionless units. These initial conditions were intended to represent a density enhancement detached from the cosmological Hubble flow, and well into the non-linear regime. In fact, the results turn out to be insensitive to the details of the initial conditions, as supplementary runs starting from a range of numbers of initial sub-clouds (from one to thirty) were carried out to test this.

We also carried out nine runs in which a collisionless dark matter component containing 90% of the mass was present (usually 2,192 dark matter and 6,000 gas particles). The results for these runs were very similar to those containing no dark matter.

The general evolution of the runs was as follows: initial collapse of sub-clouds into filamentary structure was followed by rapid fragmentation of the medium into typically $\sim$ 15 globs each containing roughly a few hundred particles. The system as a whole collapsed to a higher density phase which then re-expanded. Merging of the globs proceeded at a steady rate until about five globs remained (see Figure 1). It was typically found that the merging rate tailed off to leave $\sim$ 5 globs in a semi-stable configuration. In the cases con-

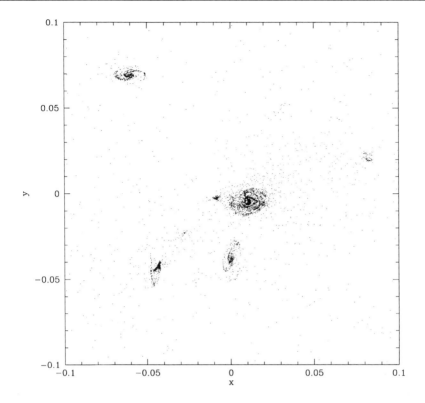

Figure 2. A simulation (after a time of 3.25 Gyr) containing 4,096 dark matter particles and 8,192 gas particles. There is little difference in the group structure between this run and the main survey runs.

taining dark matter, the baryonic globs did not contain individual dark matter haloes but instead ended up moving in a large dark matter halo encompassing the whole group. Typically the baryonic intra-group medium outside the globs contained about 20% of the baryonic matter. Twenty-seven of the gas-only simulations ended up with groups satisfying Hickson's first criterion, while all of those containing dark matter did. The usual time span of the runs was 4–5 Gyr, but a few were carried on to up to 8.5 Gyr, and in these the group remained intact (note that the dimensionless units have been scaled to 2.5 Gyr, 500 kpc, and $5 \times 10^{12} M_\odot$—the total mass was one in dimensionless units).

The globs themselves are disc-like objects supported by rotation against further collapse. They often contain transient spiral arms (see Figure 2) which are mainly material arms and which, in view of the fact that there are only a few hundred to a few thousand particles per glob, are barely credible as realistic structure. More credible are the tidal distortions and bridges which occur on scales of the inter-glob distance, and which occur during interactions and mergers (e.g., note the linear feature at time 2.2 in Figure 1, which is a tidally disrupted glob). In several cases, distinct rings (some polar) are formed around a glob because of tidal disruption of a smaller companion, and this usually ends in a merger of the two. In other cases, particularly when disruption happens to a small glob between two larger globs, the event leads to the glob material being dispersed and returned to the intra-group medium.

For about 30% of the runs, the intense interaction of the globs during the phase of highest density leads to the escape of one of the globs resulting from the three-body slingshot mechanism.

3. Discussion

These calculations demonstrate that:

- Compact groups of galaxies may form by collapse of an inhomogeneous primordial cloud, and that because of "natural selection" of the more stable orbits (*i.e.*, those orbits, where close interaction leads to merger, are removed because of that merger), several of the proto-galaxy objects formed can continue in close association without merging for a significant fraction of the age of the Universe;

- Simulations using SPH Tree-code produce plausible scenarios for galaxy formation. When greater resolution can be obtained using more particles (and improvements on SPH are implemented; see Nelson *et al.*, these proceedings) and when more detailed physics (*e.g.*, star formation) can be incorporated, it will be possible to investigate the origin of morphology and the mass spectrum of galaxies.

References

Davies, J. R. 1991, Ph.D. Thesis, (UWCC, Cardiff).
Davies, J. R., & Nelson, A. H. 1991, in *Dynamics of Disc Galaxies*, ed. B. Sundelius (Goteborg University), 405.
Hickson, P. 1982, ApJ, 255, 382.

Selected Lagrangian Weights—An Alternative to Smoothed Particle Hydrodynamics

Alistair H. Nelson, Peter R. Williams, John P. Sleath & Nicholas P. Moore

Department of Physics and Astronomy, University of Wales, College of Cardiff, Cardiff CF2 3YB, Wales, U.K.

Abstract. The solution of the equations of hydrodynamics on sets of disordered sample points is a requirement for Lagrangian calculations of self-gravitating gas. A new method for doing this is discussed.

1. Lagrangian Methods for Hydrodynamics

Recent calculations of the formation of galaxies and stars have demonstrated that Smoothed Particle Hydrodynamics (SPH) is a very powerful tool in simulating the evolution of self-gravitating gas. Its chief advantages are its geometrical flexibility, and the tendency for large numbers of particles to collect in the high density regions where more resolution is needed. The SPH method represents the fluid using particles which move with it, and calculates fluid bulk quantities and gradients by taking a weighted sum over a number of particles neigbouring the field point. Typically, the number included in the weighted sum is 50–100, and the weights are calculated by a specified weighting function which depends on the coordinate separation of the field point and the contributing particle. This algorithm has proved very robust, and delivers simulations which have considerable physical plausibility. However, the method has often been criticised as being inferior to Finite-Difference methods as far as accuracy is concerned (ignoring the computational problems associated with having a grid with sufficient adaptability to deliver the geometric flexibility). And indeed, 50–100 seems a large number of points to use in comparison to the six needed to obtain, say, a 3-D gradient on a regular grid.

There is, however, an alternative procedure which enables gradients for inclusion in numerical analogues of the field equations to be computed with a smaller number of points. If one Taylor-expands the SPH weighted sum for a gradient, then it becomes obvious that the oddness of the weights as a function of the coordinate separation is being used to minimize the truncation errors in the terms at the next order up from the gradient in the expansion. The truncation error is not exactly zero, however, since the random sampling of the points does not allow the terms at that level in the sum to exactly cancel—although the sampling error is reduced if the number of points in the sum is increased or if the positioning of the points is not completely random (*e.g.*, the artificial viscosity typically used in SPH causes the Poisson noise in an

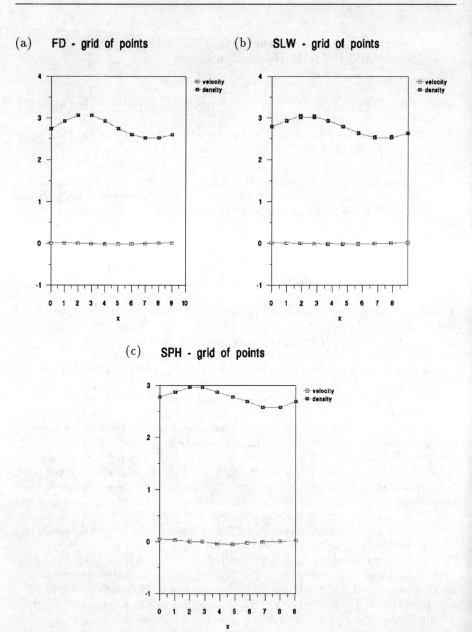

Figure 1. Sound wave solution on a regular 10 × 10 2-D grid using (a) finite-difference method, (b) SLW method, and (c) SPH method.

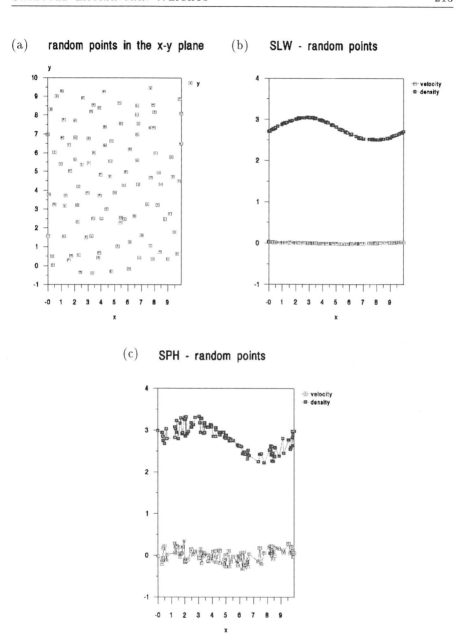

Figure 2. Sound wave solution on an irregular distribution of points. (a) shows the actual distribution used, whilst (b) and (c) show the sound wave solutions calculated by SLW and SPH respectively.

initial random distribution to be damped out, leading to a smoother distribution of particles and a smaller sampling error).

The alternative approach is to select the weights of the sample points in such a way that the truncation error is exactly zero. This is done by solving a matrix equation for the weights, and for the case of a 3-D gradient, requires the use of only nine neighbouring points to the field point. This method of Selected Lagrangian Weights (SLW) is related directly to finite-difference approximations, and in the case of a regular grid, the weights yield the standard finite-difference formulae for approximating gradients. For more details, see Nelson & Moore (1993).

Here we illustrate the use of the method for a simple 1-D sound wave propagating through a 2-D array of sample points. The initial field has the density unperturbed but the velocity is perturbed with a cosine wave in x, $i.e.$, we have two sound waves propagating in opposite directions with the same amplitude and wavelength. Initially, the density perturbations exactly cancel each other and in the subsequent evolution, the velocities will exactly cancel each other a quarter of a period later so that at that time the solution will be zero for the velocity and a sine wave for the total density perturbation.

Figure 1 shows the results of three calculations on a regular grid of sample points a quarter period after the start for the cases of (a) a standard finite-difference calculation, (b) SLW and (c) SPH. The results are very similar for all three, and close to the expected result. The error, however, is largest in the SPH case, although 50 points ($i.e.$, half of the total of 100) were used for each SPH sum to minimise the sampling error. Figure 2 shows the results of the same calculation, but this time on a disordered set of points covering the same space as for Figure 1—Figure 2a shows the actual distribution. Here, SLW performs as well as it did for the regular grid, but the sample error in SPH leads to significant errors.

2. Conclusion

The greater accuracy of SLW over SPH makes it attractive for the solution of field equations on sets of disordered points, such as in the simulation of self-gravitating gas dynamics. We plan to incorporate this algorithm into our calculations of galaxy formation.

References

Nelson, A. H., & Moore, N. P. 1993, Mem. Soc. Astron. It., 65, 1043.

Galaxy Dynamics by N-Body Simulation

J. A. Sellwood

Rutgers University, Department of Physics and Astronomy, P.O. Box 849, Piscataway, NJ 08855, U.S.A.

Abstract. I compare various popular and unpopular techniques for simulating large collisionless stellar systems. I give a quantitative comparison of the raw cpu times required for five separate codes, including tree codes and basis function expansions, which demonstrate that grid codes are most efficient for large numbers of particles. Since efficiency is only one consideration when choosing a code, I discuss other strengths and weaknesses of the various methods. While some applications may require the maximum possible number of particles, I argue that quiet start techniques can often permit reliable results to be obtained with moderate particle numbers. I suggest a quiet start procedure for spherical stellar systems and show that it leads to a significant reduction in the relaxation rate. The combination of efficient codes and quiet starts allows many galactic dynamical problems to be tackled without the need for supercomputers.

1. Introduction

In this very brief review, I discuss several techniques that attempt to mimic the gravitational dynamics of galaxies using N-body methods. The collisionless property of these stellar systems demands algorithms that try, as far as possible, to suppress the relaxation caused by $\sqrt{N}$-type fluctuations in the particle distribution. The objective, therefore, is quite different from collisional problems, discussed by Hut (these proceedings), in which the interactions between every pair of particles have to be treated as accurately as possible.

My purpose is to emphasize two quite old, but very powerful, methods which I feel are not being fully exploited in much of the current work in the field. The first is the superior efficiency of particle-mesh (PM) techniques over all others that are widely used, which I demonstrate by presenting some timing comparisons of five codes. While efficiency is important, it is not the only factor that determines the choice of code, and I discuss the strengths and weaknesses of each of these methods. Secondly, I wish to underline the advantage of using a quiet start, which I show reduces noise-driven relaxation for a fixed number of particles.

For all the tests reported here, I used a stellar polytrope of index 5, or Plummer sphere, which has the density profile

$$\rho(r) = \frac{3M}{4\pi a^3}\left(1 + \frac{r^2}{a^2}\right)^{-5/2}, \qquad (1)$$

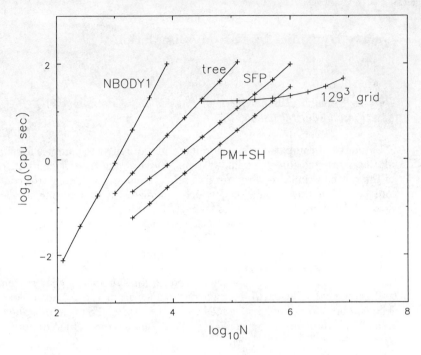

Figure 1. Comparison of the performance of five codes for collisionless N-body simulations. The lines join values obtained from sequences of runs with different numbers of particles. While the spatial resolution is kept fixed for each method, it differs from method to method.

where M is the total mass and a is a length scale. A spherical model with an isotropic distribution function of this form is stable (Binney & Tremaine, 1987). This mass distribution is not very centrally condensed, making it particularly easy to simulate. This infinite mass distribution has to be truncated to fit inside a grid; for all these tests, I therefore restrict the spatial extent of the particles to $r < 6a$ at the outset, thereby discarding just over 4% of the mass.

2. Performance Tests

Figure 1 compares the cpu time required to complete one time step using five different methods. Timings are all on a DEC Alphastation 500/333.

The methods in this comparison are two PP (particle-particle) methods (direct and tree), a basis function expansion method or SFP (smooth field-particle), and a generic PM (particle-mesh) code. The fifth method is based on an expansion in surface harmonics, which has been widely used in the past, and is a hybrid PM+SH. Of these, the SFP, PM and PM+SH codes are much easier to vectorize/parallelize than are the PP methods.

2.1. Direct method

The simplest method of all is NBODY1, the least sophisticated of the sequence of direct N-body codes devised by Aarseth (1985); the Fortran source is reproduced in an appendix of Binney & Tremaine (1987). I set the softening length $\epsilon = 0.01a$ and accuracy parameter $\eta = 0.03$. Since the individual time steps for each particle are determined by the adopted accuracy criterion, I have defined the "time per step" as one twentieth of the cpu time to reach one dynamical time $= (a^3/GM)^{1/2}$, which is the time step length I used for all the other codes.

A least-squares fit to the points in this log-log plot has a slope of 2.3, i.e, steeper than the usually quoted $\mathcal{O}(N^2)$ behavior. The steeper dependence is because the number of close neighbors rises with N, which implies that a larger fraction of the particles requires shorter steps, even though the mass of each is correspondingly less. Performance could, of course, be improved by increasing the softening parameter or relaxing the accuracy criterion. One could justifiably argue that the softening length should be reduced as N is increased, a strategy that would steepen the N dependence still more.

The rapid increase of cpu time needed with N renders this code far too expensive for desirable numbers of particles and more efficient techniques should be used. The GRAPE machines (Hut, these proceedings) implement an algorithm that is not very different; their strategy seems essential for collisional problems which require highly accurate forces, but collisionless problems allow other algorithms to be employed which yield results more efficiently, more than compensating for the advantage of custom-built chips.

2.2. Tree code

The line marked "tree" indicates timings obtained using Hernquist's (1987) public-domain tree code, which employs the original Barnes & Hut (1986) algorithm. Admittedly, this may no longer be regarded as a state-of-the-art code, but most recent improvements have been directed towards improving accuracy while running speeds do not appear to have improved significantly. In fact, the speeds reported by Dubinski (1996) for an $N = 640k$ calculation on the T3D using 128 nodes are only some ~ 20 times higher than the extrapolated line in Figure 1, which is for a single processor workstation.

The timings reported in Figure 1 are with quadrupole forces included but use the large opening angle $\theta = 1$, which leads to quite low accuracy. A more accurate calculation with a smaller opening angle would, of course, take longer. The slope of the line shown in Figure 1 is approximately 1.3, which is steeper than the predicted $\mathcal{O}(N \log N)$.

2.3. Surface harmonic method

The PM+SH method I employ has not been described elsewhere. It sits between the fully particle-based approach used by Villumsen (1982), White (1983) and McGlynn (1984) and the fully grid-based method used by van Albada (1982). I tabulate the set of active coefficients for the expansion of the potential at a number of fixed radii (a mesh) and derive the acceleration applied to each particle from linearly interpolated values of these coefficients at the radius of the particle. There is no need to sort particles and there is no gridding in the

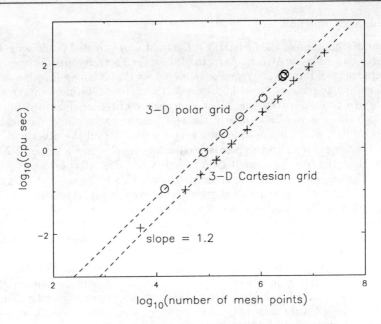

Figure 2. The cpu time requirements for the field determination on two different grids for various numbers of grid cells. These times do not include any particle movement.

non-radial directions. The radii of grid points can be more closely spaced near the center and there should be many particles between successive grid points.

With this scheme, the cpu time rises linearly with N and is virtually independent of the number of radial grid points for fixed l_{max}. The timings shown in Figure 1 are for $l_{max} = 6$ with 201 radial grid points.

2.4. Basis function expansion

The line marked SFP in Figure 1 is for a basis function expansion method. It was obtained using Hernquist's so-called SCF code, but I prefer the acronym SFP (smooth field-particle), since all N-body codes have self-consistent fields. Hernquist & Ostriker (1992) derive a new basis, but otherwise their algorithm is identically that first proposed by Clutton-Brock (1972; 1973).

As advertized, the slope of the line in this plot is precisely unity when the number of functions is held fixed. The data in Figure 1 were obtained using the Clutton-Brock (1973) basis, employing radial functions $0 \leq n \leq 10$ and an angular expansion up to $l_{max} = 6$. One could reasonably argue that more functions should be employed as N rises, which would steepen the slope. Unlike the method described in § 2.3, this method becomes more time consuming as the radial resolution is increased, since more functions need to be evaluated.

At the end of his paper, Clutton-Brock (1972) conceded that the SFP method was not competitive, in terms of raw speed, with the Fourier grid methods which were then emerging, a conclusion that has not changed in the subsequent almost two and a half decades.

2.5. Grid methods

The curve marked "129^3 grid" shown in Figure 1 results from a Cartesian PM code with this number of grid cells. It incorporates James' (1977) Poisson solver which represents the only significant algorithmic improvement since the first 3-D grid results reported by Hockney & Brownrigg (1974). It is clear from the near horizontal portion of the line at small N that the calculation time is dominated by the field determination except for very large N. In fact the slope is still well below linear at the last point which is for $N = 8M$. The dominance of the field calculation part leads to similar N-dependence for grids of other geometries.

Figure 2 illustrates how the N-independent field calculation time varies with the size of grid employed. Timings are both for James' Poisson solver and my own 3-D cylindrical polar grid (Sellwood & Valluri, 1997), which differs only in details from that described by Pfenniger & Friedli (1993). The potential returned on the Cartesian grid has the full resolution in all three dimensions afforded by each grid. The forces determined on the polar grid, on the other hand, have azimuthal harmonics restricted to $0 \leq m \leq 8$ only. The polar grid is clearly more time consuming, but has the compensating advantage of offering higher spatial resolution near the symmetry axis, where the density of particles is generally greatest.

Other grid geometries are clearly possible. Van Albada (1982) used a spherical grid to follow mildly aspherical collapses with spectacular radial resolution. To have conducted such simulations on a fixed Cartesian grid with any degree of validity would have required an impossibly large number of cells. The moral here is that when resolution demands a huge grid, then a different grid geometry would probably be more appropriate.

Adaptive mesh refinement has been shown at this meeting to be highly successful for other applications. Many adaptive PM codes have been described in the recent literature, most of them devised for the problem of cosmological structure formation in which many dense regions develop at random locations on the main grid; they seem ideally suited to the intended application (*e.g.*, Couchman, these proceedings). I have not yet found a need for such a method to follow the global dynamics of an isolated galaxy, preferring instead to tailor the grid to the mass distribution under study. Of course, an adaptive refinement strategy on a Cartesian grid would be immensely superior to a fixed grid for the collapse problems studied by van Albada, yet his well-chosen spherical grid required no adaptivity and remarkably few mesh cells.

3. Which Method to Use?

While raw cpu speed is clearly an important factor when determining which code to employ for a problem, the physical properties of the model to be studied may render a less efficient code more appropriate.

The principal disadvantage of Eulerian PM codes is that the fixed volume and geometry of the mesh make them unsuited to following wholesale rearrangements of the mass distribution, as occur during major galaxy mergers, for example. This same inflexibility means that different grids are required for different problems. As these codes are already more complex than PP codes, the need to rewrite for a new grid is a further significant handicap. Furthermore, grid codes generally require a few hundred MB of memory in addition to that needed to store the particle coordinates.

Finally, choosing an interpolation scheme that will best hide the discrete nature of the grid is a "black art"; fortunately this problem has been studied in great detail by the plasma physicists, who employ similar codes, and the theory behind the various strategies has been developed at some length; see *e.g.*, Birdsall & Langdon (1991). Nevertheless, it is not possible to hide the grid completely, and there are some delicate problems for which grid effects are intolerably large. For example, my attempts to study fully three-dimensional warped disks have been compromised by the tendency for a thin flat disk inclined at an angle to the grid planes to experience a weak torque from the grid that causes it to precess and to try to align with the grid planes. Until this tendency can be effectively removed, a grid method simply cannot be used for this problem. Fortunately, serious problems of this kind are rare.

A common, but generally misdirected, criticism of grid methods is that they lack spatial resolution. It has to be admitted that Cartesian grids cannot handle steep density gradients, but other grid geometries can (*cf.*, van Albada, 1982, and the PM+SH code described in § 2.3). Furthermore, should problems be identified for which a single fixed grid cannot resolve high density regions, one could resort to adaptive mesh refinement, at the cost of some extra software effort.

The criticism that current grid methods are inflexible and cannot follow major rearrangements of the mass distribution can be applied *a fortiori* to SFP methods. Weinberg's (1996) extension of the SFP technique to a dynamically changing basis attempts to remedy this inflexibility. Nevertheless, the form of the potential that can be represented by the few hundred terms employed in typical SFP implementations is much more restricted than can be represented on a grid of some million separate mesh points. (See also § 4.1.)

Both surface harmonic methods, and SFP codes in which the non-radial part of the basis uses an expansion in Y_l^m functions, are ill-suited to mass distributions, such as disks, that are far from spherically symmetric.

Basis function expansions are best suited to simulations in which only very minor changes in the mass distribution occur. In fact, Earn & Sellwood (1995) found the method to be ideal for following the linear growth of instabilities, where the mass distribution by definition is hardly changing, but quickly to become uncompetitive in the non-linear regime.

Lagrangian PP codes, on the other hand, are fully adaptive and can follow arbitrary changes to the mass distribution. However, the cost penalty for not using a grid is so high that PP methods should be regarded as methods of last resort. In fact, it would be worth considerable effort to develop hybrid or overlapping grids (or expansion centers) when the problem warrants, such as have been developed by Villumsen (1983), or the almost completed effort by Weeks (1988), for binary galaxies.

4. Noise

Particle noise is the major difficulty when modeling collisionless systems. In a simulation with $N \sim 100{,}000$, density fluctuations caused by shot noise are

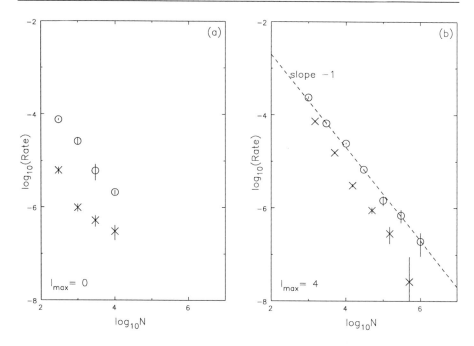

Figure 3. The relaxation rate, defined in equation (2), for various numbers of particles (a) when all non-radial terms are omitted, and (b) when terms up to $l_{max} = 4$ are included. The dashed line is not a fit, but merely indicates a slope of -1. The circles are from experiments with randomly distributed particles and the crosses are from experiments with quiet starts.

larger than those in a real elliptical galaxy, say, by factor of $\sim 1{,}000$. This problem affects all spatial scales from the importance of two-body encounters to global variations. The implication is therefore that we should try to maximize N in order to minimize the noise. Such a strategy places a still higher premium on computational efficiency.

4.1. Smoothing or bias

As simulations with sufficient spatial resolution to make meaningful use of billions of particles are still not even remotely possible, some form of smoothing of the density distribution is required. However desirable smoothing may seem, it must be borne in mind that it is a double-edged sword that yields a *biased* estimate of the potential.

The nature of the bias is to limit resolution to the (effective) particle size in PP and PM codes, which smooth only locally, and to leave larger scales unsmoothed. SFP methods, on the other hand, smooth globally but thereby introduce a much more severe form of bias, since density changes that are not represented by the truncated basis have no effect on the potential. The "*over-smoothing*" strategy advocated by Weinberg (1996) reduces the relaxation rate at the cost of further exacerbating the bias.

4.2. Quiet starts

A quiet start (Sellwood, 1983) is a strategy for reducing noise without increasing N. It is achieved by distributing the particles smoothly instead of at random. The use of such techniques in astronomy dates back at least to Hénon (1968) and probably still further in plasma physics. They have become standard in laying down the initial conditions and perturbation spectrum in cosmological simulations, as first clearly explained by Doroshkevich et al. (1980), but continue to be largely ignored for galaxies.

Figure 3 shows empirical relaxation rates, defined by

$$\text{Rate} = \frac{d}{dt}\left\langle [E_i(t) - E_i(0)]^2 \right\rangle, \qquad (2)$$

where $E_i(t)$ is the specific energy of the i-th particle at time t, and the angle brackets indicate an average over all particles. In a perfectly collisionless simulation of a stable equilibrium model, each particle would conserve its energy and this rate should be zero. If the relaxation is caused by shot noise in the particle distribution, we expect $d(\Delta E_{\text{rms}})/dt \propto 1/\sqrt{N}$; the mean square changes plotted seem consistent with a slope of -1.

Once again, the data are taken from simulations of the Plummer sphere computed, in every case, using the PM+SH method with 201 radial grid points and $l_{\max} = 0$ or 4. Simulations with particles placed at random initially (noisy starts) follow the expected $1/N$ behavior. Models begun with a quiet start, however, relax more slowly—apparently at about the rate for a noisy start with several times the number of particles.

As pointed out by Weinberg (1993), collective neutral oscillations of the system excited by the random distribution of particles are the most important source of relaxation. In the case of the Plummer sphere, experiments with purely radial forces (Figure 3a) showed that relaxation is most effectively reduced by suppressing collective radial pulsations of this spherical model. The most important part of the quiet start strategy in this case is to space several particles (ten in this case) having identical energy and angular momentum components at equal intervals of radial phase. This strategy reduces the density variations caused by the radial oscillation of randomly selected particles as they pursue their orbits, leading to an impressive reduction in the relaxation rate.

Once non-radial forces are included, however, yet more particles are required to reduce the non-radial density variations. The data in Figure 3(b) were obtained by placing replicas of each independent particle at 72° intervals around a circular ring that lies in the plane of that orbit (normal to its angular momentum vector). All the replica particles lie at a single radius and have equal radial and azimuthal components of velocity and would therefore follow exactly congruent orbits in that plane if the potential were smooth. This careful set-up procedure leads to a relaxation rate that seems to be roughly equivalent to that from a model having several times the number of particles.

The number of particles per ring should exceed $l_{\max}$, five seem to be adequate when $l_{\max} = 4$. The use of ten rings of five particles to smooth both the radial and azimuthal density variations reduces the number of independent orbits by a factor of 50. It is therefore desirable to take extra care to select a better-than-random set of orbits (E & L values) from the distribution function.

It should be noted that while it is easier to suppress noise by truncating the angular expansion at low order, a similar procedure works well on a 3-D Cartesian grid where no angular symmetries are imposed. Again, a smooth radial distribution and ten particles per ring produced a very significant reduction in the relaxation rate.

Provided the relaxation time greatly exceeds the duration of the experiment, we can regard a simulation as adequately collisionless. Thus, in simulations of mergers or collapses which may need to be followed for only a few dynamical times, relaxation caused by particle noise can be rendered unimportant even with a modest number of randomly placed particles. Quiet starts are therefore advantageous only for initially equilibrium models that are to be evolved in isolation for long periods.

5. Conclusions

While it remains true that no code is "best" for *every* problem, PM codes have continued to hold their pre-eminent position as the most efficient and should be considered first for many problems. PM methods do have a number of drawbacks, principally stemming from the finite volume enclosed by the mesh, and therefore are not appropriate for every problem. PP methods, on the other hand, are so expensive they should be regarded as an option of last resort for collisionless problems.

I have argued that the resolution limitations of PM methods are generally overstated, since it is often possible to tailor the grid geometry to the problem at hand. Cartesian grids are not the only option; both cylindrical and spherical grids have been used to excellent effect to improve resolution dramatically near the centers. The moral here is that if attaining satisfactory resolution of the mass distribution of an isolated galaxy seems to require an excessively large number of grid points, then the grid geometry is probably inappropriate. For problems I have worked on, making this change has seemed more efficient than resorting to adaptive mesh refinement, though this conclusion may not be always true.

My other principal conclusion is that it is rarely necessary to employ many millions of particles for collisionless problems. I find that $100{,}000 \leq N \leq 500{,}000$ is usually adequate and that results with more particles are generally indistinguishable. The relaxation rate, which clearly should be much slower than the rate of evolution being simulated, can be reduced without increasing N simply by employing quiet starts.

With efficient PM codes and quiet starts, I find there is a wealth of interesting problems that can be addressed quite adequately without requiring supercomputers.

Acknowledgments. I would like to thank Gerald Quinlan for making his copies of NBODY1 and Hernquist's basis function codes available for timing tests, and for his comments on a draft of this paper. Also S. Shandarin pointed out the early reference to quiet starts in cosmology. This work was supported by NSF grant AST 93/18617 and NASA Theory grant NAG 5-2803.

References

Aarseth, S. J. 1985, in *Multiple Time Scales*, eds. J. W. Brackbill & B. J. Cohen, (New York: Academic Press), 377.
Barnes, J., & Hut, P. 1986, Nature, 324, 446.
Binney, J., & Tremaine, S. 1987, *Galactic Dynamics*, (Princeton: Princeton University Press).
Birdsall, C. K., & Langdon, A. B. 1991, *Plasma Physics by Computer Simulation*, (Bristol: Adam Hilger).
Clutton-Brock, M. 1972, Ap&SS, 16, 101.
Clutton-Brock, M. 1973, Ap&SS, 23, 55.
Doroshkevich, A. G., Kotok, E. V., Novikov, I. D., Polyudov, A. N., Shandarin, S. F., & Siguv, Yu. S. 1980, MNRAS, 192, 321.
Dubinski, J. 1996, New Astron., 1, 133.
Earn, D. J. D., & Sellwood, J. A. 1995, ApJ, 451, 533.
Hénon, M. 1968, Bull. Astron. Paris, 3, 241.
Hernquist, L. 1987, ApJS, 64, 715.
Hernquist, L., & Ostriker, J. P. 1992, ApJ, 386, 375.
Hockney, R. W., & Brownrigg, D. R. K. 1974, MNRAS, 167, 351.
James, R. A. 1977, J. Comput. Phys., 25, 71.
McGlynn, T. A. 1984, ApJ, 281, 13.
Pfenniger, D., & Friedli, D. 1993, A&A, 270, 561.
Sellwood, J. A. 1983, J. Comput. Phys., 50, 337.
Sellwood, J. A., & Valluri, M. 1997, MNRAS, in press.
van Albada, T. S. 1982, MNRAS, 201, 939.
Villumsen, J. V. 1982, MNRAS, 199, 493.
Weeks, A. 1988, in *The Few Body Problem*, IAU Colloq. 96, ed. M. Valtonen (Dordrecht: Kluwer), 413.
Weinberg, M. D. 1993, ApJ, 410, 543.
Weinberg, M. D. 1996, ApJ, 470, 715.
White, S. D. M. 1983, ApJ, 274, 53.

Galaxy Splashes: The Effects of Collisions Between Gas-Rich Galaxy Disks

Curtis Struck

Department of Physics and Astronomy, Iowa State University, Ames, IA 50011, U.S.A.

Abstract. Results of three-dimensional (SPH) simulations of collisions between two model galaxies, each consisting of a rigid halo and a gas-rich disk component, are presented. I will focus on two topics: the gas bridges formed from shock impacts, and the gas infall out of these bridges. The bridges are predicted to have essentially no ongoing star formation, in contrast to tidal bridges but in agreement with observations of several systems. Which galaxy contributes the most gas to the bridge depends strongly on the companion orientation at impact. The companion disk is usually disrupted by the impact, but reforms via accretion from the bridge.

1. Introduction

Barnes & Hernquist (1992), in their review paper on interacting galaxies, remind us that "Long ago, Spitzer & Baade (1951) suggested that global shocks could sweep the gas from spiral galaxies during fast interpenetrating collisions,..." However, they go on to caution that "unless the encounter geometry is just right", not much gas will be swept up. Here, I want to consider some of these special galaxy collisions, and in particular, nearly head-on collisions between two gas-rich galaxies of roughly comparable size which produce splashes affecting a significant fraction of the gas in the two galaxies (specifically, those cases in which most of the gas in the smaller companion and less than half that in the primary is affected). While such collisions are definitely a minority among all interacting systems, they are not terribly rare. This is probably because galaxies in groups don't usually have large relative angular momenta.

Nonetheless, these collisions have not received much attention to date. This is understandable since the splash process is likely less important to galaxy evolution as a whole than gas funneling to central regions in mergers, for example. On the other hand, there are many interesting questions to be addressed here. How are the clouds affected relative to the stars? What happens to the gas? How much ends up in the centers of the galaxies? How much mass exchange between the galaxies is there? How does star formation (henceforth SF) vary as a function of time, location, and collision parameters? How is the overall structure of the galaxies affected?

Moreover, multi-waveband studies of a number of individual systems have been published recently. Because these systems are at a fairly early stage after the collision, they provide especially good opportunities for comparison to models. This is in contrast to the study of merger remnants, where multiple collisions make it impossible to figure out the details of the interaction. *Splash galaxies* include several well-known types of interaction morphology, *e.g.*, the collisional ring galaxies and some ocular galaxies (see Elmegreen *et al.*, 1991).

2. Simulations

To address some of the questions above and model some of the observed systems, I have carried a number of *splashy* galaxy collision simulations. Both galaxies in these simulations consist of a gas disk and a rigid dark matter halo, with an optional stellar disk component. A three-dimensional SPH algorithm is used with typically a total of 30,000 particles. A local self-gravity is computed for the particles on the scale of the smoothing length, and the disks are initialized such that shear exceeds self-gravity on larger scales. The gas particles obey either an isothermal equation of state or an adiabatic one with heating and cooling terms added. These terms are described in detail in Struck (1997), but in sum, while the cooling function is a simple approximation to a standard interstellar cooling curve, the heating is assumed to be the result of young star activity. Specifically, heating is turned on for a finite time in a particle whenever it exceeds a density threshold and is cool enough. Heating mainly occurs where the self-gravity is able to form large cloud "complexes".

The simulations were typically run with a 3:1 primary-to-companion galaxy mass ratio, with bound trajectories such that the companion's impact point in the primary disk was slightly off-center.

3. Splash Bridges

Splash bridges can be defined as those in which collisions between clouds belonging to the different galaxies have played a significant role in removing the gas from the disk. A number of likely examples have been mapped out in recent VLA HI studies. These include first and foremost collisional ring galaxies, where the symmetric, propagating wave provides independent evidence of a direct collision (see Appleton & Struck-Marcell, 1996). (The argument is usually run the other way, *i.e.*, the bridge provides the evidence that a collision made the ring wave, but in any case the picture is self-consistent.) Examples include the Cartwheel (Higdon, 1996), VII Zw 466 (see Figure 1 and Appleton *et al.*, 1996), and the Arp 284 (Smith & Wallin, 1992) systems. In the latter example, the models suggest a more off-center collision, and a bridge that is produced by the combined effects of splash and tidal "swing". In fact, an on-going modeling program with B. Smith has produced results suggesting that the swing is more important in this case. This is also true of the more nearly planar collisions that produce *ocular* galaxies, but HI maps suggest that even these often involve a splash in the outer disk (*e.g.*, Elmegreen *et al.*, 1995a; 1995b). A non-splash bridge is found in the *Sacred Mushroom* system, AM1724-622, where it appears that the primary was very gas-poor (*e.g.*, Wallin & Struck-Marcell, 1994).

The comparison between the observations and simulations of these systems is beginning to suggest an interesting difference between SF properties of splash

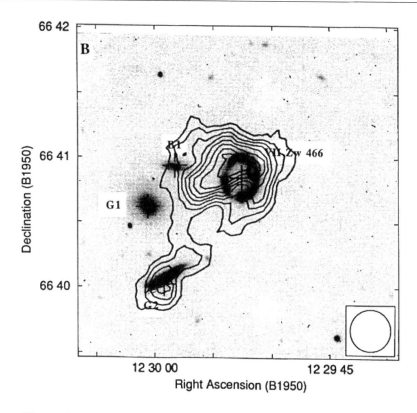

Figure 1. A grey scale optical image with HI contours from Appleton et al. (1996).

and swing bridges. It is well-known that tidal structures like the latter can generate SF, e.g., the M51 types. However, the models suggest that SF is suppressed in splash bridges, and the observations of the Cartwheel and VII Zw 466 bridges show no evidence of significant recent SF. Given the relatively crude SF (heating) algorithm in our current models, such results cannot be regarded as firm yet. However, the models suggest some general physical reasons for this suppression. Firstly, as the galaxies move apart, gas in the bridge finds itself well away from either potential center, yet it retains much of its angular momentum, so there is radial expansion (see Figure 2). Secondly, different gas elements experience different collision histories, e.g., given general orientations at impact, some gas elements from the companion follow the leading edge, and collide with gas elements (from both galaxies) that have already been collisionally accelerated forward. This is a source of vertical (kinematic) expansion. As time goes on, the bridge is stretched further. Thus, the impact can be expected to break down loose cloud structures, and the subsequent kinematic expansions prevent reformation. We would expect SF to be inhibited, unless many stars were formed in the high-velocity impact shock. The observations to date, however, suggest not.

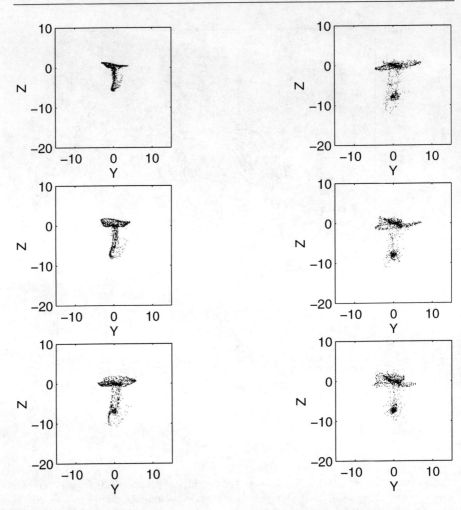

Figure 2. A sequence of cross sectional snapshots of a representative model, showing the development of the gas bridge (from Struck, 1997).

Arp 284 and some of the ocular galaxies provide especially interesting cases for further study because simulations that yield the best morphological fits suggest inclined, off-center collisions generating a combination of splash and swing. The offset between the HI bridge and the (old) stellar bridge in Arp 284 is especially intriguing (see Smith et al., 1997).

4. Companion Death and Transfiguration

Figure 2 also shows that in splash collisions, the companion gas disk is highly disrupted. This is not quite the case with the stellar disk (not shown in the figure), but it does experience large amplitude ringing. Because rigid halos were used in the simulations, this result may be conservative. That is, if the halo is also strongly perturbed, then the gravitational cohesion it provides would be less than in these models.

However, the models show that most of the bridge gas is eventually accreted onto the two potential centers. A gas disk reforms around the companion, but usually in a plane whose axis is the center-line of the bridge, rather than the initial axis. If the stellar disk survives, we would expect the companion to look like a polar ring galaxy (though, most observed polar rings do not have larger partners, so this cannot be the primary formation mechanism.)

The end result for the companion depends a lot on its disk orientation relative to the primary disk at impact (see Struck, 1997). Consider two contrasting examples. The first is where the two disks have the same orientation at impact. In this case, most of the companion gas disk is nearly "stopped" at impact, and the splash comes from the primary. In the end, the companion ends up with only 20% of its initial gas mass, and 90% of this gas originated in the primary. In the second example, the companion gas disk is in the plane of its orbital trajectory (perpendicular to the primary), and rotating in the prograde sense. Swing is important in this case, and most of the bridge material comes from the companion. The companion ends up with half of its original gas mass, and only 15% of this originates in the primary.

5. Conclusions

There is much to be learned from comparing simulations of galaxy splashes to observation. Studies of SF in ring waves (see Appleton & Struck-Marcell, 1996) have shown that high-velocity cloud collisional shocks are not needed to trigger large-scale, coherent star formation. The models above further suggest that high-velocity impact shocks are not able to trigger SF in bridges before kinematic expansion suppresses it. It may be premature to generalize, but these results provide some evidence for the conclusion that high-velocity shocks are not primarily responsible for enhanced SF in interacting systems. Gravitational instabilities and massive cloud buildup orchestrated by the interaction seem like better bets (*e.g.*, Elmegreen *et al.*, 1993).

Galaxy splashes also provide a unique environment for testing our knowledge of the global structure of the interstellar gas in gas-rich galaxies. For example, what is the porosity of the gas disks? If most of the gas mass is concentrated in dense molecular cores with small cross sections, we expect little splash. The simulations above model the diffuse components of the interstellar gas, and cannot resolve such cores. The observations and models agree that the splash mass is considerable, so the diffuse component must also be substantial. These are early days, and it seems likely that much more can be learned from this type of comparison. This should ultimately lead to substantial refinements in the models of the gas thermal physics, which is a prerequisite to understanding galaxy evolution.

Acknowledgments. I am grateful to Phil Appleton, Vassilis Charmandaris, Jim Higdon, and Bev Smith for many helpful discussions.

References

Appleton, P. N., Charmandaris, V., & Struck, C. 1996, ApJ, 468, 532.
Appleton, P. N., & Struck-Marcell, C. 1996, Fund. Cosm. Phys., 16, 111.
Barnes, J. E., & Hernquist, L. 1992, ARA&A, 30, 705.
Elmegreen, B. G., Kaufman, M., & Thomasson, M. 1993 ApJ, 412, 90.
Elmegreen, B. G., Sundin, M., Kaufman, M., Brinks, E., & Elmegreen, D. M. 1995a, ApJ, 453, 139.
Elmegreen, D. M., Kaufman, M., Brinks, E., Elmegreen, B. G., & Sundin, M. 1995b, ApJ, 453, 100.
Elmegreen, D. M., Sundin, M., Elmegreen, B. G., & Sundelius, B. 1991, A&A, 244, 52.
Higdon, J. L. 1996, ApJ, 467, 241.
Smith, B. J., & Wallin, J. F. 1992, ApJ, 393, 544.
Smith, B. J., Struck, C., & Pogge, R. W. 1997, ApJ, in press.
Spitzer Jr., L., & Baade, W. 1951, ApJ, 113, 413.
Struck, C. 1997, ApJS, submitted.
Wallin, J. F., & Struck-Marcell, C. 1994, ApJ, 433, 631.

Galaxy Formation in Dissipationless N-Body Models

Eelco van Kampen[1]

Royal Observatory Edinburgh, Blackford Hill, Edinburgh, Scotland
EH9 3HJ, U.K.

Abstract. We present a method of including galaxy formation in dissipationless N-body simulations. Galaxies that form during the evolution are identified at several epochs and replaced by single particles with increased mass and softening. This allows one to produce two-component models containing galaxies and a background dark matter distribution. We applied this technique to obtain two sets of models: one for field galaxies and one for galaxy clusters. We tested the method for the standard CDM scenario for structure formation in the universe. A direct comparison of the simulated galaxy distribution to the observed one sets the amplitude of the initial density fluctuation spectrum, and thus the present time in the simulations. The rates of formation and merging compare very well to simulations that include hydrodynamics, and are compatible with observations. We also discuss the cluster luminosity function.

1. Introduction

In order to test models for structure formation in the Universe, it is necessary to pinpoint galaxies within the modelled large-scale distribution of matter. As this distribution evolves in a non-linear fashion, the use of N-body methods is usually required. Galaxies should form, under the influence of gravity, within an N-body simulation in a self-consistent way. However, such simulations suffer from a numerical problem: small groups of particles that represent galaxies get disrupted by numerical two-body effects within clusters (Carlberg, 1994; van Kampen, 1995). This can be solved by replacing each group of particles by a single "galaxy particle" just after they have formed into a virialised system that resembles a galactic halo, thus ensuring their survival. This should also produce galaxies at the right time and the right place, with a spectrum of masses. In van Kampen (1995), such galaxy particles were only formed at a single epoch. Here we extend this scheme to "continuous" galaxy formation by applying the algorithm several times during the evolution. This means that merging of already-formed galaxies is taken into account as well, although only in a schematic fashion. An important advantage of having a galaxy formation algorithm added to the N-body integrator is that the time normalisation is now

---

[1]current address: Theoretical Astrophysics Center, Juliane Maries Vej 30, DK-2100 Copenhagen, Denmark

2. Galaxy Formation Recipe

2.1. Outline

For the identification of galaxies during the evolution of the large-scale matter distribution, we use the *local density percolation* algorithm (see also van Kampen, 1995). Instead of a fixed linking length, we modulate the linking length according to the *local density* around each particle in such a way that the linking length is shorter in high-density environments. This partly resolves the cloud-in-cloud problem. We form (and merge) galaxies several times during the evolution, so we have to consider percolation of *unequal* mass particles. In the following, **x** and **v** are *comoving* variables.

2.2. Local density percolation for unequal mass particles

In the local density percolation scheme, particles are linked together if they are separated by a certain fixed fraction p of the Poissonian average nearest neighbour distance $x_{nn}^P \equiv [4\pi\langle n\rangle/3]^{-1/3}$, where $\langle n\rangle$ is the mean number density of particles, modulated by the local number density $n^G(\mathbf{x}, s)$, which is $n(\mathbf{x})$ Gaussian-smoothed at the scale $s = x_{nn}^P$. Thus, p and s are the (dimensionless) free parameters of the algorithm. All models in this paper have $\langle n\rangle = 8.0\, h_0^3$ Mpc^{-3}, so $x_{nn}^P = 0.31 h_0^{-1}$ Mpc. For unequal mass particles, we need an extra modulation according to their masses. Initially, all particles have mass m_0. As galaxies form, particles arise with masses m_i that are integral multiples of m_0. Galaxy particles with mass m_i that are put $\sqrt{m_i/m_0}$ times further away will exert the same gravitational force as dark particles with mass m_0 and so we should take the factor $\sqrt{m_i/m_0}$ as the second modulation factor. This gives a local percolation length

$$R_p(\mathbf{x}_i, p, s) = p\, x_{nn}^P \sqrt{\frac{m_i}{m_0}} \left[\frac{n^G(\mathbf{x}_i, s)}{\langle n\rangle}\right]^{-1/3}. \qquad (1)$$

Since each particle has its own linking length, we use their mean to test pairs of particles. Furthermore, to prevent excessive percolation lengths, we adopt an absolute maximum of $x_{nn}^P/2$ for the pair linking lengths (after taking the mean of the individual ones), *i.e.*, a lower limit for the local particle density which is equal to eight times the background particle density.

In addition, we require pairs to have a relative pairwise velocity

$$v_{\|} \equiv (\mathbf{v}_j - \mathbf{v}_i) \cdot (\mathbf{x}_j - \mathbf{x}_i)/|\mathbf{x}_j - \mathbf{x}_i| \qquad (2)$$

of less than 800 km s^{-1}. This is twice the relative pairwise velocity dispersion at separations around x_{nn}^P for the field, and somewhat smaller than found for a sample including the Coma cluster (Mo *et al.*, 1993). It is also twice the maximum internal velocity dispersion we allow for a group. We include this velocity linking length to exclude fast-moving particles which are geometrically

GALAXY FORMATION

linked to a group. This often occurs within the potential wells of galaxy clusters. One should see this criterion as the velocity equivalent of a (constant) spatial linking length, so that we actually find groups in phase-space. However, the velocity linking length is less restrictive than the spatial linking length since it serves a different purpose, as said.

2.3. Virial equilibrium criterion for unequal mass particles

A group of particles should only be transformed into a single, soft galaxy particle if it forms a physical system roughly in virial equilibrium. This "virial criterion" is a necessary addition to the local density percolation algorithm for the purpose of defining galaxies. We will use a virial-equilibrium criterion in a simplified form using the half-mass radius R_h, motivated by Spitzer (1969) who found that for many equilibrium systems, the virial equilibrium equation can be written as

$$\sigma_v \equiv \langle v^2 \rangle \approx 0.4 \frac{GM}{R_h}, \qquad (3)$$

where v is now the proper velocity. If galaxies are identified only once, one has to deal with equal-mass particles, and R_h can simply be calculated by obtaining the median of the distances of all particles with respect to the centre of the group being tested. For groups where the masses of the member particles can differ by a few orders of magnitude, the half-mass radius will often exactly coincide with a galaxy particle. This makes the median (*i.e.*, the half-mass radius) a rather noisy estimator for the total gravitational energy of a group of unequal mass particles. Because for many probability distributions the mean and the median are almost identical, we use the mass-weighted mean distance from the centre of the group as an estimator for R_h. This is a more smoothly-defined and well-behaved quantity than the median distance. The new galaxy particle has a softening parameter corresponding to the R_m of the original group, which certifies a reasonable conservation of energy (see van Kampen, 1995).

Discreteness noise will cause some scatter in the group quantities, so we should allow for some tolerance in the difference between the estimated virial mass and the true mass of the group. The allowed tolerance determines the "reach" the criterion has in time: larger permitted deviations from virial equilibrium result in the acceptance of groups that are still collapsing. We accept groups as real when the virial mass is within 25 per cent of the true mass.

2.4. Choice of the galaxy formation parameters

For the (dimensionless) local density percolation parameters, we choose $p = 1$ and $s = 1/2$. The maximum percolation length is $x_{nn}^P/2$, which corresponds to $0.16\, h_0^{-1}$ Mpc for our simulations. We set the upper mass limit for galaxy particles to be $1.4 \times 10^{13}\, M_\odot$. This ensures that possible cD galaxies are *not* modelled by single galaxy particles, since that would produce undesirable numerical problems and galaxies that massive are not (numerically) disrupted anyway. We add an extra limit on the internal galaxy velocity dispersion of $\sigma_v < 400$ km s^{-1}, the maximum value found for typical ellipticals (de Zeeuw & Franx, 1991) and comfortably within the velocity dispersion of halos around spirals given their typical circular velocities of 200–300 km s^{-1}, Finally, we need to adopt a lower limit of seven particles in a group because of discreteness noise that causes an artificially large scatter in the virial mass estimate.

3. Description and Timing of the Simulations

We have run eight simulations of average patches of the universe, and 99 cluster models. This latter set forms a catalogue of galaxy clusters, and is discussed extensively in van Kampen & Katgert (1997). The actual N-body code we use is the Barnes & Hut (1986) tree code, slightly adapted for our purposes and supplemented with the galaxy-formation algorithm. We ran the models up to $\sigma_8 = 1$, which is sufficiently beyond the time that is expected to be the present epoch for the $\Omega_0 = 1$ CDM scenario adopted: σ_8 was found to be significantly smaller than unity in most earlier work (*e.g.*, Davis *et al.*, 1985; Frenk *et al.*, 1990; Bertschinger & Gelb, 1991). From a comparison of the galaxy-galaxy autocorrelation function obtained for the field models to that observed, we find σ_8 to be in the range 0.46 to 0.56 (van Kampen, 1997), while a similar comparison of the statistical properties of clusters gives roughly the same range (van Kampen & Katgert, 1997).

4. Galaxy Properties

4.1. Galaxy formation and merging rates

As a first check on how our modelling of the formation and merging of galaxies compares to other techniques, notably hydrodynamical simulations, we look at the galaxy formation and merger number density rates as a function of time. These are plotted in Figure 1 for $\sigma_8 = 0.46$, where t_0 is the present epoch. The formation rate peaks at $z \approx 1.3$, whereas the merger rate does not show a clear peak. The merging of small objects into galaxies with masses that are included in the formation rate is not included in the merger rate.

The shapes and amplitudes of both rates compare remarkably well with those found from hydrodynamical simulations performed by Summers (1993), also shown in Figure 1, if we triple his time scale. This can be justified quantitatively as follows: Summers (1993) has a higher mass resolution and forms galaxies down to a lower mass cut-off. The smallest galaxy masses in his simulations are roughly a hundred times smaller than our lowest mass galaxies. The CDM spectrum on galactic scales is a power-law with index -2. We can then use the scaling law, $t_{\text{form}} \sim M^{1/4}$, which applies for such a spectrum, to find that this mass difference gives a factor of three difference in the formation time.

Summers (1993) used a full-fledged hydro code (and identified galaxies with the ordinary friends-of-friends algorithm). The fact that we find similar shapes and amplitudes for the rates means that the use of a galaxy formation recipe with an ordinary collisionless N-body code can give comparable results to more advanced simulation techniques that incorporate more (but certainly not all) physical processes.

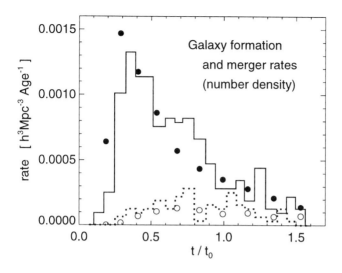

Figure 1. Galaxy formation and merger rates for our models (symbols) and the simulations of Summers (1993), rescaled to our units and galaxy masses (see text). Filled symbols and solid lines represent galaxy formation, open symbols and dotted lines represent merging.

4.2. Cluster luminosity function

For our cluster simulations, we study the luminosity function within the projected Abell radius. Since we know only the masses of the galaxies, we need to assume a constant mass-to-light ratio, Υ, to obtain a luminosity function. For the B_J magnitude, $\Upsilon_J \approx 1{,}200$ for an $\Omega_0 = 1$ universe. Because, on average, 25 per cent of the mass is locked into our galaxies (including dark halos), $\Upsilon_J \approx 300$ for the galaxies. The joint luminosity function for our cluster models is plotted in Figure 2 for all galaxies (all symbols), and for the $M > 1.5 \times 10^{12}\ M_\odot$ ones (filled symbols only), along with a fitted Schechter (1976) luminosity function for each of them (dashed line for all galaxies, solid line for the limited set), with slope α and characteristic magnitude B_J^* as free parameters.

The fit for all galaxies is quite good but not perfect: $\alpha = -1.5$. We find that the fit gets better for the limited set of massive galaxies: $\alpha = -1.25$. This means that we either do not model low-mass galaxies very well, or that the mass-to-light ratio is not constant. The first option is probably true anyway since we cannot model merging of galaxies with masses below our lower limit towards the low-mass end of the mass function that we try to fit.

With this in mind, it is fair to say that the Schechter function does fit rather well for the limited set. We find $B_J^* = -20.3$ for this set, which corresponds remarkably well with the value that Colless (1989) found for a sample of 14 observed clusters. It compares less well with $B_R^* = -22.6$ found by Vink & Katgert (1994) for a sample of 80 clusters, corresponding to $B_J^* \approx -20.8$. Still we can say that the modelling performs reasonably well given the uncertainties in both the fits and the assumption of a constant mass-to-light ratio.

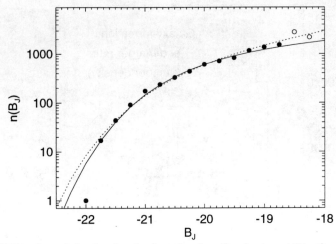

Figure 2. Joint luminosity function for all galaxies within the projected Abell radius. Filled symbols represent galaxies with masses larger than $1.5 \times 10^{12}\ M_\odot$. The dashed line indicates a fit of a Schechter (1976) function to all galaxies. The solid line is a similar fit to the massive ones only (filled symbols). Both fits were made with α and B_J^* as free parameters, and assume a constant mass-to-light ratio $\Upsilon_J = 300$.

Acknowledgments. Joshua Barnes and Piet Hut are gratefully acknowledged for allowing use of their tree code, Edmund Bertschinger and Rien van de Weygaert for their code to generate initial conditions, and Eric Deul for allowing me to use the computer systems that are part of the DENIS project. I acknowledge EelcoSoft Software Services for partial financial support during the early stages of the project, and a European Community Research Fellowship as part of the HCM programme during its final stages.

References

Barnes, J. E., & Hut, P. 1986, Nature, 324, 446.
Bertschinger, E., & Gelb, J. M. 1991, Computers in Physics, 5, 164.
Carlberg, R. G. 1994, ApJ, 433, 468.
Colless, M. 1989, MNRAS, 237, 799.
Davis, M., Efstathiou, G., Frenk, C. S., & White, S. D. M. 1985, ApJ, 292, 371.
de Zeeuw, P. T., & Franx, M. 1991, ARA&A, 29, 239.
Frenk, C. S., White, S. D. M., Efstathiou, G., & Davis, M. 1990, ApJ, 351, 10.
Mo, H. J., Jing, Y. P., & Börner, G. 1993, MNRAS, 264, 825.
Schechter, P. L. 1976, ApJ, 203, 297.
Spitzer Jr., L. 1969, ApJ, 158, L139.
Summers, F. J. 1993, Ph.D. Thesis, (University of California at Berkeley).
van Kampen, E. 1995, MNRAS, 273, 295.
van Kampen, E. 1997, MNRAS, submitted.
van Kampen, E., & Katgert, P. 1997, MNRAS, in press.
Vink, J., & Katgert, P. 1994, preprint.

Global Error Measures for Large N-Body Simulations

Wayne B. Hayes & Kenneth R. Jackson
Department of Computer Science, University of Toronto, Toronto, ON M5S 3G4, Canada

> "Prediction can be difficult, particularly about the future."
> — Mark Twain.

1. Introduction and Motivation

N-body systems are chaotic, which implies that small perturbations to a solution, such as numerical errors, are exponentially magnified with the passage of time. Although this is widely recognized, its impact on qualitative properties of numerical N-body simulations is not well understood. Animated movies of large N-body simulations, like spiral galaxies or cosmological systems, are very exciting to watch and often look quite reasonable, but it is little more than an "article of faith" that the results are qualitatively correct (Heggie, 1988).

Can these simulation results really be trusted? What conditions must a simulation meet for its accuracy to be assured? Is there a limit on the length of time a system can be followed accurately? What measures can be used to ascertain the accuracy of a simulation? More fundamentally, what do we *mean* by "accuracy" and "error" in these simulations, given that numerical errors are exponentially magnified with the passage of time? How do we know if a simulation is unreasonable, unless it is *grossly* so? In summary, *how badly are we allowed to integrate?*

There are several formal methods of measuring the accuracy of an ODE integration. *Forward* error measures the divergence of the numerical solution from the true solution with the same initial conditions; *backward* or *defect* error asks about the existence of a true solution to a slightly different problem, *i.e.*, the field is allowed small time-dependent perturbations. Finally, although we know that a numerical solution diverges exponentially away from the exact solution with the same initial conditions in the exact field, there remains the (unintuitive) possibility that there exists another exact solution, in the exact field, but with slightly different *initial conditions*, that remains close to the numerical solution. Such an exact solution, if it exists, is called a *shadow*, which we discuss below.

At the least stringent end of the global error spectrum, quantities like energy and momentum should be well conserved, but it is unclear if this is enough because there are an infinite number of solutions that have exactly the same energy and momentum but vastly different phase space trajectories. Symplectic integrators (*e.g.*, Sanz-Serna, 1992) may provide better error control, but it is far from proven that this is enough. A less rigorous, but almost certainly satisfactory requirement, might simply be that a slightly "out-of-focus" animation of the N-body simulation be indistinguishable from an "out-of-focus" animation of the exact solution. Even less stringent, we might require that statistical "clumpiness" be reproduced correctly, *e.g.*, if we are simulating a spiral galaxy, we may require that the time evolution of the Fourier spectrum of the density distribution in cylindrical coordinates be reproduced reasonably accurately. This is one of the least stringent properties with which we can imagine one might be satisfied. If such a statistic is not reproduced by a simulation, then the numerical results are probably worthless.

2. Shadowing N-Body Systems

A true trajectory of an ODE system lying near a numerical one is called a *shadow*. In this paper, we consider shadows of numerical solutions to the N-body problem. If you are interested only in general dynamics of solutions chosen at random, and not in the behaviour of a particular solution, then the existence of shadows implies that the numerical simulation may be dynamically valid in a very strong sense. (However, there still remains the possibility that shadows are not typical of true solutions chosen at random.)

A shadow found by our algorithm (and, most likely, shadows in general) typically stays quite close to the numerical solution for some period of time, and then it abruptly diverges at a point called a *glitch*. The relevant measures of error are the proximity of the shadow to the numerical solution, and the length of time before a glitch occurs. In gravitational N-body systems, glitches seem to occur most often at close encounters between particles (Quinlan & Tremaine, 1992). Glitches do not necessarily imply no shadow exists, only that our algorithm failed to find one.

Shadowing was initially studied in the context of gravitational N-body systems by Quinlan & Tremaine (1992), who considered a simplified system with one particle moving "pinball fashion" amongst 99 fixed ones. Not surprisingly, they found that shadows generally lasted longer, and usually remained closer to the numerical trajectory, as the local accuracy of the numerical solution increased. For a typical integration accuracy, they found that they could shadow the single particle for about 10 crossing times before a glitch occurred. They also found that the shadow length was moderately extended when the gravitational potential was "softened".

However, real N-body systems have more than one moving particle! Our work extends the results of Quinlan & Tremaine (1992) by studying shadow lengths for systems with more moving particles. To this end, we first spent several months speeding up Quinlan and Tremaine's shadowing algorithm by about a factor of 100 (Hayes, 1995). We then considered M particles moving amongst $N - M$ fixed ones.

Figure 1 plots various shadow lengths as a function of M, for systems with $N = 100$ particles. We use the standardized units of Heggie & Mathieu (1986). As can be seen in the first plot, the shadow length for an unsoftened potential seems to scale as $1/M^2$, which is not encouraging. However, perhaps we are not scaling the problem in a realistic way. The second plot shows that shadow length for a softened potential stays roughly constant, or very slowly decreases, with M. A still better way to scale the problem is to leave the potential unsoftened but to increase N, the total number of particles. This makes the problem extremely computationally intensive, and so we have only attempted this up to $M = 10$

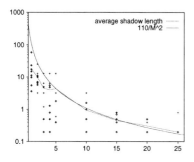 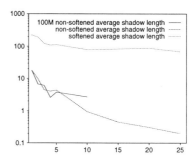

Figure 1. On the left, shadow lengths in crossing times as a function of M, the number of moving particles, while $100 - M$ particles remain fixed. The potential is unsoftened. Each point represents a shadow length for a particular orbit, while a line joins their averages. $110/M^2$ is plotted for comparison. On the right, we re-plot the average from the left, and add the average for a softened potential and the average for an unsoftened system that is scaled so that the total number of particles is $100M$ while the total mass of each system remains constant.

and $N = 1,000$, but it appears that the shadow length is longer in this case than for the smaller $N = 100$ particle system.

3. Further Work on Shadowing

There are many directions to pursue. First, the algorithm used here does not allow time step perturbations. During a close encounter between two stars, two solutions that differ in a direction along the orbit will temporarily separate greatly as the stars speed past each other. The difference leaves the "linear regime", and shadowing fails. Van Vleck (1995) and others have shown that, for various problems, allowing time step perturbations can increase shadow lengths *tremendously*. Other possibilities include extending our computations to larger N and M, shadowing of simpler equations that don't require ODE integrations, and testing the dependence of shadowing on the choice of integration method. Still more possibilities are listed in Hayes (1995).

References

Hayes, W. 1995, M. Sc. Thesis (Department of Computer Science, University of Toronto).
Heggie, D. C. 1988, in *Long-Term Dynamical Behaviour of Natural and Artificial N-Body Systems*, ed. A. E. Roy, (Dordrecht: Klewer), 329.
Heggie, D. C., & Mathieu, R. D. 1986, in *The Use of Supercomputers in Stellar Dynamics*, eds. P. Hut & S. L. W. McMillan, (Berlin: Springer-Verlag), 233.
Quinlan, G., & Tremaine, S. 1992, MNRAS, 259, 505.
Sanz-Serna, J. M. 1992, in *Acta Numerica*, ed. A. Iserles, (Cambridge: Cambridge University Press), 243.
Van Vleck, E. 1995, SIAM J. Sci. Comp., 16, 1172.

Part V
AGN/ACCRETION/ OUTFLOW

Jet Stability: Numerical Simulations Confront Analytical Theory

Philip E. Hardee

Department of Physics and Astronomy, University of Alabama, Tuscaloosa, AL 35487, U.S.A.

Abstract. Observations of extragalactic radio sources, protostellar systems, and the galactic superluminals have provided a strong motivation for studying the dynamics of highly-collimated outflows. All such outflows are subject to the Kelvin-Helmholtz instability. In this article, we review the basics of stability theory and, in particular, the synergistic confrontation between theory and numerical simulation relevant to these highly-collimated jet flows.

1. Introduction

The highly-collimated outflows observed to emanate from the centers of galaxies and quasars, from neutron star and black hole binary star systems, and from protostellar systems are subject to the Kelvin-Helmholtz instability. These jets, some inferred to be less dense and some denser than the surrounding medium, can propagate to distances of tens to hundreds of times their diameter. The observed jet structures can be related to jet velocity, density, and magnetic fields via stability theory and modeling. Modeling is of particular importance to extragalactic jets or to the galactic jets where the continuum nature of synchrotron or inverse Compton emission provides no direct measure of velocity, density and magnetic field strength, and whose apparent motions may reflect a wave pattern or shock speed different from the speed of the underlying flow. The line emission associated with protostellar jets provides detailed kinematical information on jet velocity and density structures which can be used to confront the adequacy of theory and numerical simulations.

In general, two theoretical approaches to an analysis of jet stability have been taken involving a choice of either slab jet or cylindrical jet geometry. A slab jet is spatially resolved along two Cartesian axes and is infinite in extent in the third dimension. In the original development of the theory, the choice of geometry was largely made as a result of the complexity of the situation being analyzed. Cylindrical geometry was used to study various aspects of the temporal growth of the instability on fluid jets with a sharp discontinuity between a uniform jet and a uniform external environment (Ferrari *et al.*, 1978; Hardee, 1979) or with magnetic fields (Ferrari *et al.*, 1981; Ray, 1981; Cohn, 1983; Fiedler & Jones, 1984; Bodo *et al.*, 1989; 1996). Slab jet geometry was used to study the evolution of velocity, density, and magnetic field across a shear layer of finite width between a jet and the external environment (Ferrari *et al.*, 1982). Further

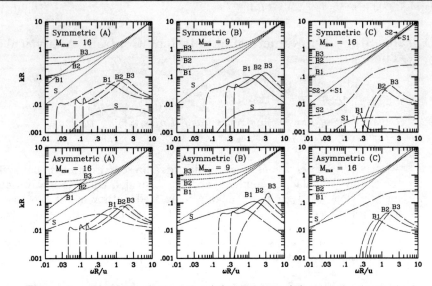

Figure 1. Typical solutions for (A) adiabatic, (B) strongly magnetized and (C) radiative cooling slab jets of half-width R. Dotted (solid/dashed) lines give the real (imaginary) part of the wavenumber.

development considered the spatial rather than the temporal growth of instability on cylindrical jets (Birkinshaw, 1984; Payne & Cohn, 1985; Hardee, 1987a) and on slab jets (Hardee & Norman, 1988; Hardee et al., 1992) as study of spatial growth is more relevant to astrophysical jets. More recently, the stability properties of more complex cylindrical configurations including velocity shear (Birkinshaw, 1991a) and helically twisted magnetic fields (Appl & Camenzind, 1992) have been investigated. A fine review can be found in an article by Birkinshaw (1991b). Finally, the dramatic observations of protostellar jets in which energy loss via radiative cooling is important have spurred consideration of the effects of radiative cooling on the stability properties of jets (Massaglia et al., 1992; Hardee & Stone, 1997).

The theory was relatively well-developed when it became possible to conduct numerical jet simulations. The first simulations were of relatively stable 2-D axisymmetric jets (e.g., Norman et al., 1984). These simulations were soon followed by 2-D slab jet simulations and much more recently by fully 3-D jet simulations which allow the much more unstable non-axisymmetric modes to develop. In this review let us concentrate on 2-D slab jets and 3-D cylindrical jets where the theory has been confronted by numerical simulations.

2. Stability Theory—2-D Slab Jets

Jet distortion can be considered to arise from Fourier components whose wave propagation and growth (damping) is governed by a dispersion relation. On a supermagnetosonic slab jet, coupling between the jet boundaries results in symmetric (pinch) and asymmetric (sinusoidal) normal modes of jet distortion. Each

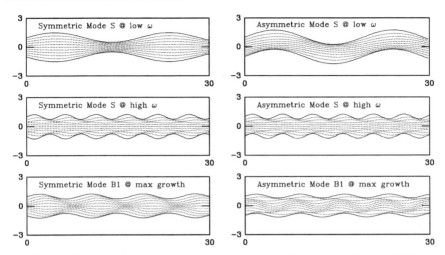

Figure 2. Maximum displacements for surface (S) modes at low and high frequencies and for the first body (B1) mode at the maximally-growing frequency for a 2-D slab jet.

normal mode contains surface (S) and multiple body (Bm) wave solutions. The mathematical behavior of the normal modes is illustrated in Figure 1. In general, solutions have a maximum growth rate (imaginary part of the wavenumber) at a "resonant" frequency that increases for higher-order body waves. The resonant frequency and maximum growth rate decrease and the growth length increases as the magnetosonic Mach number increases. The magnetosonic Mach number is $M_{\mathrm{ms}} \equiv u/(a^2+V_{\mathrm{A}}^2)^{1/2}$, where u, a and V_{A} are the jet velocity, sound speed and axial Alfvén speed, respectively. Note the increase in maximum growth rates that accompanies the decrease in Mach number for case B ($M_{\mathrm{ms}} = 9$) relative to case A ($M_{\mathrm{ms}} = 16$) even though the magnetic field strength has increased. On supermagnetosonic jets, the pinch mode surface (S) wave has a plateau in the growth rate which is much less than the maximum growth rate for other wave modes and is reduced as the magnetic field strength is increased.

Case C in Figure 1 illustrates the changes that can result from the inclusion of radiative cooling. When radiative cooling is important, a new pinch mode surface (S2) wave can have relatively large growth rates. The low (high) frequency growth rate of S2 increases as cooling internal (external) to the jet increases. The high-frequency growth rate of the sinusoidal mode surface wave is similarly affected by radiative cooling. The growth rate of body waves is decreased somewhat by radiative cooling. These results are sensitive to the temperature dependence and magnitude of the cooling function (Hardee & Stone, 1997).

Supermagnetosonic adiabatic or cooling jets are only mildly influenced by magnetic fields. However, on transmagnetosonic jets, the growth rate of the sinusoidal surface wave mode is large while the growth rate of all other surface and body wave modes is severely depressed. On the other hand, when the jet is sub-Alfvénic, the jet is stabilized to all fast-growing modes of instability.

The distortions induced by pinch and sinusoidal surface and body wave modes can be computed from the theory and are illustrated in Figure 2. The

Figure 3. Density image of a Mach 3, 2-D slab jet one-tenth the ambient density. Axial image dimension is about 65 jet radii.

maximum displacement amplitude should be less than that leading to the overlap of displacement surfaces because overlap implies the development of shocks. The surface modes are capable of large-scale fluid displacements at frequencies below the maximally-growing frequency. At higher frequencies, fluid displacements are reduced with much less internal displacement out of phase with the surface displacement. The body modes produce distortions in which interior fluid displacements remain relatively large. Along a transverse cut through the jet perpendicular to the axis, the axial phase shift means that there are reversals in fluid displacement at null surfaces within the jet. The number of null surfaces increases and the maximum amplitude decreases for higher-order body modes.

3. Numerical Simulations—2-D Slab Jets

An image from one of the first simulations (Norman & Hardee, 1988) performed to confirm theoretical predictions is shown in Figure 3. In this simulation and those to follow, a jet that is established completely across the computational grid in pressure equilibrium with the surrounding medium is sinusoidally perturbed at the inlet. In this simulation, the perturbation is at a frequency near the resonant frequency predicted by stability theory. Random perturbation leads to a similar result (Zhao et al., 1992). The growing sinusoidal oscillation seen in Figure 3 has wavelength and spatial growth rate in accordance with predictions made by the linear theory and is seen to grow to sufficient amplitude, on the order of the half-width of the slab, to disrupt the initial continuous flow.

In other simulations of constantly expanding "wedge" jets initialized in equilibrium with a suitable atmosphere (Hardee et al., 1994), the jet is sinusoidaliy perturbed at different frequencies and the results of these numerical experiments are shown in Figure 4. The top panels show that perturbation at frequencies above the resonant frequency leads to reduced surface amplitudes.

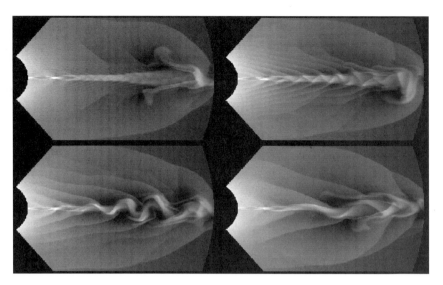

Figure 4. Density images from simulations which show jets perturbed from high (top left) to low (bottom right) frequencies.

Internal structures predicted by the theory result from the high-frequency response of the sinusoidal surface or first body wave mode (see Figure 2). The bottom two panels of Figure 4 show the predicted response of the jet to perturbations below the resonant frequency where the jet can respond bodily to the oscillation. The long wavelength of the lowest-frequency perturbation allows a relatively large amplitude oscillation before transverse motions approach the magnetosonic speed.

Somewhat counter-intuitively, the theory predicts that a stronger magnetic field can be destabilizing through reduction in the magnetosonic Mach number. Note, however, that a sub-Alfvénic flow is predicted to be almost totally stabilized. These predicted effects have been explored numerically (Hardee et al., 1992; Hardee & Clarke, 1995) and some results from these studies are shown in Figure 5. The left-most image is from the same simulation shown in the lower left panel of Figure 4. In the middle panel, the jet contains a relatively strong axial magnetic field which has reduced the magnetosonic Mach number by about a factor of 2 relative to the jet in the left-most panel. It is readily apparent that the jet with reduced magnetosonic Mach number develops large amplitude transverse oscillations much closer to the origin and in agreement with the theory. However, the simulation also revealed that non-linear magnetic tension effects reduce somewhat the disruptive effects of large transverse motions. The right-most image shows that the strongly magnetized sub-Alfvénic jet remains stable until the jet becomes super-Alfvénic. At the Alfvén point, the now submagnetosonic jet destabilizes abruptly as would be expected from the high growth rates predicted by the theory to accompany the sinusoidal surface mode. However, non-linear effects associated with magnetic tension are now ca-

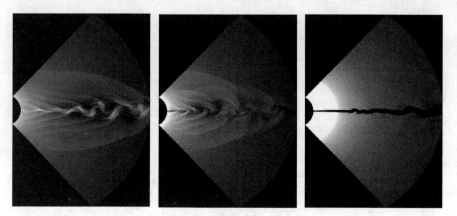

Figure 5. Density images from three simulations showing from left to right the effect of increasing the axial magnetic field strength.

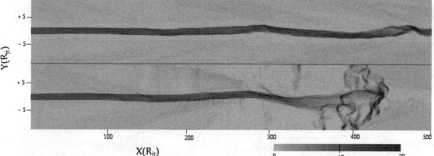

Figure 6. Density images from Mach 63, 2-D slab jet simulations with density ten times the ambient density showing (top) an adiabatic jet and (bottom) a radiatively cooled jet.

pable of preventing disruptive effects associated with large amplitude transverse motions.

The effects of radiative cooling are illustrated in Figure 6. In contrast with the low-density, lower-Mach-number jet simulations discussed above, these dense, high Mach number jet simulations have parameters appropriate to protostellar jet outflows. Both adiabatic and cooling jets develop internal sinusoidal shocks at a spacing indicative of the first sinusoidal body wave mode predicted by the theory. In the non-adiabatic jet, the shocks can produce dense knots of cooling jet gas. Note that the internal body mode grows to lesser amplitude than the surface wave. The surface wave on the radiatively cooling jet has a higher predicted growth rate than the adiabatic jet (see Figure 1) and the jet disrupts into high-density cooling knots. The wavelength and spatial transverse amplitude growth observed in adiabatic and cooling jet simulations at different perturbation frequencies has shown remarkable agreement with the theoretical predictions (Stone et al., 1997).

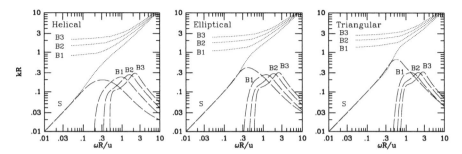

Figure 7. Typical solutions to the dispersion relation for helical, elliptical and triangular normal modes on a 3-D supermagnetosonic jet. Dotted (solid/dashed) lines give the real (imaginary) part of the wavenumber.

In the simulations shown in the figures, typically 40 grid zones spanned the jet diameter. The use of periodic boundary conditions in numerical simulations and reduction in the number of axial grid zones required has allowed resolution studies of 2-D slab jets with from 36 to 292 zones across the jet (Bassett & Woodward, 1995). These studies reveal increasingly small-scale structures and enhanced mixing between jet and ambient material on the higher resolution grids. However, the large-scale structures remain essentially unchanged.

4. Stability Theory—3-D Cylindrical Jets

In 3-D, a single dispersion relation can be found describing the propagation and growth of the Fourier components used to describe an arbitrary perturbation. In addition to a pinching ($n = 0$) mode (almost identical physically and mathematically to the pinch mode on the slab jet), the dispersion relation admits a helical ($n = 1$) mode similar to the sinusoidal mode on the slab jet, as well as elliptical ($n = 2$), triangular ($n = 3$), rectangular ($n = 4$), etc., normal modes that have no analog on a slab jet. Each normal mode contains surface (S) and multiple body (Bm) wave solutions. The mathematical behavior of these solutions is indicated in Figure 7. The normal mode solutions have a maximum growth rate at a resonant frequency both of which increase for the higher-order normal modes. The resonant frequency and maximum growth rate decrease and the growth length increases as the magnetosonic Mach number increases just as for the slab jet. The largest growth rate at a given frequency is always associated with a surface (S) wave. The body waves grow faster than the associated surface wave only for the pinch (see Figure 1) and helical modes.

The distortion induced by the pinch mode surface and body waves is identical to that shown in Figure 2 for the slab jet symmetric pinch mode. The distortion induced by the asymmetric surface and body normal wave modes can be computed from the theory (Hardee et al., 1997) and is illustrated in Figure 8. The maximum displacement amplitude must be less than that leading to the overlap of displacement surfaces because overlap implies the development of shocks. In the axial direction all distortions twist through 360° over a distance $2\pi n/k_R$ where n is the mode number and k_R is the real part of the wavenum-

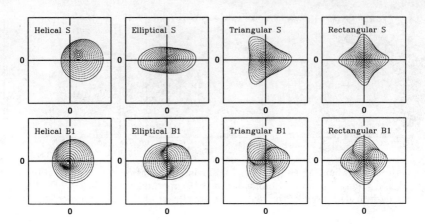

Figure 8. Cross section displacements for helical, elliptical, triangular and rectangular surface (S) and first body (B1) modes at the maximally-growing frequency for a 3-D cylindrical jet.

ber. As found in the slab jet, the surface wave modes are capable of larger fluid displacements at frequencies below the resonant frequency. At higher frequencies, fluid displacements are reduced with much less internal displacement out of phase with the surface displacement. The body wave modes produce distortions in which interior fluid displacements remain relatively large. Along a radial cut through the jet, the phase shift means that there are reversals in fluid displacement at null surfaces within the jet. For higher-order body wave modes, the phase shift increases, the number of null surfaces increases, and the maximum amplitude decreases. Only the helical mode produces non-zero displacement at the jet center. The largest amplitudes are associated with the lowest-order surface and accompanying body wave modes.

5. Numerical Simulations—3-D Cylindrical Jets

Given the additional fast growing asymmetric normal modes on a 3-D cylindrical jet predicted by the theory, it is not surprising that numerical simulations find the 3-D jet to be much more unstable than the 2-D slab jet (Hardee & Clarke, 1992; Bassett & Woodward, 1995). The difference between 2-D slab and 3-D cylindrical geometry is illustrated by comparing Figures 9 and 3. Both 2-D and 3-D simulations involve a low-density Mach 3 unmagnetized jet with identical initial conditions. In this 3-D simulation, a jet is established completely across the computational grid in pressure equilibrium with the surrounding medium and is perturbed by precession at the inlet. The precessional frequency is near the resonant frequency of the surface helical mode predicted by the theory. Continuous flow is disrupted about 20 jet radii from the origin in 3-D compared to 30 jet radii from the origin in 2-D. Filamentation, which promotes entrainment and mixing, is largely responsible for the rapidity of disruption when compared to the 2-D slab jet. Filamentation occurs as low-order modes, *e.g.*, the elliptical mode, induce large distortions in the initially circular cross section.

Figure 9. Density slice image of a Mach 3, 3-D cylindrical jet with one-tenth the ambient density. Axial image dimension is about 45 jet radii.

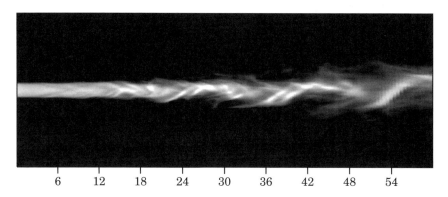

Figure 10. Image formed by line-of-sight integration of the magnetic pressure from a dense, 3-D cylindrical jet of radius R. Markings in the axial direction are at $6R$ intervals.

Three-dimensional simulations of low-density unmagnetized jets reveal a very rapid breakup through filamentation and mixing. Thus, low density and/or unmagnetized jets are a poor choice for studying jet structures predicted by the theory and for producing long stable jets such as those observed in high power radio sources (Hardee et al., 1995) or in protostellar systems. The most recent studies have considered dense magnetized jets and found that overdense jets (i.e., jets whose density is greater than the medium immediately surrounding the jet) show complex internal structures over many tens of jet diameters (Hardee et al., 1997). Figure 10 is an image from a supermagnetosonic Mach 7.7 jet whose density is four times the ambient density and whose axial magnetic pressure is comparable to the jet thermal pressure. The image reveals complex twisted internal structures, apparent transverse oscillation of the jet, and narrowing and broadening of the jet. The transverse oscillation is produced by helical twisting resulting from jet precession, while the narrowing and broadening is produced by elliptical distortion, both as predicted by the theory. Figure 10 and axial

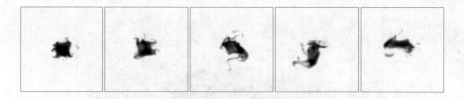

Figure 11. Magnetic pressure cross sections at axial distances of 18, 24, 30, 36, and 42R (left to right).

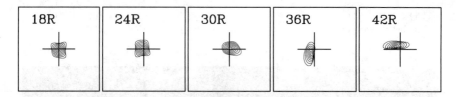

Figure 12. Fits to the magnetic pressure cross section using helical, elliptical, triangular and rectangular surface modes from the theory.

velocity plots (not shown) provide evidence for the existence of a few of the lower-order pinch, helical, and elliptical body waves. The velocity plots also indicate that the internal body waves grow to less than 25% of their potential maximum amplitude, as predicted by the theory.

Jet distortions arising from the K-H instability can be studied by examining magnetic pressure cross sections. Figure 11 shows cross sections from the jet imaged in Figure 10. Note the clear evidence for elliptical and rectangular distortion. At 18R, a triangular distortion in addition to a rectangular distortion is evident. The cross sections reveal the existence of the helical, elliptical, triangular and rectangular K-H surface waves predicted by the theory. The higher-order rectangular and triangular distortions grow most rapidly but then decline in amplitude as the lower-order elliptical and helical distortions grow to large amplitude. The observed behavior is in qualitative agreement with the spatial growth rates computed from the theory.

The cross section distortions shown in Figure 11 can be fit reasonably well by a linear sum of displacements associated with the helical, elliptical, triangular and rectangular surface wave modes and the result of this fitting is shown in Figure 12. The fits imply that helical and elliptical distortions can reach the maximum displacements indicated by the linear theory while the observed triangular and rectangular distortions can be fit by displacements that are considerably less than the maximum displacements. In any event, the simulations confirm the presence of the surface waves predicted by the linear theory. Perhaps the most surprising result is that non-linear displacements can be fit reasonably well using displacements computed from the linear theory.

6. Conclusion

Confrontation between theory and numerical simulations has provided considerable insight into the behavior of jet flows. The unstable modes indicated by the theory have been verified by the numerical simulations. Structures observed in the simulations can be interpreted in terms of structures predicted by the theory. The numerical simulations can be used to ascertain the non-linear behavior and amplitudes to which the various unstable modes develop. The numerical simulations indicate which and under what circumstances the normal modes are destructive to continuous jet flow.

Future work in this area requires that 3-D numerical simulations be performed to study the stability properties of jets expanding in response to gradients in the external medium. This is needed as most jets in astrophysical systems are observed to expand and the theory (Hardee, 1987b) and 2-D slab jet simulations (Hardee et al., 1991) show that expansion is stabilizing. An additional area for future work involves comparison between fully-relativistic K-H instability theory and numerical simulations of relativistic jets.

One of the more interesting results of the present confrontation between 3-D numerical simulation and theory is that organized structures seen in extragalactic and galactic jets may be modeled using structures predicted by the linear theory but with amplitudes suggested by the numerical simulations. For example, it should prove possible to use the present theoretical and numerical results to model twisted dynamical structures in extragalactic and galactic jets, and also to compute the expected emission from such features.

In this review I have restricted the discussion to structures in the numerical simulations that can be directly compared to the linear stability theory. Of equal interest astrophysically is the non-linear mixing and entrainment process that also governs the morphology of jet outflows. The initial development of the mixing and entrainment process is intimately related to the linear growth rates, scale sizes and ultimate amplitudes of the normal modes of jet instability. Tom Jones, Chris Loken, and Alex Rosen (in these proceedings) present some numerical simulation results that address these issues.

Acknowledgments. I would like to thank my collaborators for their contributions to the work reviewed here: David Clarke, Michael Norman, Alex Rosen and Jim Stone. This research was supported in part by NSF grants AST-8611511, AST-8919180 and AST-9318397, by NASA grant NAG-4202, and by NSERC of Canada. The images shown here are from simulations performed at the National Center for Supercomputing Applications and at the Pittsburgh Supercomputing Center.

References

Appl, S., & Camenzind, M. 1992, A&A, 256, 354.
Bassett, G. M., & Woodward, P. R. 1995, ApJ, 441, 582.
Birkinshaw, M. 1984, MNRAS, 208, 887.
Birkinshaw, M. 1991a, MNRAS, 252, 505.
Birkinshaw, M. 1991b, in *Beams and Jets in Astrophysics*, ed. P. A. Hughes, (Cambridge: Cambridge University Press), 278.
Bodo, G., Rosner, R., Ferrari, A., & Knobloch, E. 1989, ApJ, 341, 631.

Bodo, G., Rosner, R., Ferrari, A., & Knobloch, E. 1996, ApJ, 470, 797.
Cohn, H. 1983, ApJ, 269, 500.
Ferrari, A., Massaglia, S., & Trussoni, E. 1982, MNRAS, 198, 1065.
Ferrari, A., Trussoni, E., & Zaninetti, L. 1978, A&A, 64, 43.
Ferrari, A., Trussoni, E., & Zaninetti, L. 1981, MNRAS, 196, 1051.
Fiedler, R., & Jones, T. W. 1984, ApJ, 283, 532.
Hardee, P. E. 1979, ApJ, 234, 47.
Hardee, P. E. 1987a, ApJ, 313, 607.
Hardee, P. E. 1987b, ApJ, 318, 78.
Hardee, P. E., & Clarke, D. A. 1992, ApJ, 400, L9.
Hardee, P. E., & Clarke, D. A. 1995, ApJ, 449, 119.
Hardee, P. E., Clarke, D. A., & Howell, D. A. 1995, ApJ, 441, 644.
Hardee, P. E., Clarke, D. A., & Rosen, A. 1997, ApJ, 485 (August 20).
Hardee, P. E., Cooper, M. L., & Clarke, D. A. 1994, ApJ, 424, 126.
Hardee, P. E., Cooper, M. A., Norman, M. L., & Stone, J. M. 1992, ApJ, 399, 478.
Hardee, P. E., & Norman, M. L. 1988, ApJ, 334, 70.
Hardee, P. E., Norman, M. L., Koupelis, T., & Clarke, D. A. 1991, ApJ, 373, 8.
Hardee, P. E., & Stone, J. M. 1997, ApJ, 483 (July 1).
Massaglia, S., Trussoni, E., Bodo, G., Rossi, P., & Ferrari, A. 1992, A&A, 260, 243.
Norman, M. L., & Hardee, P. E. 1988, ApJ, 334, 80.
Norman, M. L., Winkler, K.-H. A., & Smarr, L. L. 1984, in *NRAO Workshop No. 9: Physics of Energy Transport in Extragalactic Radio Sources*, eds. A. H. Bridle & J. A. Eilek, (Green Bank: NRAO), 150.
Payne, D. G., & Cohn, H. 1985, ApJ, 291, 655.
Ray, T. 1981, MNRAS, 196, 195.
Stone, J. M., Xu, J., & Hardee, P. E. 1997, ApJ, 483 (July 1).
Zhao, J.-H., Burns, J. O., Norman, M. L., & Sulkanen, M. 1992, ApJ, 387, 83.

A Restarting Jet Revisited: MHD Computations in 3-D

David A. Clarke
Department of Astronomy and Physics, Saint Mary's University, Halifax, NS B3H 3C3, Canada

Abstract. That VLBI-scale jets are variable and indeed episodic has long been established. More recently, extensive VLA observations of M87 show that extragalactic jets can be variable even on kpc scales. It is a natural extension, therefore, to expect that some VLA-scale jets are episodic to the point where jets may be completely turned off for a while, only to resume flow at a later time into the "relic" radio lobe.

3-D simulations of a "restarting jet" allow for more realistic comparisons with candidates such as 3C 219 than previous 2-D simulations. Line-of-sight integrations of the synchrotron emissivity accounting for Doppler boosting are performed to determine the morphological signatures of a restarting jet. While the restarting jet picture accounts for many of the observations, there remain some profound discrepancies. For example, in the simulation, a prominent bow shock is excited in the relic radio lobe by the passage of the restarting jet, whereas no such feature has been observed in Nature. This difference is difficult to reconcile, and possible implications for the model are discussed.

1. Introduction

Variability and episodic behaviour in VLBI jets have been known for some time. Examples are numerous, and include 3C 120 (Walker *et al.*, 1987), 3C 273 (Davis *et al.*, 1991), 3C 345 (Zensus *et al.*, 1995), *etc.* Direct evidence for variability of VLA jets (kpc scale) has been reported in the nearby galaxy M87 (Biretta *et al.*, 1995) and there is significant *indirect* evidence that VLA-scale jets are episodic in a number of radio sources.

The spectral index image of the lobes in 3C 388 (Roettiger *et al.*, 1994) show there is, in fact, a younger (flatter spectrum) region embedded within an older (steeper spectrum) region, and that the transition between these two spectral regions is quite sharp. This is naturally explained by an episodic jet model in which the older region, or lobe, formed from a previous epoch of jet activity, is currently being displaced by the younger lobe inflated by the present-day jet.

A nearby example is Centaurus A, in which the "inner lobes" (Clarke *et al.*, 1992*b*) are, in fact, embedded in a "middle lobe" which in turn is embedded in the "outer lobes" spanning 8° in the southern sky (Haynes *et al.*, 1983). This may be evidence for at least three distinct phases of jet activity—the most recent being spawned by the current state of cannibalism in NGC 5128 (Graham, 1979).

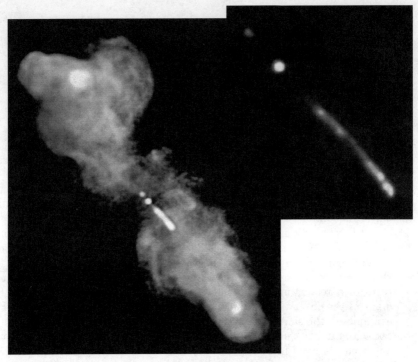

Figure 1. Grey-scale, 22 cm total intensity image of 3C 219 at $1\rlap{.}''4$ ($2.8h_{100}^{-1}$) resolution. The core is the central bright spot, the counter jet is the spot to the NE of the core, and the main jet is the extended feature to the SW of the core. The inset shows the counter-jet, core, and main jet at 6 cm and $0\rlap{.}''4$ resolution (from CBBPN).

3C 219 (Figure 1, Clarke et al., 1992a, hereafter CBBPN) is one of several known radio galaxies possessing a "partial jet"—one that appears to stop well before the end of the lobe. Other partial jets are found in 3C 277.3 (van Bruegel et al., 1985), 3C 33.1 (Rudnick, 1985), 3C 288 (Bridle et al., 1989), 3C 228, and 3C 445. These may be examples of restarting jets "caught in the act", and provide a useful data base for performing careful tests on the model.

2. The Case of 3C 219

The "born-again, relativistic jet" model (Bridle et al., 1986; CBBPN) posits that the jet in 3C 219 is episodic, relativistic, and oriented well out of the plane of the sky. This model predicts that the "main jet" (SW) is inclined toward the observer and is longer than the "counter jet" (NE) because of time-of-flight effects (assuming both jets are launched from the core at the same time). The model also predicts that the main jet is brighter (Doppler enhancement) and has a flatter spectrum (Doppler shifting) than the counter jet. Each of these predictions is borne out by the observations (CBBPN), at least qualitatively.

In detail, however, the model leaves some questions unanswered. Assuming the two jets are launched simultaneously with the same velocity (βc), the time-of-flight effects should result in a counter jet foreshortened relative to the main jet by a factor of $q \equiv (1+\beta \sin\theta)/(1-\beta \sin\theta)$, where θ is the inclination angle of the jet out of the plane of the sky. For 3C 219, $q \sim 3.8$ (Figure 1). In addition, if the emission from the two jets is intrinsically symmetric, the ratio of jet to counter jet brightnesses is given by $r = q^{2+\alpha}$, where α is the spectral index of the radio emission ($I_\nu \propto \nu^{-\alpha}$). For $\alpha = 0.7$, $r \sim 37$, or about a factor of seven higher than the observed brightness ratio (CBBPN). This problem may be circumvented by dropping the intrinsic symmetry assumptions, in which case the model begins to take on an *ad hoc* flavour.

Slightly more troublesome is the observed brightness contrast between the main jet and the cocoon. Given that the jet and cocoon "stuff" are essentially the same, and given the greater line of sight through the cocoon than the jet, one might reasonably ask 'Why do we see the jet at all?' (*e.g.*, Scheuer, 1989). This question is usually answered in terms of Doppler boosting by the relativistic flow of the jet. One can easily show that the *maximum* Doppler boosting occurs when $\beta = \sin\theta$ in which case, the jet emission is enhanced by a factor of $D = (\sin\theta)^{2+\alpha}$. If we use the value of q determined above, $\sin\theta \sim 0.76$ [just slightly higher than the value radio galaxies are supposed to have according to Barthel (1989)], and $D \sim 2.1$. This is lower by *two orders of magnitude* than the observed ratio of jet and cocoon emissivities. Clearly, something other than Doppler favouritism must be invoked to explain the observed brightness of the jet relative to the cocoon (*e.g.*, shearing magnetic field at the jet surface, particle aging in the cocoon, *etc.*).

To proceed further, it is necessary to determine a test for the model independent of particle aging, magnetic reconnection, and the intrinsic symmetry, if any, of the two jets. Fortunately, fluid dynamics provides such a test. Leading the reborn jet as it propagates through the relic radio lobe should be a powerful bow shock compressing the local magnetic field and re-energising the electron population in its wake, thus generating a sharp brightness contrast across the shock. The question then becomes: "Would such a bow shock dominate the relic lobe enough to render the brightness contrast observable?". To help answer this question, a 3-D numerical simulation of a restarting jet was performed.

3. The Numerics

The simulations were performed using ZEUS-3D, a two-fluid MHD solver described in Clarke (1996), on the Cray C-90 at the PSC.

A Mach 6, low-density jet ($\eta \equiv \rho_j/\rho_a = 0.02$, where the subscripts j and a refer to quantities in the jet and ambient medium, respectively) with a weak magnetic field is launched into a 3-D cube (48 kpc on a side) of uniform, quiescent, unmagnetised gas. The cube is resolved into $280 \times 172 \times 172$ (8.3×10^6) zones on a Cartesian grid. There are 20 zones across the jet diameter ($r_j = 0.8$ kpc).

The jet is wiggled with a 1% perturbation at the orifice to break the azimuthal symmetry. At $t = 7.5 \times 10^6 \, (r_j/800 \, \text{pc})(1,000 \, \text{km s}^{-1}/c_a)$ yr (where c_a

is the ambient sound speed), outflow is stopped. At $t = 1.2 \times 10^7$ yr, outflow resumes and continues until $t = 1.6 \times 10^7$ yr.

4. The Simulation

In previous 2-D axisymmetric simulations, Clarke & Burns (1991, hereafter CB) showed that there are two "partial jet" phases when the jet is episodic. These are the "vestigial jet" once outflow has stopped, and the "reborn jet" once outflow has resumed. For light jets ($\eta \ll 1$), both phases are extremely short-lived. The vestigial jet is consumed by a rarefaction wave which propagates along the jet axis at a speed $v_j + c_j$ once the flow is terminated. Now, since the hot spot depends upon the momentum flux of the jet for confinement, and since the surface brightness of the hot spot decreases strongly with radius, it too disappears very soon after the vestigial jet is spent. The reborn jet then races down the length of the relic lobe (essentially an evacuated cavity), slowing only when it encounters the contact discontinuity separating the old lobe and the original ambient medium. Thus, it too is short-lived and the chances of "catching" a restarting jet as a partial jet is rather low (consistent with the relative rarity of such jets).

CB also noted that during the period of jet quiescence, the lobe gradually collapses upon itself since it is no longer supported by momentum (ram) pressure. Because the simulation was axisymmetric, the condensate fell precisely to the jet axis, forming what CB call a "false jet". This is an elongated feature aligned with the lobe axis resembling a jet but without the usual outflow. However, the most important aspect of the restarting jet—the observability of a bow shock—was not addressed in these 2-D simulations.

The 3-D simulation presented herein (Figures 2 and 3) confirms some of the 2-D results, modifies others, and addresses the issue of the bow shock. Certainly, the two partial-jet phases are found in the 3-D simulations as well, and the hot spot does not survive for any appreciable time once the vestigial jet has disappeared. Because the hot spot in the southern lobe of 3C 219 is the most compact, brightest feature in the 22 cm image of Figure 1, the vestigial jet must still be dumping into the hot spot (despite the fact that no convincing evidence of a vestigial jet can be seen in the data). Thus, the reborn jet must have been launched during the brief vestigial jet phase, implying a very short (compared to the lifetime of the radio source) period of jet intermittency.

Unlike 2-D, a "false jet" does not condense out of the relic lobe in 3-D because there is no strict symmetry to force material to converge *exactly* on axis. However, as the relic lobe ages, it does become more centrally condensed.

Figure 2 shows a 2-D slice of density (ρ) and velocity divergence ($\nabla \cdot \mathbf{v}$) containing the jet axis shortly after the reborn jet is launched. Propagation of the jet is from left to right and the bow shock leading the reborn jet is clear in both images. Figure 3 shows the surface brightness of the jet depicted in Figure 2 at various angles relative to the plane of the sky. The synchrotron emissivity is taken as proportional to $p(B \sin \psi)^{3/2}$ (Clarke et al., 1989). For orientations consistent with radio galaxies ($< 45°$), the bow shock remains a distinctive feature in the emission image. Only when orientations are consistent with QSR's does the bow shock get confused with the tangle of filaments in the

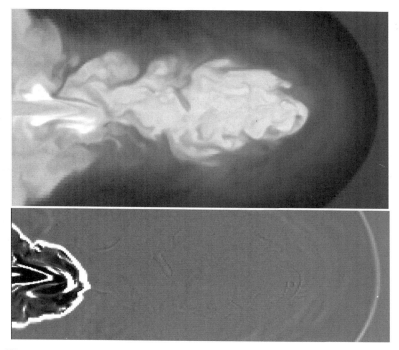

Figure 2. Grey-scale images of density (top) and $\nabla \cdot \mathbf{v}$ (bottom) along 2-D slices through the data cube containing the jet axis. Jet flow is from left to right. Dark tones indicate high values, light tones indicate low values for density and large negative values for $\nabla \cdot \mathbf{v}$. The reborn jet can be seen propagating into the relic lobe (top) while shocks are most visible as strong negative $\nabla \cdot \mathbf{v}$ features (bottom).

radio lobe. This suggests, therefore, that a reborn jet in a radio galaxy should be preceded by an observable bow shock.

Notes and Caveats:

- Regardless of initial conditions, the cocoon will end up being less dense and hotter (by a factor of a few) than the jet material itself. The reason is simple: the cocoon material expands (adiabatically) from the hot spot, and will reach rough pressure balance with the jet and shocked ambient medium (since it is separated by contact discontinuities from each). But since the jet shock near the working surface generates entropy, the cocoon must be hotter (and less dense to maintain pressure balance) than the jet. Thus, assuming the reborn jet is similar (η, M) to the original jet, it will always see an environment hotter and less dense than itself, and will therefore propagate somewhat ballistically into the relic lobe.

- The nature of the bow shock is primarily dependent on the Mach number. High Mach-number flows will tend to have narrower bow shocks than low Mach-number flows. It has been suggested (A. H. Bridle, private communi-

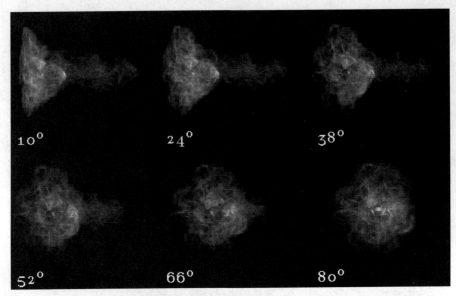

Figure 3. Simulated total intensity images generated shortly after outflow is resumed taken at various angles with respect to the plane of the sky. Outflow velocity is taken as $0.97\,c$ and still the jet is not sufficiently Doppler boosted to compare in brightness with the jet in 3C 219. The brightest feature is the apex of the bow shock which can be distinguished as such at angles up to 45°.

cation) that the Mach number may be high enough to force the bow shock to hug the jet all along its length, rendering it *indistinguishable* from the jet itself. In this scenario, a Mach number greater than 1,000 is implicated. However, it is unlikely Mach numbers much over 100 can be sustained by collimated outflow since slight disturbances along the jet surface will excite strong oblique shocks into the interior of the jet thereby heating the jet, increasing the internal sound speed, and lowering the Mach number.

- Even with optimal Doppler boosting, it is difficult to make a jet stand out against the background emission of the cocoon (Figure 3). One must invoke particle aging in the cocoon and/or poorly understood processes at the contact between the cocoon and the jet (related to the strong shear layer there) to account for some observed brightness ratios. Currently, simulations cannot reproduce this effect.

5. Conclusions

Our present understanding of the dynamics of extragalactic radio jets and lobes is predicated on the fluid model. Magnetic fields and particle acceleration are obviously two fundamental ingredients to this picture, but can only modify some very basic predictions. One of these predictions—that a bow shock should lead a supersonic jet—is disturbingly unaccounted for in the observations, and at this stage, one can only speculate as to the possible reason. What could hide the bow shock from the sensitive observations depicted in Figure 1? Could the jet, in fact, be uninterrupted all the way to the southern hotspot, and only *appear* to stop part way (CBBPN)? It is difficult to construct a process capable of shutting off the emission so abruptly, other than a working surface itself. If 3C 219 were an isolated case, one might hope to discount it as an unlikely exception. However, there are several such sources known and it remains a challenge to the fluid model for extragalactic radio sources to account for such exceptions. If an explanation within the fluid model cannot be found, modifications to the current paradigm, or an altogether new model, may have to be devised.

Acknowledgments. This work is supported, in part, by operating and equipment grants from NSERC of Canada, and by the Pittsburgh Supercomputer Center through NSF grant AST930010P.

References

Barthel, P. D. 1989, ApJ, 336, 606.
Biretta, J. A., Zhou, F., & Owen, F. N. 1995, ApJ, 447, 582.
Bridle, A. H., Fomalont, E. B., Byrd, G. G., & Valtonen, M. J. 1989, AJ, 97, 674.
Bridle, A. H., Perley, R. A., & Henriksen, R. N. 1986, AJ, 92, 534.
Clarke, D. A. 1996, ApJ, 457, 291.
Clarke, D. A., Bridle, A. H., Burns, J. O., Perley, R. A., & Norman, M. L. 1992a, ApJ, 385, 173 (CBBPN).
Clarke, D. A., & Burns, J. O. 1991, ApJ, 369, 308 (CB).
Clarke, D. A., Burns, J. O., & Norman, M. L. 1992b, ApJ, 395, 444.
Clarke, D. A., Norman, M. L., & Burns, J. O. 1989, ApJ, 342, 700.
Davis, R. J., Unwin, S. C., & Muxlow, T. W. B. 1991, Nature, 354, 374.
Graham, J. A. 1979, ApJ, 232, 60.
Haynes, R. F., Cannon, R. D., & Ekers, R. D. 1983, Proc. Astron. Soc. Australia, 5, 241.
Roettiger, K., Burns, J. O., Clarke, D. A., & Christiansen, W. A. 1994, ApJ, 421, L23.
Rudnick, L. 1985, in *Physics of Energy Transport in Extragalactic Radio Sources* (Proc. of NRAO Workshop 9), eds. A. H. Bridle & J. A. Eilek (Green Bank: NRAO), 35.
Scheuer, P. A. G. 1989, in *Hot Spots in Extragalactic Radio Sources* (Lecture Notes in Physics, 327), eds. K. Meisenheimer & H.-J. Röser (Berlin: Springer-Verlag), 159.
van Breugel, W., Miley, G., Heckman, T., Butcher, H., & Bridle, A. H. 1985, ApJ, 29, 496.
Walker, R. C., Benson, J. M., & Unwin, S. C. 1987, ApJ, 316, 546.
Zensus, J. A., Cohen, M. H., & Unwin, S. C. 1995, ApJ, 443, 35.

Effect of Magnetic Fields on Mass Entrainment in Extragalactic Jets

Alexander Rosen, Philip E. Hardee

Department of Physics and Astronomy, University of Alabama, Tuscaloosa, AL 35487, U.S.A.

David A. Clarke

Department of Astronomy and Physics, Saint Mary's University, Halifax, NS B3H 3C3, Canada

Audress Johnson

Astronomy Department, University of Texas, Austin, TX 78712, U.S.A.

Abstract. We have investigated the spatial mass-entrainment rate in a set of jet simulations that vary the jet-to-ambient density ratio, and the magnetic field strength and orientation. The lateral expansion of simulated synchrotron emission appears to be related to the entrainment process. The entrainment process exhibits three spatial growth stages.

1. Introduction

The plume-like appearance of jets in FR I radio sources, when contrasted with the highly structured appearance of jets in FR II radio sources, suggests that mass entrainment can be an important process in understanding the propagation of extragalactic jets. Additionally, the assumption of jet deceleration resulting from mass entrainment within a suitable galactic atmosphere when combined with the assumption of a low Mach number, weakly-relativistic flow, has led to a correct prediction of the FR I–FR II break in the radio-optical plane (Bicknell, 1994). These facts have motivated investigations by a number of researchers. For example, Bowman et al. (1996) have studied the possible effect of entrainment of stellar wind material on relativistic jets. In other work, a temporal mass-entrainment rate involving mixing at a contact discontinuity has been estimated (DeYoung, 1996), and other numerical simulations have been performed in an attempt to investigate the temporal rate of mixing between the jet and ambient media (Loken et al., 1996; Bodo et al., 1994; Bodo et al., 1995). These studies have not included the effects of magnetic fields. We have performed a set of three-dimensional MHD jet simulations with the aim of understanding the mixing process in a spatial sense and of understanding the effect of magnetic field strength and orientation on this process.

2. Simulations

The numerical simulations were performed using ZEUS-3D including the Consistent Method of Characteristics, which solves the transverse momentum transport and magnetic induction equations simultaneously and in a planar-split fashion (Clarke, 1996). Seven simulations have been run and a complete analysis is underway (Rosen et al., in preparation). Here we restrict our attention to simulations of a *light* jet with a strong *axial* magnetic field (magnetic pressure is approximately equal to the jet thermal pressure for the strong field), a *dense* jet with a strong *axial* magnetic field, and a *dense* jet with a strong *toroidal* magnetic field. The density of both the light and dense jets differs from the ambient density by a factor of four. Our "equilibrium" jet simulations represent the jet flow well behind the jet front and terminal shock, and the ambient medium in these simulations corresponds to the cocoon or shocked ambient material in a "propagating" jet simulation. The grid of 325 × 130 × 130 zones spans 60 jet radii along the jet axis and 30 jet radii transverse to the jet axis, with a transverse resolution of 30 zones across the jet diameter. In all of the simulations, the jets are supermagnetosonic, with magnetosonic Mach numbers $M_{\text{jet}}^{\text{ms}} = 6$ (light jet) and 8 (dense jet), where $M_{\text{jet}}^{\text{ms}} \equiv u/(v_A^2 + a_{\text{jet}}^2)^{1/2}$, v_A is the axial Alfvén speed, and a_{jet} is the jet sound speed. The input velocity was precessed at the origin in order to stimulate the helical modes of the Kelvin-Helmholtz (K-H) instability.

3. Results

We display line-of-sight integrations through the data cube of the (simulated) total synchrotron intensity and fractional polarization from the light jet with the strong axial field in Figures 1 and 2. These integrations reveal three regions with different lateral expansion rates: a slow linear initial expansion, a subsequent faster expansion, and a final region with no expansion. The appearance is consistent with the K-H instability growing from the linear regime through the non-linear regime to saturation. The region closest to the inlet is dominated by short-wavelength, high-order modes. Farther down the jet, longer wavelength, lower-order modes (specifically the helical and elliptical modes) dominate (see Hardee, these proceedings). Note that a high-intensity jet in the left half of Figure 1, which corresponds to fast organized flow, disappears in the right half of the figure. This high-intensity jet has a higher fractional polarization than the portion farther down the jet (see Figure 2), and this is a result of mixing, entrainment, and disruption of the highly ordered flow.

A dense jet with a strong axial field remains organized for a longer distance (see Figures 3 and 4) than the light jet. In this case, the high-intensity region persists out to nearly 50 jet radii. The high-intensity core is surrounded by a low-intensity sheath. The dense jets shows three regions with different lateral expansion rates, although the three regions are displaced axially farther down the jet than for the light jet. Note that the polarization vectors are aligned with the brighter filamentary regions, even when these regions are not aligned with the jet axis. This result seems to be a general one for other MHD jet simulations (Clarke et al., 1989, although the polarization is from both the jet

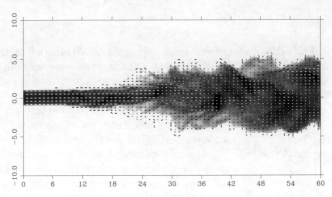

Figure 1. Simulated total intensity image for a light jet with a strong axial magnetic field. The length unit is a jet radius. The lines overlayed on the image indicate B-field polarization vectors, which have a length proportional to the fractional polarization.

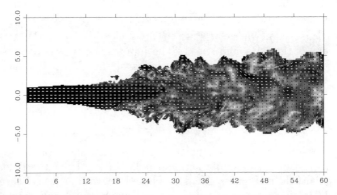

Figure 2. Simulated fractional polarization image for a light jet with a strong axial magnetic field. As in Figure 1, the lines overlayed on the image indicate B-field polarization vectors.

and magnetized cocoon) as well as observations (NGC 6251: Perley *et al.*, 1984; M87: Owen *et al.*, 1989).

The effect of a strong toroidal magnetic field is to restrict jet expansion and to maintain a high-intensity core with very little sheath across the entire grid. Note that the fractional polarization (Figure 6) diminishes near the edge of the jet, where the polarization changes from perpendicular to parallel to the jet axis.

Since only the jet is initialized with a magnetic field, we can estimate spatial mass-entrainment rates by plotting the linear mass-density of magnetized material as a function of the axial position. Results for these three simulations are shown in Figure 7. In this plot, the behavior of the simulations is similar, showing an initial region that slowly increases, a subsequent one with a faster increase, and a final one that remains roughly constant. Conceptually, this corresponds well with the K-H instability growing from the linear regime through the non-linear regime, and then saturating. Suppression of instability and mixing in the toroidal field simulation is suggested by the reduced linear mass density in

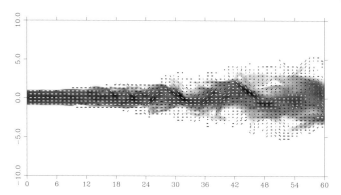

Figure 3. Same as for Figure 1 but for a dense jet with a strong axial magnetic field. Note the more ordered structure and magnetic vectors compared to Figure 1.

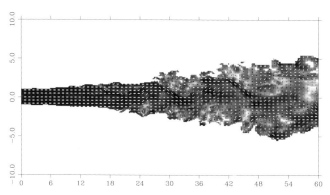

Figure 4. Same as for Figure 2 but for a dense jet with a strong axial magnetic field. Note the higher central fractional polarization compared to Figure 2.

the mass-entrainment plot. The dense jets shown here ultimately have a total linear mass density approximately six times the initial jet linear mass density. Other simulations with either a weak axial or toroidal magnetic field show approximately the same entrained mass as the strong axial magnetic field case shown here. On the other hand, the light jet shown here ultimately has a total linear mass density that is about 50 times the initial jet linear mass density. In general, for each simulated jet, the total mass of jet plus entrained material seems correlated to the width of the simulated emission. In addition, there is a weak anti-correlation between emission width and jet-to-ambient density ratio.

4. Conclusions

Dense jets maintain organized structure and exhibit a core-sheath morphology, whereas light jets lose their organized structure and exhibit a plume-like morphology. The light jet entrains much more external material as a fraction of

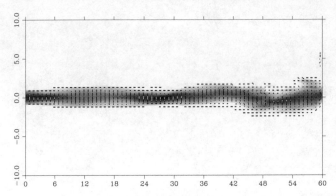

Figure 5. Same as for Figure 3 but for a dense jet with a strong toroidal magnetic field. Note the lack of extended emission, and the different vector orientation relative to the strong axial magnetic field.

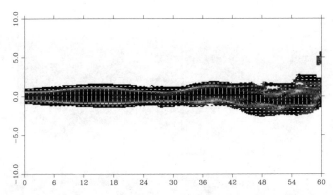

Figure 6. Same as for Figure 4 but for a dense jet with a strong toroidal magnetic field. Note the relatively high fractional polarization except where the vectors rotate by 90°.

the initial jet material than the dense jets. The strong toroidal field has clearly slowed mixing of jet and ambient material.

These simulations indicate that a reduction in spatial mixing is achieved by spatially extending the linear-growth regime and reducing mixing in the non-linear regime. This can be accomplished by an increase in jet density relative to the ambient medium and by inclusion of a strong toroidal magnetic field. Note, however, that the present magnetic fields are stronger than might be expected in an astrophysical jet if a large lobe is to form (Clarke, 1994). Although the magnetosonic Mach numbers in these simulations are comparable, the velocity of the light jet is about three times higher than the dense jets. Previous simulations (Hardee *et al.*, 1995) showed that a higher jet velocity leads to more rapid heating and disruption as jet and external fluids mix. Thus, it is likely that the rapid disruption of the light jet is partially attributable to the higher velocity. However, the most stable combination (linearly and non-linearly) for a given magnetosonic Mach number is always associated with the denser, lower-velocity jet.

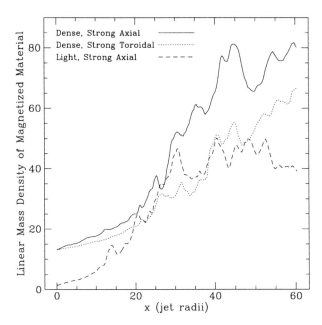

Figure 7. Plot of jet and entrained mass *vs.* jet axis distance.

Acknowledgments. PEH and AR acknowledge support from National Science Foundation grant AST-9318397. DAC is supported in part by the Natural Sciences and Engineering Research Council of Canada (NSERC). The simulations were run on the Cray C90 at the Pittsburgh Supercomputing Center through grant AST930010P.

References

Bicknell, G. V. 1994, ApJ, 422, 542.
Bodo, G., Massaglia, S., Ferrari, A., & Trussoni, E. 1994, A&A, 283, 655.
Bodo, G., Massaglia, S., Rossi, P., Rosner, R., Malagoli, A., & Ferrari, A. 1995, A&A, 303, 281.
Bowman, M., Leahy, J. P., & Komissarov, S. S. 1996, MNRAS, 279, 899.
Clarke, D. A. 1994, in *Jets in Extragalactic Radio Sources*, eds. H.-J. Röser & K. Meisenheimer (Heidelberg: Springer-Verlag), 243.
Clarke, D. A. 1996, ApJ, 457, 291.
Clarke, D. A., Norman, M. L., & Burns, J. O. 1989, ApJ, 342, 700.
DeYoung, D. S. 1996 in *ASP Conf. Ser. 100: Energy Transport in Radio Galaxies and Quasars*, eds. P. E. Hardee, A. H. Bridle, & J. A. Zensus, (San Francisco: ASP), 261.
Hardee, P. E., Clarke, D. A., & Howell, D. A. 1995, ApJ, 441, 644.
Loken, C., Burns, J. O., Bryan, G., & Norman, M. 1996 in *ASP Conf. Ser. 100: Energy Transport in Radio Galaxies and Quasars*, eds. P. E. Hardee, A. H. Bridle, & J. A. Zensus, (San Francisco: ASP), 267.
Owen, F. N., Hardee, P. E., & Cornwell, T. J. 1989, ApJ, 340, 698.
Perley, R. A., Bridle, A. H., & Willis, A. G. 1984, ApJS, 54, 291.

Jet Turbulence and High-Resolution 3-D Simulations

C. Loken

Astronomy Department, New Mexico State University, Las Cruces, NM 88003, U.S.A.

Abstract. The morphology of the edge-darkened FR I radio sources, particularly in their outer regions, is generally taken as evidence for a turbulent subsonic or transonic flow. Despite the great importance of understanding the role of turbulence in extragalactic jets, the vast majority of numerical jet simulations are relevant only to supersonic FR II-type jets. Only recently has numerical resolution improved to the point where simulations can yield meaningful results concerning turbulent entrainment. Here we present initial results from a series of simulations of disrupting, 3-D equilibrium jets at various resolutions. A passive scalar is employed to quantify the degree and rate of entrainment and we conclude that a resolution of 20 to 25 zones across the jet radius is adequate to converge on entrainment characteristics.

1. Introduction

As outlined by Bicknell (1984) and Canto & Raga (1991), an initially laminar supersonic jet is expected to develop a turbulent boundary layer or mixing layer at its surface which expands outward and inward. The jet becomes fully turbulent when the mixing layer manages to work its way in to the jet axis. Detailed experimental results concerning subsonic, turbulently entraining jets have been available for some years (*e.g.*, Dimotakis *et al.*, 1983; Liepmann, 1991; Panchapakesan & Lumley, 1993; Schefer *et al.*, 1994; Hussein *et al.*, 1994). More recently, direct observational results have been obtained for the astrophysically more interesting case of supersonic jets. For example, Fourguette *et al.* (1991) studied the temporal evolution of the mixing layer of a round, pressure-matched, supersonic ($M = 1.5$) air jet (Re = 4.6×10^5). Data were collected over a streamwise extent of 12 jet radii allowing the authors to plot the streamwise velocity along the axis, radial pressure profiles, and shear layer growth rates as a function of convective Mach number. They also present time-lapse photographs (using laser-based Rayleigh scattering techniques) of the development of the mixing layer and azimuthal views of the jet. The jet spreads linearly and the authors recorded the presence of large mixing regions or structures spanning the width of the mixing layer which rotate and move downstream (see also Mungal & Hollingsworth, 1989) with velocities greater than that corresponding to the expected convective Mach number (Bogdanoff, 1983).

A common numerical approach to modeling high Reynolds number turbulence is provided by Large Eddy Simulations (LES) in which the largest eddies

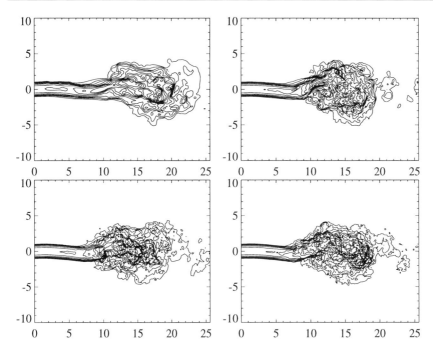

Figure 1. Self-convergence study. Line-of-sight integrations of the passive scalar are shown at dynamical time $\tau = 12$ for an $M_j = 2.5$ jet computed with resolutions of 6, 10, 15 and 20 zones per jet radius, respectively (top left to bottom right).

are directly simulated but the effect of unresolved eddies (*i.e.*, those smaller than a grid zone) is included by prescribing a form for the sub-grid dissipation (*e.g.*, Rogallo & Moin, 1984; Speziale *et al.*, 1988; Yoshizawa, 1991). Boris *et al.* (1992) have argued that any monotonic, finite-difference scheme (such as FCT or PPM) inherently includes a sub-grid turbulence model and that in using such a code one is effectively performing an LES. Direct numerical simulations of subsonic ($M = 0.6$), underdense ($\eta = 0.7$) axisymmetric jets were carried out by Grinstein *et al.* (1987) using FCT while Gathmann *et al.* (1993) used a PPM scheme to investigate the transition to turbulence in a supersonic shear layer. In the astrophysical context, high-resolution direct numerical simulations of turbulent jets have been performed by Bassett & Woodward (1995) and Hardee *et al.* (1995).

2. Simulations

All the simulations discussed here were performed with CMHOG (Bryan *et al.*, 1995), a PPM hydrodynamics code implemented on the CM5 at NCSA. Simulations were initialized with an equilibrium jet stretching the length of the computational volume in the x-direction. The initial jet is parameterized by its density with respect to the ambient medium, $\eta = \rho_j/\rho_{\rm amb}$, and its internal Mach

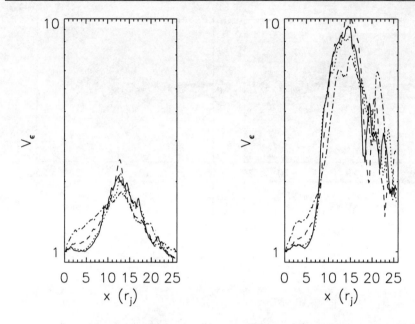

Figure 2. Self-convergence study. The volume of entrained material (V_ϵ, $\epsilon = 0.01$) is plotted vs. radius at dynamical times $\tau = 5$ (left) and $\tau = 9$ (right) for the $M_j = 2.5$ jet. The dash-dot, dashed, dotted and solid lines correspond to the simulations with resolutions of 6, 10, 15 and 20 zones per jet radius, respectively.

number, $M_j = v_j/c_j$ where c_j is the sound speed in the jet. The jet was perturbed by applying a small component of velocity to the jet material at the inlet. This perturbing velocity component was perpendicular to the jet axis with a position angle varying sinusoidally in time with $\omega \sim 1.3 c_{\mathrm{amb}}/r_j$.

A passive scalar, Φ, was injected at the inlet along with the jet fluid and advected in order to trace the amount of jet material present in any zone on the grid. Initially, $\Phi = 1$ in the jet and 0 elsewhere. Following Boris et al. (1992), we define the *entrainment volume*, V_ϵ, as the volume of fluid on the grid which has $\Phi \geq \epsilon$ where $0 < \epsilon < 1$.

2.1. Resolution study

The effects of resolution were investigated with a series of simulations of an equilibrium jet with $M_j = 2.5$ and $\eta = 0.1$ on a grid that extends $26\,r_j$ in the axial direction and $\pm 11\,r_j$ from the jet axis in the transverse direction. Grid zones were of uniform size everywhere within $5\,r_j$ of the jet axis and the simulations had resolutions of 6, 10, 15 and 20 zones per r_j. The perturbation amplitude was 5%. Figure 1 shows line-of-sight integrations through the volume of Φ at the final epoch in the four simulations. We calculated V_ϵ (normalized to the value at $t = 0$) in planes that were one zone thick and perpendicular to the jet axis in order to plot the entrained volume as a function of radius (Figure 2) at

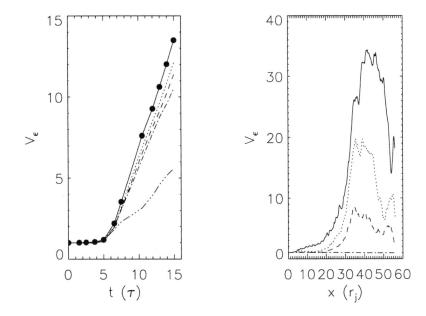

Figure 3. High-resolution simulation. The left panel shows the volume of entrained material (V_ϵ) plotted vs. time for the $M_j = 5.0$ jet. Descending from the top, the lines correspond to $\epsilon = 0.001$, 0.01, 0.02, 0.05, and 0.2, respectively. The right panel shows the entrainment volume (V_ϵ, $\epsilon = 0.001$) plotted vs. radius at dynamical times $\tau = 15.0$, 10.5, 7.5 and 3.0 (going from top to bottom).

two different epochs. Figure 2 demonstrates self-convergence on V_ϵ at the higher resolutions.

2.2. Mach 5 jet

Our high-resolution simulation of a jet with $M_j = 5$ and $\eta = 0.1$ has a resolution of 25 zones per r_j in the transverse direction (within $2\,r_j$ of the axis) and 16 zones per r_j along the jet axis. The perturbation amplitude is 1% and the grid is $56\,r_j$ long and $18\,r_j$ wide ($896 \times 192 \times 192$ zones). The jet parameters were chosen to match those of Hardee et al. (1995) but the resolution is significantly better. This jet was evolved for 15 dynamical times and the evolution of Φ in the central plane is shown in Figure 3. The total entrained volume as a function of time and the entrained volume at 3 specific epochs as a function of radius are plotted in Figure 4. By the end of the simulation, there is clearly detectable jet material in a volume which is more than twelve times greater than the initial jet volume.

Acknowledgments. Thanks to Greg Bryan, Mike Norman and Jim Stone for access to, and help with, the CMHOG code. This work was funded in part by a grant from the National Science Foundation (AST-9317596).

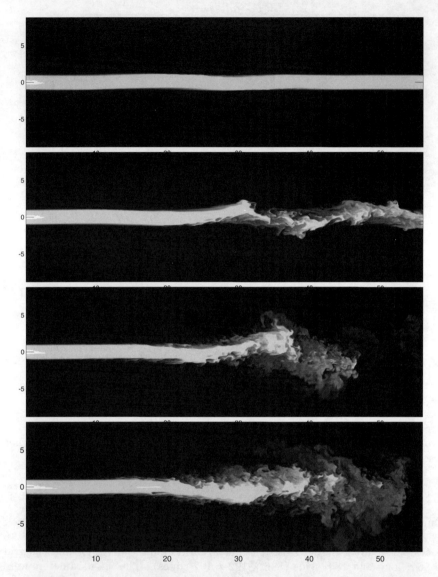

Figure 4. High-resolution simulation. Grey-scale images of the passive scalar (Φ) in the central plane of the jet at dynamical times $\tau = 3.5$, 6, 10.5 and 15.

References

Basset, G. M., & Woodward, P. R. 1995, ApJ, 441, 582.
Bicknell, G. V. 1984, ApJ, 286, 68.
Bogdanoff, D. W. 1983, AIAAJ, 21, 926.
Boris, J. P., Grinstein, F. F., Oran, E. S., & Kolbe, R. L. 1992, Fluid Dyn. Res., 10, 199.

Bryan, G. L., Norman, M. L., Stone, J. M., Cen, R., & Ostriker, J. P. 1995, Comp. Phys. Comm., 89, 149.
Canto, J., & Raga, A. C. 1991, ApJ, 372, 646.
Dimotakis, P. E., Miake-Lye, R. C., & Papantoniou D. A. 1983, Phys. Fluids, 26, 3185.
Fourguette, D. C., Mungal, M. G., & Dibble, R. W. 1991, AIAAJ, 29, 1123.
Gathmann, R. J., Si-Ameur, M., & Mathey, F. 1993, Phys. Fluids A, 5, 2946.
Grinstein, F. F., Oran, E. S., & Boris, J. P. 1987, AIAAJ, 25, 92.
Hardee, P. E., Clarke, D. A., & Howell, D. A. 1995, ApJ, 441, 644.
Hussein, H. J., Capp, S. P., & George, W. K. 1994, J. Fluid Mech., 258, 31.
Liepmann, D. 1991, Phys. Fluids A, 3, 1179.
Mungal, M. G., & Hollingsworth, D. K. 1989, Phys. Fluids A, 1, 1615.
Panchapakesan, N. R., & Lumley, J. L. 1993, J. Fluid Mech., 246, 225.
Rogallo, R. S., & Moin, P. 1984, Ann. Rev. Fluid Mech., 16, 99.
Schefer, R. W., Kerstein, A. R., Namazian, M., & Kelly, J. 1994, Phys. Fluids, 6, 652.
Speziale, C. G., Erlebacher, G., Zang, T. A., & Hussaini, M. Y. 1988, Phys. Fluids, 31, 940.
Yoshizawa, A. 1991, Phys. Fluids, 3, 714.

Modern Schemes for Solving Hyperbolic Conservation Laws of Interest in Computational Astrophysics on Parallel Machines

Dinshaw S. Balsara

NCSA, University of Illinois at Urbana-Champaign, Urbana, IL 61801, U.S.A.

Abstract. In this review, I describe several facets involved in the computational solution of hyperbolic systems of conservation laws of interest in numerical astrophysics. After taking the reader through some of the theoretical issues that are central to the design of modern numerical schemes, I present the schemes themselves including modern-day TVD and ENO schemes. I then choose MHD and relativistic hydrodynamics as two examples of hyperbolic systems that have been solved using these techniques. I then briefly catalog some of the recent advances in parallel computing, especially as they apply to parallel adaptive-mesh solution techniques for these systems.

1. Introduction

Computational solutions to astrophysical problems have taken on an increasing importance in recent years. In this paper I review modern techniques for solving hyperbolic systems of partial differential equations, especially those that are in conservation form. Several such systems of equations exist. The ones that are of great importance to computational astrophysics include the equations of non-relativistic ideal hydrodynamics and magnetohydrodynamics (MHD). Their relativistic counterparts also obey similar hyperbolic conservation laws. In most astrophysical problems, one is interested in time-dependent problems. For such problems, a certain class of solution techniques, known as higher-order Godunov schemes, have shown themselves to be very robust, accurate and efficient. I therefore focus on these schemes and their role in solving hyperbolic systems of interest to computational astrophysics.

In an almost concurrently written set of lecture notes, LeVeque (1997) has focussed on the theory of hyperbolic systems and their solution using total-variation-diminishing (TVD) schemes. Here, I focus on more recent developments including essentially non-oscillatory (ENO) schemes. Both TVD and ENO schemes are Godunov schemes. Furthermore, since LeVeque has already treated the underlying theory, I focus on practical aspects of solving those conservation laws that are of interest to computational astrophysics. This enables the two reviews to complement each other.

Space limitations allow this review to to treat all topics in only the briefest fashion. In § 2, I describe general features of hyperbolic systems. In § 3, I describe issues in the numerical solution of hyperbolic problems. In § 4 and 5,

I describe issues in numerical MHD and relativistic flow, respectively. In § 6, I talk about the use of parallel computers to solve these hyperbolic systems.

2. Hyperbolic Systems in Conservation Form

Systems of conservation laws can be formally written as:

$$\partial_t U + \partial_x F(U) + \partial_y G(U) + \partial_z H(U) = 0 \tag{1}$$

where U is a vector of n conserved variables and F, G and H are fluxes in the x-, y- and z-directions. The fluxes are functions of the conserved variables. It is easy to see from Gauss' Law that for a closed system the volumetric integral of the vector of conserved variables is indeed conserved. A local version of the conservation law approach would require that if the equations were discretized on a grid, then the volume integral of a conserved variable in each zone after a time step should equal the volume integral of the same before the time step except for the fluxes integrated over the zone boundaries. As an example, for the Euler equations, the vector U would consist of the mass density, the three momentum densities and the total energy. A much more detailed development of the theory presented in this section is given in Lax (1972).

It is often helpful to take a dimension-by-dimension view of equation (1). It also helps to write the system in terms of its primitive variables. The primitive variables are basically any set of variables in which the mathematical or computational manipulation of the conservation laws becomes more convenient. We denote such a vector of n primitive variables as V. Again as an example, for the Euler equations, the vector V might consist of the density, the flow speeds in the three directions and the pressure. For variations that are restricted to the x-direction, the system can be written as

$$\partial_t V + A(V)\partial_x V = 0 \tag{2}$$

where $A(V)$ is an $n \times n$ matrix. The behavior of the system in equation (2) is strongly determined by the properties of the matrix $A(V)$. If the matrix has n real eigenvalues, then the system is hyperbolic. If the eigenvalues form an ordered set, then it is known as strictly hyperbolic. Several important systems, say for example the Euler or MHD systems, are hyperbolic but not strictly hyperbolic. Corresponding to the real eigenvalues, the matrix $A(V)$ has real right and left eigenvectors. The advantage of viewing the system in equation (2) in terms of its eigenvalues and eigenvectors is that infinitesimal variations of the primitive variables V when projected onto the space of the right eigenvectors propagate in the space-time plane with speeds given by the eigenvalues. While variations in a numerical scheme are seldom infinitesimal, this view is very helpful in the construction of numerical schemes. The reason is that for short enough time steps, we can pretend that the system evolves in that fashion as long as we also factor in the fact that shocks can emerge in the computation. These variations are formally known as characteristic variables. Thus, to establish some notations, we denote the eigenvalues by $\lambda^1, \lambda^2, ..., \lambda^n$. The right eigenvectors will be denoted by $r^1, r^2, ..., r^n$ and the left eigenvectors by $l^1, l^2, ..., l^n$. The eigenvalues and eigenvectors will in general be functions of the primitive variables.

Traditionally, one likes to ensure that the left eigenvectors are orthonormal to the right eigenvectors. The characteristic variable for the kth eigenvalue is denoted by $w^k = l^k \cdot V$. When the spatial variations are smooth, left multiplying equation (2) by l^k shows us that the evolution equations for the characteristic variables satisfy simple wave equations

$$\partial_t w^k + \lambda^k \partial_x w^k = 0 \qquad (3)$$

Understanding the nature of these waves yields a good understanding of the hyperbolic system. The space-time lines along which the waves in equation (3) propagate are known as characteristics.

I motivate this analysis of waves by reminding the reader of a few simple results for the Euler equations from Landau & Lifshitz (1987). A one-dimensional analysis of sound waves shows that there are two families of sound waves. Both have a self-steepening property so that a finite amplitude sound wave traveling in either direction steepens into a shock. This is a consequence of the non-linearity of the Euler equations. The non-linearity is also manifested by the fact that the characteristics of a given shock family flow into the shock. The Euler equations also admit another kind of wave, the entropy wave. The discontinuity associated with that wave family is known as a contact discontinuity and it moves with the speed of the flow. Unlike shocks, the characteristics associated with a contact discontinuity do not flow into a contact discontinuity. Thus, entropy waves do not have a self-steepening property. This discussion serves to motivate the fact that different waves in a hyperbolic system have different properties. Thus their computational treatment necessarily needs to be different. I will draw out these differences further in the section on interpolation. Here it helps to explain to the reader that the distinction between self-steepening waves and waves that are not self-steepening has a mathematical codification. Thus waves with a self-steepening property are called genuinely non-linear waves. The kth wave field of a hyperbolic system is said to be genuinely non-linear if it satisfies

$$[\nabla_V \lambda^k(V)] \cdot r^k(V) \neq 0 \quad \forall V \qquad (4)$$

On the other hand if the quantity in equation (4) is exactly zero for all V, then that wave field is said to be linearly degenerate. Such a wave field does not have a self-steepening property.

The above discussion makes it clear that singular solutions can emerge in the flow. Thus the computational strategy must be able to handle them robustly. Shocks are not the only family of simple waves associated with a genuinely non-linear characteristic field. The system can also admit rarefaction waves.

A central issue associated with computational solution of hyperbolic systems is the resolution of the cell-break problem, also known as the Riemann problem. The Riemann problem starts with assuming that two slabs of fluid are separated by an infinitesimally-thin planar membrane. The slabs can have arbitrary values for the fluid variables. The Riemann problem describes the time evolution of the fluid once the membrane is broken. The result of this removal is that the discontinuity is resolved into a sequence of simple waves. In particular, there is one simple wave associated with each wave family in the problem. This is not strictly true when the hyperbolic system can admit compound waves, as is the case with MHD. However, as realized by Roe (1981), only a small portion of the

Hyperbolic Conservation Laws

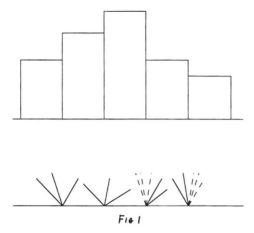

Figure 1.

information generated in the Riemann problem is actually used in the numerical solution of hyperbolic systems. This has given rise to the development of various approximate Riemann solver strategies and such strategies have even shown their worth in numerical MHD.

3. Higher-Order Godunov Schemes

The original Godunov scheme (Godunov, 1959) consisted of discretizing the fluid variables in each zone as discrete slabs of fluid (Figure 1). At each time step, the Riemann problem was solved at the zone boundaries. A numerical flux was obtained by using those fluid variables in the Riemann problem's solution that lie along the zone boundary. Finite differencing the fluxes yielded a conservative scheme that could handle shock formation with little or no post-shock oscillations. The utilization of the Riemann solver was central to this physically correct handling of shocks. The original Godunov scheme was extremely dissipative and had only first order of accuracy. Thus, despite its stability, it was seldom used.

van Leer (1979) showed that Godunov's original scheme could be made second-order accurate. He did this by endowing the slabs of fluid with internal structure (Figure 2). Thus, by fitting piecewise linear profiles to the slabs of fluid, he was able to arrive at a scheme that was spatially second-order accurate. van Leer was able to show that not every linear profile that is consistent with second-order accuracy is acceptable and that the linear profiles had to satisfy a constraint known as the monotonicity constraint. Thus, the slopes that were fit had to be limited using a slope limiter. This slope limiter as applied to each zone is essentially a highly non-linear function of the slopes in the adjoining zones.

It is easy to see why fitting linear profiles increases the accuracy. In Godunov's original scheme, the dissipation was provided by the Riemann solver; the larger the jump in variables on either side of the Riemann problem, the higher the dissipation. But the fact that Godunov's original scheme put large

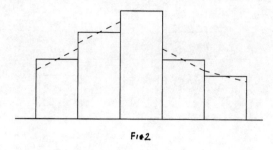

Figure 2.

jumps at the zone boundaries meant that it put large amounts of dissipation all over. Endowing the fluid variables with linear profiles brings the fluid variables on either side of the Riemann problem closer together when the flow is smooth and hence reduces the dissipation. The monotonicity constraint was essential for having oscillation-free behavior at shocks. In intuitively simple terms, the monotonicity constraint implies that the interpolation process puts no further maxima or minima in the interpolated profiles of Figure 2 than are present in the original slabs of fluid in Figure 1. The monotonicity constraint also implies that around shocks, the formal order of accuracy is relinquished in favor of stability, *i.e.*, at a shock, van Leer's scheme behaves much like a Godunov scheme. Notice too from Figure 2 that strict application of van Leer's monotonicity switch would have the extremum in Figure 2 be interpolated by a flat profile. Thus even though the extremum in Figure 2 might not be a shock it will be dissipated as if the flow around the extremum needed to be treated with first-order accuracy. Subsequently, Osher & Chakravarthy (1984) and Harten (1982) designed schemes where the limiters were applied directly to the fluxes. Such schemes are said to be based on flux limiters as opposed to the original van Leer scheme which limited the slopes. The rigorous mathematical connections between monotonicity, accuracy, stability and dissipation were made later by Sweby (1984) and Tadmor (1987). For reasons of pedagogy, we keep our focus on schemes that are based on slope limiters.

The original higher-order schemes still had an ambiguity associated with them. They fitted linear profiles to any set of "good" variables where such variables were taken to be variables that varied rapidly in a shock. Glimm & Lax (1970) had already shown that for a 2×2 convex hyperbolic system, the variation in the characteristic variables decreases as a function of time. Harten (1982) was able to show that to codify this property for numerical schemes, one has to apply the monotonicity constraint to the characteristic variables. These are schemes expressly designed to make the total variation decrease with time, hence the nomenclature TVD schemes. Using the characteristic variables as interpolants also provides three further advantages. First, it enables one to increase the temporal accuracy with little additional computational effort. Second, it enables linearly-degenerate characteristic fields to be treated differently from genuinely non-linear characteristic fields. In particular, left to their own devices, linearly-degenerate fields have a tendency to disperse, a property first noted by Harten (1977) and subsequently developed by Yang (1990). This can

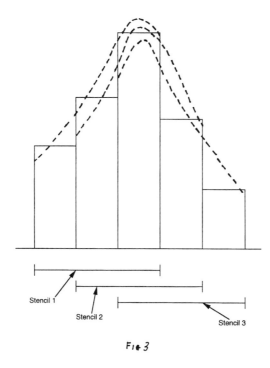

Figure 3.

be selectively controlled; for an example drawn from MHD, see Balsara (1997a). Third, it allows one to formulate very refined numerical dissipation strategies to eliminate slight, but undesirable post-shock oscillations in genuinely non-linear characteristic fields (see Balsara, 1997a). All of this brings us to a viewpoint, first suggested by Roe (1985), that given any hyperbolic system, one can analyze its wave structure. Having understood its wave structure, one can design a strategy for treating each wave field. Different wave fields might be treated differently. Having brought together one's understanding of the wave structure, one, therefore, puts together the solution strategy for that hyperbolic system by incorporating several of the insights catalogued in this section.

Osher & Chakravarthy (1984) tried to design TVD schemes of ever-increasing order of accuracy. The attempt proved futile though its real value lay in the fact that it pointed out an important limitation of TVD schemes in general. The limitation stems from the fact that the monotonicity property implicit in TVD schemes causes all extrema, such as the one in Figure 2, to be fit with a flat profile, i.e., extrema get clipped. This causes the flow solution around the extremum to degenerate to first-order accuracy, as was pointed out in the previous paragraph. The net result is a decrease in the scheme's order of accuracy. Harten et al. (1987) found a way out of this impasse. They realised that all that was required was to find an interpolation strategy that put no further oscillations in the interpolating function than are there in the original slabs of fluid. Hence they named the scheme Essentially Non-Oscillatory (ENO). Notice that this requirement does not require that a flat profile be fitted in a region having an extremum, thus alleviating the problem pointed out in Figure 1. In order to accomplish this they realized that there always exist several stencils covering a zone that allow one to interpolate in that zone with the desired degree of accuracy. Figure 3 provides an example of this for interpolating the flow profile in the central zone to third-order accuracy. The stencils that allow one to do this interpolation will be three zones wide and will all cover the central zone. There are three such stencils. In Figure 3, we show the three stencils and their interpolating functions shown with dashed lines having slight vertical offsets from each other. While all three stencils interpolate with third order, if Stencil 1 were used everywhere in a numerical scheme, it would be numerically unstable. The same is true for Stencil 3. Only Stencil 2, being centered, is numerically stable. The trick, as Harten et al. (1987) discovered, lay in selectively choosing between the stencils based on the local nature of the solution.

Because the ENO schemes of Harten et al. (1987) always chose the smoothest stencil, they always picked up that stencil which introduced the least oscillation in the interpolated profile. In regions containing shocks, picking a stencil that introduces the least oscillation in the interpolated profile also causes one to perform selectively the interpolation from the direction that is most stable, i.e., one-sided stencils can take precedence over centered stencils provided they contribute to stability. Moreover, a little reflection shows that in the vicinity of a shock, that direction is the upwinded direction. It is by these two devices that ENO schemes achieve numerical stability. Notice that the issue of what constitutes the smoothest stencil was intentionally left open above. Thus, in the original ENO schemes, Harten et al. (1987) resolve that issue based on constructing a table of divided differences. Later it was found that this strategy of choosing just one stencil could destabilize the solution by introducing too rapid a switching of the choice of smoothest stencil from one zone to another (Shu, 1990). Thus, it was deemed better to use a convex combination of the smoothest and second from smoothest stencil. This strategy is known as biasing. Later, Liu et al. (1994) and Jiang & Shu (1996) showed that if all the stencils were ascribed weights based on their smoothness measures, one could achieve an increase in the scheme's accuracy. Because of the weighting process, such ENO schemes were referred to as weighted ENO (WENO) schemes. Balsara & Shu (in preparation) showed that such WENO strategies can be constructed stably for several higher orders and that their order of accuracy can be formally proved. While the discussion in this paragraph has been couched in the language of finite-volume interpolation, Shu & Osher (1988; 1989) showed that the most efficient versions of ENO arose from finite-difference interpolation. Casper & Atkins (1994) compared the two different formulations. It has been the experience of several of the authors mentioned in this paragraph that ENO schemes offer a factor of two more resolving power than the well-designed PPM scheme of Colella & Woodward (1984).

A lot of insight has also been gained in the design of Riemann solvers. Many of these derive from the same quest for simplicity initially expressed in Roe (1981). Thus, Bell et al. (1989), Marquina & Donat (1996), and Balsara & Shu (in preparation) have all suggested Riemann solvers that are based on some form of linearization. Einfeldt et al. (1991) have shown the shortcomings of linearization strategies but they have also shown how these shortcomings can be rectified. Quirk (1994) has also shown how multiple strategies for Riemann solvers can be used with good effect.

4. Numerical Magnetohydrodynamics

The eigenvector analysis for this system was done by Zachary & Colella (1992) and with much more precision by Roe & Balsara (1996). Roe & Balsara (1996) also showed that the breakdown of strict hyperbolicity for this system does not cause a computational breakdown because a complete set of eigenvectors is always available. Further, they showed that the flux computation based on linearization strategies remains valid even at the triple-umbilic point. For an analysis of compound waves in MHD, see Brio & Wu (1988) and for MHD test problems that include compound waves see Ryu & Jones (1995).

Zachary et al. (1994), Balsara (1997a) and Balsara & Shu (in preparation) have used characteristics-based interpolation strategies for the solution of the MHD system. Balsara (1997a) and Balsara & Shu (in preparation) have shown how artificial compression techniques can be used to improve the quality of linearly degenerate characteristic fields including Alfvén waves. PPM interpolation schemes have been used for MHD by Dai & Woodward (1994a). Balsara & Shu (in preparation) have developed ENO schemes for numerical MHD. A linearized Riemann solver for MHD based on the strategy of Bell et al. (1989) was developed by Zachary et al. (1994). A linearized Riemann solver for MHD based on the strategy of Roe (1981) was developed by Balsara (1997b). Non-linear Riemann solvers prove to be extremely cumbersome, especially in numerical MHD, but such Riemann solvers have been developed by Dai & Woodward (1994b).

No other issue in numerical MHD seems to generate as much controversy as the divergence of the magnetic field. The controversy stems from the fact that most numerical discretizations of the MHD equations that are based on higher-order Godunov schemes allow the divergence of the magnetic field to build up over time. Based on a few short-time scale simulations, several authors have opined that it is acceptable for the divergence of the magnetic field to build up in the course of a computation. An extreme version of this viewpoint (Powell, 1994) relies on the generation and advection of divergence and incorporates extra terms into the MHD equations to accommodate this process. The fact that searches for magnetic monopoles have consistently failed would, therefore, have to be considered irrelevant. Balsara (1997a) has run several test problems with a variety of schemes, some of which incorporate divergence cleaning and some which do not. It was found that the removal of magnetic divergence was essential for guaranteeing good results.

5. Relativistic Hydrodynamics

Balsara (1994) and Marti & Muller (1994) designed solution strategies for the Riemann problem for ultrarelativistic hydrodynamics. Marquina et al. (1992) designed a scheme based on the ideas of Marquina & Donat (1996). Balsara (1996) showed that the ENO schemes developed in Balsara & Shu (in preparation) could be extended to relativistic hydrodynamics. He also reviewed the role of relativistic hydrodynamics in extragalactic jet simulations. In non-relativistic hydrodynamics, the pressure and density can be derived directly from the conserved variables. In relativistic hydrodynamics, this entails solving a transcendental equation iteratively. This does not pose a severe challenge but is one of several intricacies that makes relativistic hydrodynamics computationally costly.

6. Hyperbolic Systems Solved on Parallel Computers

The methods catalogued here are accurate, stable and robust. This makes them attractive for general purpose use. They are, nevertheless, rather float point intensive and so require powerful computers to solve problems of reasonable size. The highest levels of computer processing power and memory size are available on massively parallel machines. Thus, it becomes interesting to want to run such codes on parallel machines. Initially such effort relied on making vendor-specific implementations. With the advent of the Message Passing Interface (MPI; see Gropp *et al.*, 1994) and High Performance Fortran (HPF; see Koelbel *et al.*, 1994) there is a drive towards standardization. The author's RIEMANN code has been shown to have optimal speedup with increasing number of processors on a variety of massively parallel machines (see Balsara & Melhem, 1997). The float point intensive aspect of these methods can be used to good advantage to increase the on-processor computing and reduce the off-processor communication.

The quest for parallelism comes hand-in-hand with the need to do parallel adaptive-mesh refinement. Here the solution is adaptively refined in selective regions of the flow where particularly complicated flow topologies arise. The basics that underlie this strategy were laid down by Berger & Colella (1989). The complexity of the problem necessitates the use of object-oriented languages such as C++ and its library extensions on parallel machines. Recent work in that direction has been reported in Balsara & Quinlan (1997) and Parashar & Browne (1997).

References

Balsara, D. S. 1994, J. Comput. Phys., 114, 284.
Balsara, D. S. 1996, in *Pune AGN Workshop*, eds. J. J. Perry & S. A. Lamb.
Balsara, D. S. 1997a, ApJ, in press.
Balsara, D. S. 1997b, ApJ, in press.
Balsara, D. S., & Quinlan, D. 1997, in *Eighth SIAM Conference on Parallel Processing for Scientific Computing*,
Balsara, D. S., & Melhem, R. 1997, Comp. in Phys., (submitted)
Bell, J. B., Colella, P., & Trangenstein, J. A. 1989, J. Comput. Phys., 82, 362.
Berger, M., & Colella, P. 1989, J. Comput. Phys., 82, 64.
Brio, M., & Wu, C. C. 1988, J. Comput. Phys., 75, 400.
Casper, J., & Atkins, H. A. 1994, AIAA J., 32, 1970.
Colella, P., & Woodward, P. R. 1984, J. Comput. Phys., 54, 174.
Dai, W., & Woodward, P. R. 1994a, J. Comput. Phys., 115, 485.
Dai, W., & Woodward, P. R. 1994b, J. Comput. Phys., 111, 354.
Einfeldt, B., Munz, C. D., Roe, P. L., & Sjogreen, B. 1991 J. Comput. Phys., 92, 273.
Glimm J., & Lax, P. D. 1970, Mem. Amer. Math. Soc., 101, 1.

Godunov, S. K. 1959, Mat. Sb., 47, 271.
Gropp, W., Lusk, E., & Skjellum, A., 1994, *Using MPI*, (Cambridge: MIT Press).
Harten, A. 1977, Comm. Pure and Appl. Math., 30, 661.
Harten, A. 1982, *On Second Order Accurate Godunov Type Schemes*, (unpublished).
Harten, A., Engquist, B., Osher, S., & Chakravarthy, S. 1987, J. Comput. Phys., 71, 231.
Jiang, G.-S., & Shu, C.-W. 1996, J. Comput. Phys.
Koelbel, C. H., et al. 1994, *The High Performance Fortran Handbook*, (Cambridge: MIT Press).
Landau, L. D., & Lifshitz, E. M. 1987, *Fluid Mechanics*, (Oxford: Pergamon Press).
Lax, P. D. 1972, Amer. Math. Monthly, 79, 227.
LeVeque, R. J. 1997, in *SAAS-Fee Lectures on Computational Methods for Astrophysical Flow*, eds. O. Steiner & A. Gautschy.
Liu, X.-D., Osher, S., & Chan, T., 1994, J. Comput. Phys., 115, 200.
Marquina, A., et al. 1992, A&A258, 566.
Parashar, M., & Browne, J. C. 1997, SAMR Workshop held in Minnesota
Powell, K. G. 1994, *ICASE Report 94-24*
Quirk, J. J. 1994, Int. J. Num. Meth. Fluids, 18, 555.
Roe, P. L. 1981, J. Comput. Phys., 53, 357.
Roe, P. L. 1985, in *Lectures in Applied Mathematics, volume 22*, eds. B. Engquist. & R. J. Osher, (Somerville: American Mathematical Society).
Roe, P. L., & Balsara, D. S. 1996, SIAM J. Num. Anal., 56, 57.
Ryu, D., & Jones, T. W. 1995, ApJ, 442, 228.
Shu, C.-W. 1990, J. Sci. Comput., 5, 127.
Shu, C.-W., & Osher, S., 1988, J. Comput. Phys., 77, 439.
Shu, C.-W., & Osher, S., 1989, J. Comput. Phys., 83, 32.
Sweby, P. K. 1984, SINUM, 21, 995.
van Leer, B. 1979, J. Comput. Phys., 32, 101.
Yang, H. 1990, J. Comput. Phys., 89, 125.
Zachary, A. L., & Colella, P. 1992, J. Comput. Phys., 99, 341.
Zachary, A. L., Malagoli, A., & Colella, P. 1994, SIAM J. Sci. Comput., 15, 263.

Relativistic Jet Simulations and VLBI Maps

Philip Hughes

Astronomy Department, University of Michigan, Ann Arbor, MI 48109-1090, U.S.A.

Amy Mioduszewski

JIVE/NRAO, P.O. Box 0, Socorro, NM 87801, U.S.A.

Comer Duncan

Department of Physics and Astronomy, Bowling Green State University, Bowling Green, OH 43403, U.S.A.

Abstract. We describe recent radiation transfer calculations whose aim is to explore the appearance of extragalactic jets simulated with a relativistic hydrodynamic code, and which include Doppler boost and time-delay effects. We note the role of these effects in influencing the appearance of such flows, and assess the viability of performing an extensive set of such computations on workstation-class machines for comparison with VLBI maps.

1. Introduction

Apparent superluminal motion and brightness temperatures well in excess of the inverse-Compton limit of $\sim 10^{12}$ K (Hughes & Miller, 1991) provide compelling evidence that at least on the parsec-scale, extragalactic jets have relativistic bulk speeds. These highly-collimated flows of synchrotron-emitting plasma have been modeled using computational fluid dynamics for several decades (Norman, 1993), but until recently all such simulations have ignored the relativistic character of the flow. This is a major deficiency because not only may relativistic flows display dramatically different character (*e.g.*, narrow, very high density shocked domains), but a detailed map of the *relativistic* velocity field is essential in order to compute the radiative properties of these flows for comparison with VLBI/P maps.

In this contribution we present results from both recent relativistic fluid dynamic simulations, and radiation transfer calculations through the computed flows. We emphasize the latter, and in particular discuss the practical difficulties associated with the need to account for "time delay" effects when producing simulated VLBI maps.

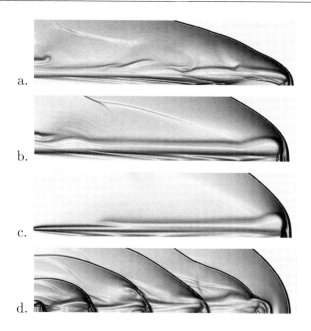

Figure 1. A series of progressively more relativistic jet simulations, showing increasing stability: a) $\gamma = 5$, $\mathcal{M} = 17$, $\Gamma = 5/3$; b) $\gamma = 10$, $\mathcal{M} = 16$, $\Gamma = 5/3$; c) $\gamma = 10$, $\mathcal{M} = 15$, $\Gamma = 4/3$; d) a sinusoidally modulated run of the previous simulation.

2. Computational Fluid Dynamics

Our calculations assume axisymmetry, an inviscid and compressible gas, and an ideal equation of state with constant adiabatic index. We use a Godunov-type solver which is a relativistic generalization of the method described by Harten et al. (1983) and Einfeldt (1988), in which the full solution to the Riemann problem is approximated by two waves separated by a piecewise constant state. We evolve mass density R, two components of the momentum density M_ρ and M_z, and the total energy density E relative to the laboratory frame. The rest frame energy density, e, and mass density, n, needed for the estimation of the sound speed, are determined via a Lorentz transformation. The scheme is second order in both time and space.

We utilize an *Adaptive Mesh Refinement* scheme following Quirk (1991) and Berger & Colella (1989). This uses a hierarchical collection of grids consisting of embedded meshes to discretize the flow domain. The algorithm orchestrates: *i*) the flagging of cells for refinement/collection into meshes; *ii*) the construction of boundary zones for each mesh; *iii*) a sweep over all levels/each mesh in a given level to update the solution; *iv*) the transfer of data between the meshes in the hierarchy. This reduces computational time and memory requirements to the extent that we can run extensive simulations on workstation-class machines.

Figure 1 shows the results of a set of simulations spanning various values of Lorentz factor (γ), Mach number ($\mathcal{M}$) and adiabatic index (Γ) for a jet with rest frame density one-tenth ambient, and initially in pressure balance with its surroundings. The images are "schlieren-type", and display the gradient of the laboratory frame density on a non-linear scale. This facilitates the display of structures covering a large range of spatial and density scales. The last panel of the sequence shows the result of modulating the inflow Lorentz factor between ~ 1 and 10 for the case shown in panel c. The set of nested bow shocks that results is probably a poor approximation to the much weaker shocks thought to be responsible for VLBI knots (Hughes *et al.*, 1989), but is expedient for our first exploration of the radiative properties of flows, and may be pertinent to the study of sources which display intermittent behavior, *i.e.*, restarting jets (*e.g.*, Clarke, these proceedings).

3. Radiation Transfer

We estimate the rest frame emission and absorption coefficients by assuming that both the number density of synchrotron-emitting particles and the energy density in the magnetic field (which is not computed by our hydrodynamic code) are proportional to the energy density, and thus the pressure, of the thermal particle distribution. The magnetic field is assumed to be tangled on a scale much less than our computational cell size, whence the effects of relativistic aberration can be ignored. Radiation transfer calculations are performed using the standard theory (Rybicki & Lightman, 1979), allowing for both Doppler frequency shift and Doppler boost, and for time delay effects: *i.e.*, that the flow evolves significantly within the time taken by a light ray to cross the flow.

The latter point makes radiation transfer a great challenge: in the nonrelativistic domain, a simulated flow at some epoch may be used as the basis of such a calculation, but here the history of the flow is required, thus requiring us to store the results from the computational fluid dynamics at a large number of times. In fact, the radiation transfer *may* be performed in conjunction with the fluid dynamics, obviating the need to store the latter, and that approach has been adopted by Komissarov & Falle (1996). However, we feel that such a method is restrictive, as it precludes much of the exploration that is so essential for understanding the radiative properties of the simulated flows. In particular, divorcing the radiation transfer from the fluid dynamics admits *i*) an exploration of different maps between the thermal particle pressure and the particle-field values that determine transfer coefficients; *ii*) recomputation with a different set of angles of view; *iii*) recomputation with different flow attributes such as opacity; *iv*) a full use of diagnostics (*e.g.*, a coding of the contributions by location and epoch to the total intensity along a line of sight); *v*) the application of the radiation transfer to the flows independently computed by other workers (for comparison purposes). Our goal here is to consider the practical implications of "after the fact" radiation transfer, and to demonstrate that this approach *is* viable on the machines used for the original fluid dynamics.

The key issue is storage: the radiation transfer calculations are quite simple, and as long as a local storage medium is used, avoiding transfer of large data sets over a network, the requisite data can be read rapidly. One cycle of the full 2-D solutions shown in Figure 1 requires about 10 ± 1Mb of storage. The CFL number for the fluid dynamic calculations depends on scale, as we employ Adaptive

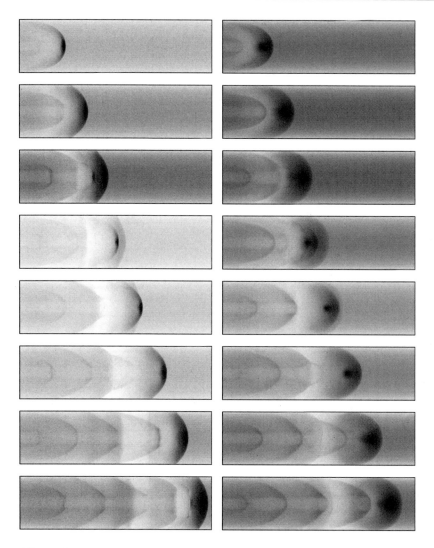

Figure 2. A sequence of time slices for the modulated flow of Figure 1d seen at 90° to the flow axis. Left: without time delays; Right: with time delays.

Mesh Refinement; however it may be characterized by that number adopted on the coarsest scale: $\mathrm{CFL}|_{\max}$. By definition of this number, there is only slight evolution of the flow within $N_* \sim 1/\mathrm{CFL}|_{\max}$ cycles, and thus we do not need to save the solution more frequently than every $\mathcal{O}(\lesssim 1) \times N_*$ cycles to get adequate time resolution. The simulations shown above were run for typically 3,000–5,000 cycles, with $\mathrm{CFL}|_{\max} \sim 0.1$, implying that it would be reasonable to save the solution every ~ 5 cycles, leading to a storage requirement of 10Gb at most. Multigigabyte disks are now quite cheap and commonly available,

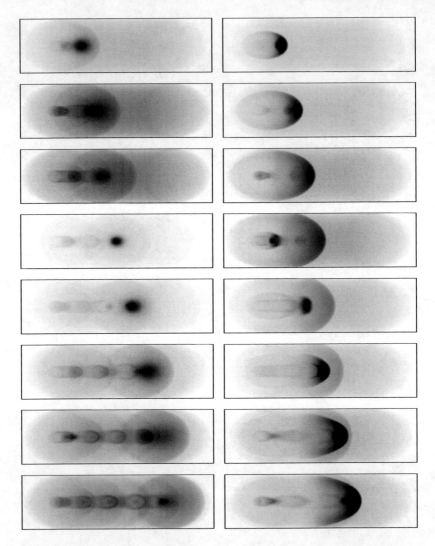

Figure 3. A sequence of time slices for the modulated flow of Figure 1d seen at 30° to the flow axis. Left: without time delays; Right: with time delays.

and so this is not an impractical amount of storage. In fact, storing a subset of the hydrodynamic solution with limited accuracy and saving somewhat less frequently—we have found no discernible difference in intensity structure when saving every $\sim 1.5/\text{CFL}|_{\max}$ cycles—reduces storage requirements to $\sim$ 2Gb or even less.

Figures 2 and 3 show the results of such a radiation transfer calculation performed on the same workstation as was used for the underlying hydrodynamics. Because the bow shocks are strong, they propagate significantly more slowly than the underlying flow, and thus the overall morphology does not change dramatically within one light crossing time. Nevertheless, the role of time delays is evident: from "front" to "back" of the source in the side-on case, and along the flow in the 30° case. Understanding the location and epoch of the main contribution to flux along any line of sight in the time delay cases is important for the interpretation of VLBI maps and is an aspect of the simulations currently under study. We note that sources such as 3C 345, which are believed to be viewed at only a few degrees to the flow axis, will require less frequently modulated, longer, simulations, in order to keep the components distinct even in projection. This will be an even greater challenge, in terms of computer memory and storage, but one which we believe to be viable with presently available resources.

References

Berger, M., & Colella, P. 1989, J. Comput. Phys., 82, 67.
Einfeldt, B. 1988, SIAM J. Numerical Analysis, 25, 294.
Harten, A., Lax, P., & van Leer, B. 1983, SIAM Rev., 25, 35.
Hughes, P. A., Aller, H. D., & Aller, M. F. 1989, ApJ, 341, 68.
Hughes, P. A., & Miller, L. 1991, in *Beams and Jets in Astrophysics*, ed. P. A. Hughes (Cambridge: Cambridge University Press), 1.
Komissarov, S. S., & Falle, S. A. E. G. 1996, in *Energy Transport in Radio Galaxies and Quasars (ASP Conference Series, Vol. 100)*, eds. P. E. Hardee, A. H. Bridle, & J. A. Zensus, (San Francisco: ASP), 165.
Norman, M. L. 1993, in *Astrophysical Jets*, eds. D. Burgarella, M. Livio, & C. P. O'Dea (Cambridge: Cambridge University Press), 211.
Quirk, J. J. 1991, Ph.D. Dissertation (Cranfield Institute of Technology).
Rybicki, G. B., & Lightman, A. 1979, *Radiative Processes in Astrophysics*, (New York: Wiley).

Simulations of Jet Production in Magnetized Accretion Disk Coronae

David L. Meier, Samantha Edgington, Patrick Godon

Jet Propulsion Laboratory, California Institute of Technology, Pasadena, CA 91109, U.S.A.

David G. Payne

Intel Scalable Systems Division, Beaverton, OR 97006, U.S.A.

Kevin R. Lind

Cray Research Corporation, Livermore, CA 94550, U.S.A.

Abstract. We describe the results of over 40 MHD simulations, performed on Caltech/JPL parallel supercomputers, of the coronae of magnetized accretion disks around compact objects. All produce some type of outflow from the disk. Our parameter study investigated the character of the outflow as a function of the strength of the initial poloidal magnetic field and its angle with respect to the disk rotation axis. When the radial component of the field is significant, this outflow takes the form of a collimated jet ejected from the center of the accretion disk. The jet velocity is a strong function of the strength of the initial magnetic field: for Alfvén velocities (V_A) below the escape speed (V_{esc}), the jet velocity is of order V_A, but for V_A only slightly above V_{esc}, the jet velocity is an order of magnitude or more greater.

This "magnetic switch" behaves similarly for a broad range of magnetic field polar angle. However, when the initial coronal magnetic field is nearly completely dominated by an axial component, the importance of the central jet diminishes and the outflow becomes dominated by a poorly-collimated wind from a broader region of the accretion disk.

The magnetic switch may have applications to galactic and extragalactic radio sources and other objects.

1. Introduction

Jets are observed to be ejected from a variety of astrophysical objects: the centers of radio galaxies and quasars, very old stars (planetary nebulae, accreting neutron star and black hole binary systems), and very young stars (*e.g.*, primitive solar system nebulae).

One of the most promising models for producing jets in accreting systems is the Blandford-Payne magnetohydrodynamic (MHD) process in the coronae of thin accretion disks (Blandford & Payne, 1982). In this model, a magneto-centrifugally-driven wind develops if the angle θ between the poloidal magnetic field component B_p [$\equiv (B_R^2 + B_Z^2)^{1/2}$] and the disk rotation axis is greater than 30°. Because of differential rotation, the magnetic field lines wrap around the Z axis, providing a Lorentz hoop stress that collimates the flow into a jet.

A variety of approaches have been used to study this mechanism in more detail. Most analytical or semi-analytical approaches assume open magnetic field lines anchored in an infinitely thin and dense Keplerian disk and then look for steady-state, collimated wind solutions in the disk corona (Lovelace et al., 1986; Li et al., 1992, hereafter LCB; Kudoh & Shibata, 1996). These studies have been successful in obtaining jet flow speeds with Lorentz factors $\Gamma \equiv (1 - v^2/c^2)^{-1/2} > 10$ (LCB), typical of the fastest quasar jets observed. Most numerical approaches, on the other hand, assume a disk of finite thickness, initially threaded by open magnetic field lines, and use angular momentum loss and accretion in the disk to produce a $\theta > 30°$ condition which then ejects some of the disk material itself in a jet (Uchida & Shibata, 1985; Stone & Norman, 1994; Bell & Lucek, 1995; Matsumoto, 1997; Koide et al., 1997). However, the jets produced in these simulations appear to have a low Γ factor. A few investigators have studied the anchored field line/ejected disk corona case numerically (Ustyugova et al., 1995; Ouyed, these proceedings). However, none of these have obtained jet flow speeds approaching the highly relativistic ones predicted by the semi-analytic studies either.

2. "Weakly" Magnetized Accretion Disks and Coronal Flow

Assuming that the magnetic field is anchored in a dense disk and that the disk corona is the source of the jet is more than a convenient mathematical simplification. A good parameter for measuring the importance of magnetic forces is the ratio of the Alfvén velocity [$V_A \equiv B/(4\pi\rho)^{1/2}$] to the escape velocity. (Here B is the magnetic field strength in Gauss and ρ is the mass density of the plasma.) Standard accretion disk models like the α-model (Shakura & Sunyaev, 1973) naturally have a low Alfvén velocity in the disk:

$$V_{A,\,\mathrm{disk}} \equiv \frac{B}{(4\pi\rho_\mathrm{disk})^{1/2}} \sim \alpha^{1/2}\frac{H}{R} V_\mathrm{esc} \ll V_\mathrm{esc}, \qquad (1)$$

where $V_{A,\,\mathrm{disk}}$ and V_esc are the local Alfvén and escape velocities respectively in the disk, $H \ll R$ is the disk half-thickness at some radius R, and $\alpha < 1$ is the usual viscosity parameter. While even weak disk fields can have significant consequences for the internal structure of the disk and probably play an important role in the viscosity itself, the effective α from local shear instabilities is expected to be small ($\alpha \sim 0.01$; Brandenburg et al., 1996; Stone et al., 1996). Since the inward accretion drift time scale is several α^{-1} orbital times, the assumption of a coronal magnetic field anchored at a fixed radius is a good approximation for dynamical calculations.

Furthermore, although the Alfvén velocity is low in the disk, it can still be high in the corona because of the low density there. (Such conditions are not unlike those in the solar atmosphere and corona.) In fact, there exists a critical coronal density $\rho_\mathrm{crit} \equiv \alpha (H/R)^2 \rho_\mathrm{disk}$ such that if $\rho_\mathrm{corona} \lesssim \rho_\mathrm{crit}$, then $V_{A,\,\mathrm{corona}} \gtrsim V_\mathrm{esc}$. For coronae of sufficiently low density then, magnetic forces can dominate gravitational forces and significantly affect the dynamics of the corona, even though they are not strong enough to disrupt the disk itself.

3. The MHD Simulations

We performed formally non-relativistic, two-dimensional axisymmetric magnetohydrodynamic simulations of magnetized accretion disk corona, which include toroidal, as well as poloidal, magnetic field and velocity components and gravity.

3.1. Initial and boundary conditions

For initial conditions we assume that open poloidal magnetic field lines ($B_{\phi_0} = 0$) are anchored in a dense, cold, Keplerian rotating accretion disk and thread a warm, also rotating, corona whose temperature is a significant fraction of virial but is still bound in a relatively thin structure. Permeating the entire region is a tenuous, hot halo ($\rho_{\text{halo}}/\rho_{\text{corona}} = 0.1$) in pressure equilibrium with the corona. The strength of all parameters varies with cylindrical radius as a Shakura & Sunyaev (1973) accretion disk, such that the ratio of local velocities in the corona (V_A, V_{sound}, and V_{wind}) to the local escape velocity V_{esc} is a constant. The parameter $V_{A,\text{corona}}/V_{\text{esc}}$ is allowed to vary between 0.2 and 4.0 and the initial poloidal field angle θ_0 at the base of the corona ($Z = 0$) is allowed to vary between 8° and 83° [$\log(\tan\theta) = -0.861$ to 0.889 in steps of 0.25].

For boundary conditions, we assume "flow through" conditions on the $Z = Z_{\max}$ and $R = R_{\max}$ outer boundaries and cylindrical symmetry at $R = 0$. The $Z = 0$ boundary condition perpetuates the initial conditions by continually injecting magnetized coronal material in a low velocity wind ($V_{\text{wind}}/V_{\text{esc}} \sim 0.05$). The simulations were dimensionless ($G = M = 1.0$). To re-introduce scales, one must specify a measure of the compact object radius R_g, and the Keplerian velocity at R_g [$V_g = (GM/R_g)^{1/2}$], with the unit of time being their ratio.

3.2. The code

The simulations were performed using the code FLOW (Lind et al., 1989) on the Caltech Intel Touchstone Delta parallel supercomputer using 150 radial by 300 axial zones. Tests of HD and MHD against analytic solutions indicate that the code maintains second-order accuracy in space and time. The main limitations are boundary reflection effects, which limit the total evolution time, and the cylindrical $R = 0$ condition, which generates a small error in flow near the axis. Both affect very low V_A/V_{esc} flows. Nevertheless, in this regime our simulations obtain results similar to others who computed low to moderate Alfvén velocity flows (Ustyugova et al., 1995; Ouyed, these proceedings).

3.3. Relativistic flow

As long as $V_A^2 < c^2$ and $V_{\text{sound}}^2 < c^2$, even a non-relativistic code can be used to investigate relativistic flow. If, in the relativistic fluid equations, the inertia of the magnetic and thermal energy densities are ignored and the equations are written in terms of the three spatial components of the four-velocity ($\mathbf{u} \equiv \Gamma\mathbf{V}$), then their form is *identical* to that of the non-relativistic equations, save for one term—the electric force normal to the magnetic field lines—but this also does not become important until $V_A^2/c^2 \sim 1$. Note that, even when this term is important, it does not change the velocity or angular momentum along a field line, but does determine the exact angle and shape of the field lines (LCB). In addition, even when $V_A^2/c^2 \sim 1$, including the inertia terms would make the fluid only twice as heavy as in our simulations and the electric force would modify the tangent of the magnetic field by, at most, about a factor of 2. Since we investigate a wide range of magnetic field angle, and find similar effects throughout, we conclude that our non-relativistic MHD code, with suitable transformations and care in analyzing the results, can be used to investigate flow for $V_A/c \sim 1$ and $\Gamma_{\text{jet}} \to \infty$.

4. Results

Our results are reported and discussed in more detail elsewhere (Meier et al., 1997). Here we give a brief account of them.

4.1. Typical model

Most thoroughly studied was the $\theta_0 = 54°$ branch, with V_A/V_{esc} ranging from 0.2 to 1.5. In addition, we computed a $V_A/V_{\text{esc}} = 0$ case as a control experiment, which produced no outflow of any kind. The $V_A/V_{\text{esc}} = 1.2$ case, which we studied most, produced a strong, well-collimated jet after only two inner disk rotation periods. The internal flow velocity of the jet was $\sim 19\, V_{\text{esc}_0}$, where V_{esc_0} is the escape velocity measured at the point in the disk where the coronal field has a maximum value ($R_0 = 7.2$). Because of the high velocity, with the mass flux limited to that injected at the base of the corona, the resulting jet density was very light ($\rho_{\text{jet}}/\rho_{\text{ext}} \equiv \eta \sim 10^{-3}$), so its actual propagation velocity across the grid was only $\sim 1\, V_{\text{esc}_0}$. The calculation was continued to late times (10 inner disk rotation periods), well after the jet propagated off the $Z = Z_{\text{max}}$ boundary and reached a steady state.

4.2. Variations with magnetic field strength: The magnetic switch

Additional calculations were performed with higher and lower magnetic field strengths. Increasing V_A to 1.5 V_{esc} increased the jet speed by about 70%, but *decreasing* V_A to 0.7 V_{esc} decreased the internal jet speed by a factor of ~ 40 to 0.5 V_{esc_0}! Further decreases in the V_A/V_{esc} parameter resulted in jets with speeds of a few tenths of V_{esc_0}. There apparently is a boundary at $V_A/V_{\text{esc}} \sim 1$, below which jets have velocities $\sim V_A$, but above which the jets attain high speeds. For classification purposes, we shall call the low-speed jets Type 1 flows and the high-speed jets Type 2 flows.

A simple physical explanation of this "magnetic switch" is as follows. When $V_A/V_{\text{esc}} < 1$, the flow is dominated by gravitational, not magnetic, forces. The jet is accelerated out of the potential well by the recoil of the toroidal magnetic pressure, similar to a Parker wind. When $V_A > V_{\text{esc}}$, magnetic forces dominate and gravitational effects are unimportant. The jet accelerates via the magneto-centrifugal effect, as occurs in pulsar wind solutions.

4.3. Variations with initial poloidal field angle

We investigated the magnetic switch effect further for a wide range of poloidal field angles. We found that the sharp transition from Type 1 to Type 2 jets occurs for all magnetic field angles investigated. Even the $\theta < 30°$ cases produced Type 2 jets if the magnetic field strength was great enough, and Type 1 jets otherwise. Therefore, while the magnetic switch is a strong function of field strength, it is not a strong function of the angle of the field at the base of the corona.

Why is this the case? Steady-state solutions predict no outflow for $\theta < 30°$. The answer is in the non-steady development of the jets at early times. Initially, the field is twisted by differential rotation, yielding a strong toroidal component. The hoop stress from this component drives an inward flow just above the disk, pinching the field lines inward, enhancing a $\theta > 30°$ condition for those with $\theta_0 > 30°$ *and producing such a condition even when $\theta_0 < 30°$ initially.* Having achieved this critical Blandford-Payne condition dynamically, the system proceeds to drive an outflow which eventually collimates. Following the flow along a field line, we find it first bends inward toward the axis just above the disk and then turns outward to produce the outflow and jet. The importance of the angle θ_0 in non-steady simulations, then, is not that it is above or below 30°, but that it determines how much of the magnetic field is radial and hence can be wound into a toroidal field to initiate this process.

The very small angle cases ($\theta_0 \sim 8°$–$14°$) show an additional effect. As θ_0 decreases and keeping the magnetic field strength constant, the strength of the jet diminishes and an uncollimated or poorly-collimated wind appears from much of the disk. The wind velocity also seems to exhibit the magnetic switch effect as well, with its velocity low when $V_A/V_{esc} < 1$ and high otherwise.

5. Discussion

From our simulations we find that jets are ubiquitous, being produced by magnetized accretion disks in a wide range of parameter space, and that the magnetic switch is a robust effect as well. Scaling our results to the black hole and protostellar cases, we find the following. For the black hole case, making the transformation $\mathbf{V} \to \Gamma \mathbf{V}$, we find jet velocities of $V_{jet} \sim 0.2$–$0.5\,c$ when the switch is off, and $\Gamma_{jet} \sim 5$–20 when the switch is turned on. For the protostar case, $V_{jet} \sim 80$–$200\,\mathrm{km\,s^{-1}}$ when the switch is off and $V_{jet} \sim 2{,}000$–$8{,}000\,\mathrm{km\,s^{-1}}$ when turned on. Whether or not such high velocity protostellar jets occur will depend on their accretion disks developing hot and readily-replenishable coronae.

Our results are consistent with both the semi-analytic results of LCB (when suitable relativistic transformations are taken into account) and of Ustyugova *et al.* (1995), and Ouyed (these proceedings) in their respective regions of parameter space.

Applications of the magnetic switch may be considerable. The galactic superluminal source GRO J1655-40 displayed such a bimodal behavior, remaining a weak radio source until ~ 12 days after the X-ray outburst (Harmon *et al.*, 1995; Tingay *et al.*, 1995), and only then producing a strong radio jet. The ejection of the jet appeared to coincide with the formation of a hot corona and a drop in the X-ray flux, both indicators that the switch may have turned on because of processes occurring when accreting near the Eddington limit (Meier, 1997).

The strength of jets in some active galactic nuclei may also be explained by this mechanism. The properties of FR I and FR II radio sources closely resemble those of Type 1 and Type 2 jets, respectively (Meier *et al.*, 1997). We are working on a more detailed analysis to see if the mechanism can be applied to the Ledlow & Owen (1996) relation, for example. Finally, such a mechanism also may provide an alternative model for the broad absorption line (BAL) quasars—another class of radio weak objects with strong winds (Stocke *et al.*, 1992). The small incidence of BALs in the quasar population would not be caused by covering factor but by the low incidence of super-Eddington accretion in quasars. The anomalous radio weakness of the BAL class may be because of the presence of a super-Eddington wind, which would shut off the magnetic switch.

Acknowledgments. Part of this research was carried out at the Jet Propulsion Laboratory, under contract to NASA. S. E. acknowledges the support of an Associated Western Universities Summer Fellowship. During this research, P. G. held a National Research Council Research Associateship at the NASA Jet Propulsion Laboratory.

References

Bell, A. R., & Lucek, S. G. 1995, MNRAS, 277, 1327.
Blandford, R. D., & Payne, D. G. 1982, MNRAS, 199, 883.
Brandenburg, A., Nordlund, A., Stein, R. F., & Torkelsson, U. 1996, ApJ, 458, L45.
Harmon, B. A., Wilson, C. A., Zhang, S. N., Paciesas, W. S., Fishman, G. J., Hjellming, R. M., Rupen, M. P., Scott, D. M., Briggs, M. S., & Rubin, B. C. 1995, Nature, 374, 703.
Koide, S., Shibata, K., & Kudoh, T. 1997, in *Accretion Phenomena and Related Outflows*, IAU Colloq. 163, in press.
Kudoh, T., & Shibata, K. 1996, ApJ, 452, L41.
Ledlow, M. J., & Owen, F. N. 1996, AJ, 112, 9.
Li, Z.-Y., Chiueh, T., & Begelman, M. C. 1992, ApJ, 394, 459 (LCB).
Lind, K. R., Payne, D. G., Meier, D. L., & Blandford, R. D. 1989, ApJ, 344, 89.
Lovelace, R. V. E., Mehanian, C., Mobarry, C. M., & Sulkanen, M. E. 1986, ApJS, 62, 1.
Matsumoto, R. 1997, in *Accretion Phenomena and Related Outflows*, IAU Colloq. 163, in press.
Meier, D. L. 1997, in *Accretion Phenomena and Related Outflows*, IAU Colloq. 163, in press.
Meier, D. L., Edgington, S. F., Godon, P., Payne, D. G., & Lind, K. R. 1997, Nature, submitted.
Shakura, N. I., & Sunyaev, R. A. 1973, A&A, 24, 337.
Stocke, J. T., Morris, S. L., Weymann, R. J., & Foltz, C. B. 1992, ApJ, 396, 487.
Stone, J. M., Hawley, J. F., Gammie, C. F., & Balbus, S. A. 1996, ApJ, 463, 656.
Stone, J. M., & Norman, M. L. 1994, ApJ, 433, 746.
Tingay, S. J., *et al.* 1995, Nature, 374, 141.
Uchida, Y., & Shibata, K. 1985, PASJ, 37, 515.
Ustyugova, G. V., Koldoba, M. M., Romanova, V. M., Chechetkin, V. M., & Lovelace, R. V. E. 1995, ApJ, 439, L39.

Numerical Simulations of Rotating Accretion Flows near a Black Hole

Dongsu Ryu[1]

Department of Astronomy, University of Washington, Seattle, WA 98195, U.S.A.

Sandip K. Chakrabarti

S. N. Bose Center for Basic Sciences, Calcutta 700091, India

Abstract. We present time-dependent solutions of thin, supersonic accretion flows near a black hole and compare them with analytical solutions. Such flows of inviscid, adiabatic gas are characterized by the specific angular momentum and the specific energy. We confirm that for a wide range of the above parameters, a stable standing shock wave with a vortex inside it forms close to the black hole. Apart from steady-state solutions, we show the existence of non-steady solutions for thin accretion flows where the accretion shock is destroyed and regenerated periodically. The unstable behavior should be caused by dynamically induced instabilities, since inviscid, adiabatic gas is considered. We discuss the possible relevance of the periodic behavior on quasi-periodic oscillations observed in galactic and extragalactic black hole candidates.

1. Introduction

Rotating accretion flows are important ingredients in many astrophysical systems containing a black hole, which involve mass transfer from one object to another (such as in a binary system), or from one set of objects to another (such as in a galactic center). The standard disk model of such accretion flows by Shakura & Sunyaev (1973) assumes a Keplerian distribution of accreting matter. The inner edge of the disk is chosen to coincide with the marginally stable orbit located at three Schwarzschild radii, $r_i = 3R_g$, where the Schwarzschild radius, $R_g = 2GM_{\rm BH}/c^2$, is the horizon of a black hole of mass $M_{\rm BH}$. This disk model is clearly incomplete, since the inner boundary condition on the horizon is not taken care of. As an accretion flow approaches the horizon, its radial velocity reaches the velocity of light. Therefore, a black hole accretion flow is necessarily supersonic and must pass through a sonic point where the flow has to be sub-Keplerian. Thus, independent of heating and cooling processes, a black hole accretion has to deviate from a standard Shakura-Sunyaev type Keplerian disk. The disk with realistic accretion flows is called the advective disk (Chakrabarti,

---

[1] also, Department of Astronomy and Space Science, Chungnam National University, Korea

1996). Here, we present the results of numerical study of accretion flows near a black hole by assuming they are thin, axisymmetric, and inviscid.

2. Analytic Consideration

We choose cylindrical coordinates (r, θ, z) and place a black hole at the center. We assume that the gravitational field of the black hole can be described in terms of the potential introduced by Paczyński & Wiita (1980)

$$\phi(r,z) = -\frac{GM_{\text{BH}}}{R - R_g}, \quad (1)$$

where $R = \sqrt{r^2 + z^2}$. The accreting matter is assumed to be adiabatic gas without cooling and dissipation and described by $P = K\rho^\gamma$, where γ is the adiabatic index which is considered to be constant with $\gamma = 4/3$ throughout the flow. K is related to the specific entropy of the flow, s, and varies only at shocks, if present. Since we consider only the weak viscosity limit, the specific angular momentum of the accretion flow, $\lambda = rv_\theta$, is assumed to be conserved. Thus, unlike a Bondi flow (Bondi, 1952) which is described by a single parameter (say, specific energy), the one-dimensional accretion flows are described by two parameters which are the specific energy, $\mathcal{E}$, and the specific angular momentum, λ. Figure 1 shows the classification in parameter space (Chakrabarti, 1989; Ryu & Chakrabarti, in preparation), as tabulated below:

N: No sonic points. Shock only if supersonic injection.
O: Outer sonic point only as in a Bondi solution. No shock.
I: Inner sonic point only. Shock only if supersonic injection.
O*: Outer and center sonic points. Solution does not extend to the horizon.
I*: Inner and center sonic points. Solution does not extend to large distance.
SA: Two (outer and inner) sonic points. Shock in accretion solutions but not in wind solutions.
SW: Two sonic points. Shock in wind solutions but not in accretion solutions.
NSA: Two sonic points. No shock condition satisfied in accretion solutions.
NSW: Two sonic points. No shock condition satisfied in wind solutions.

Although these solutions are strictly valid for inviscid flows, given that the viscous time scale is likely to be much larger compared to the infall time scale even when viscosity is high, the inviscid solutions are likely to remain important.

3. One-Dimensional Numerical Solutions

Figure 2 shows an example numerical solution from the "SA" region. We superposed analytical solutions (solid lines) and numerical solutions with TVD (dashed lines) and SPH (dotted lines) codes (Molteni et al., 1996). The TVD

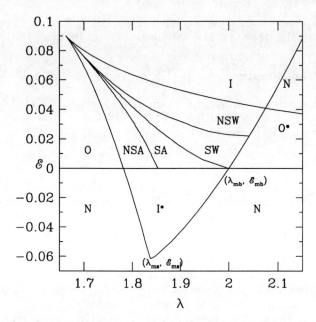

Figure 1. Classification of one-dimensional accretion flows in the parameter space of specific energy, $\mathcal{E}$, and specific angular momentum, λ. $\mathcal{E}$ is given in units of c^2 and λ in units of $R_g c$.

calculation was done with 512 zones and the SPH calculation was done with ~ 560 particles of size $h = 0.3 R_g$. Matter was injected at the outer boundary located at $x = 50 R_g$. An absorption condition was used to mimic the black hole horizon at the inner boundary which is located either at $x = 1.5 R_g$ for the TVD calculation or at $x = 1.25 R_g$ for the SPH calculation. The adiabatic index $\gamma = 4/3$ was used.

Figure 2 shows excellent agreement between the analytic and numerical solutions. Here, the flow starts out subsonically, presumably from a Keplerian disk. Then, it enters through the outer sonic point (located at $x = 27.97 R_g$), passes through the shock (located at $x = 7.98 R_g$), and subsequently passes through the inner sonic point (located at $x = 2.57 R_g$) before entering the black hole.

4. Two-Dimensional Numerical Solutions

In multi-dimensional accretion flows, non-steady solutions, as well as steady solutions, exist. Here, we discuss some examples of accretion flows with zero specific energy ($\mathcal{E} = 0$). More extensive discussion on accretion flows with $\mathcal{E} = 0$ was reported in Ryu et al. (1997). Discussion of extensive calculations for accretion flows with non-zero specific energy ($\mathcal{E} \neq 0$) will be reported elsewhere (Ryu & Chakrabarti, in preparation).

In these simulations, we inject supersonic matter (with a radial Mach number $M = v_r/a = 10$) at the outer boundary, $r_b = 50 R_g$. The inflow at the

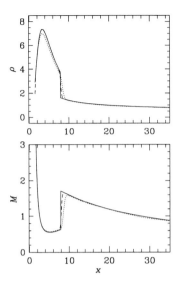

Figure 2. One-dimensional accretion flow with a standing shock in the region of "SA". $\mathcal{E} = 0.036$ and $\lambda = 1.80$ are used. The solid line is the analytical solution, and the long and short dashed lines are the solutions of the TVD and SPH simulations, respectively. Upper panel is the mass density in arbitrary units and the lower panel is the Mach number of the flow.

outer boundary is assumed to have a small thickness, h_{in}, or a small arc angle, $\theta_{in} = \arctan(h_{in}/r_b) \ll 1$. If zero-energy accretion flows occur in the region of "O", most of the material is accreted into the black hole, forming a stable quasi-spherical flow or a simple disk-like structure around it (just as a Bondi flow). If accretion flows belong to the region of "SA" and "O*", the incoming material produces a stable standing shock with one or more vortices behind it and some of it is deflected away at the shock as a conical outgoing wind of higher entropy.

Accretion flows with parameters in the region "NSA" with $1.782 R_g c < \lambda < 1.854 R_g c$ show an unstable behavior. Figure 3 shows an example numerical solution with $\lambda = 1.85 R_g c$. In this case, structure with an accretion shock and a generally subsonic high-density disk is established around the black hole. However, the structure is not stable. At the accretion shock, the incoming flow is deflected but some of the post-shock flow, which is further accelerated by the pressure gradient behind the shock, goes through a second shock where the flow is deflected once more downwards. The downward flow squeezes the incoming material, and the accretion shock starts collapsing ($t = 1.4 \times 10^4 R_g/c$). In the process of the collapse, some of the post-shock material escapes as wind but most is absorbed into the black hole. After the collapse, the rebuilding of the accretion shock starts with the incoming material bouncing back from the centrifugal barrier. The subsonic post-shock region becomes a reservoir of material, so the material is accumulated behind the shock. With the accumulated material, a giant vortex is formed, which in turn supports the accretion shock ($t = 2 \times 10^4 R_g/c$). This continues until the incoming flow is squeezed enough so that the accretion shock collapses, and the cycle continues.

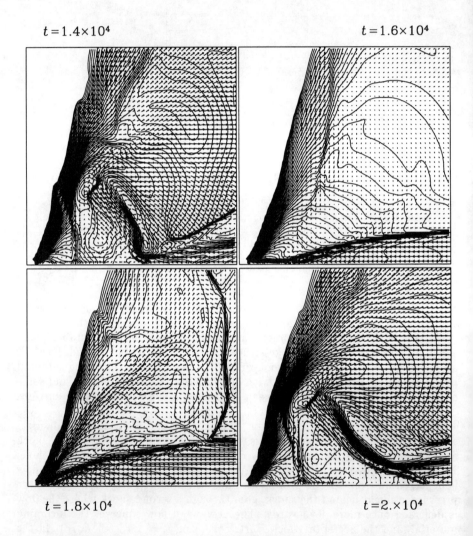

Figure 3. Two-Dimensional accretion flow showing an unstable behavior in the region of "NSA". $\mathcal{E} = 0$ and $\lambda = 1.85$ are used. Contours are for density and arrows are for velocity vector. Time is given in units of R_g/c.

5. Discussion

The time scale of the periodicity of unstable flows is interesting. It is in the range of

$$\tau \approx 4 - 6 \times 10^3 \frac{R_g}{c} = 4 - 6 \times 10^{-2} \left(\frac{M_{\rm BH}}{M_\odot}\right) \text{ s.} \quad (2)$$

The modulation of amplitude is also very significant, and could be as much as 100%, depending on detailed processes. Oscillations with these characteristics have been observed in black hole candidates and are called Quasi-Periodic Oscillations (QPO). For instance, in the QPO from the low-mass X-ray binaries, the oscillation frequency has been found to lie typically between 5 and 60 Hz (Van der Klis, 1989). Thus, compact objects with mass $M \approx 0.3$–$5 M_\odot$ could generate oscillations of the right frequencies because of the instability discussed in this paper. Similar oscillations of period on the order of a few hours to a few days are expected in soft X-rays and UV emissions from galactic centers. However, by considering simplified physics which we have assumed here, it may be premature to assume that the presented mechanism would explain all the QPO observed. In some cases, the oscillation may be caused by the dynamic instability considered in Ryu et al. (1995), or it may be caused by resonance of the cooling time scale (bremsstrahlung or Comptonization, whatever the case may be) and the infall time scale in the enhanced density region near the centrifugal barrier, as shown by Molteni et al. (1996). In detailed works, radiative processes as well as viscosity should be included in the accretion calculations to examine the observational consequences of the present instability.

Acknowledgments. The work by DR was supported in part by Seoam Scholarship Foundation.

References

Bondi, H. 1952, MNRAS, 112, 195.
Chakrabarti, S. K. 1989, ApJ, 347, 365.
Chakrabarti, S. K. 1996, Phys. Rep., 266, 229.
Molteni, D., Ryu, D., & Chakrabarti, S. K. 1996, ApJ, 470, 460.
Molteni, D., Sponholz, H., & Chakrabarti, S. K. 1996, ApJ, 457, 805.
Paczyński, B., & Wiita, P. J. 1980, A&A, 88, 23.
Ryu, D., Brown, G. L., Ostriker, J. P., & Loeb, A. 1995, ApJ, 452, 364.
Ryu, D., Chakrabarti, S. K., & Molteni, D. 1997, ApJ, 474, 378.
Shakura, N. I., & Sunyaev, R. A. 1973, A&A, 24, 337.
Van der Klis, M. 1989, ARA&A, 27, 517.

Part VI

COSMOLOGY AND NUMERICAL RELATIVITY

The Binary Black Hole Grand Challenge Project

M. W. Choptuik
Center for Relativity, Department of Physics, University of Texas at Austin, Austin, TX 78712-1081, U.S.A.

Abstract. The status of the Binary Black Hole Grand Challenge project is reviewed. This effort, which has involved over 40 researchers at 10 institutions, aims to simulate the in-spiral and merger of black hole binaries, and to provide predicted gravitational waveforms from such events. Conceivably, radiation from black-hole mergers could be detected within the next decade by the new generation of large scale gravitational wave detectors currently under construction. This report summarizes advances which have been made on physical, mathematical and computational aspects of the problem, and outlines some of the key hurdles which remain.

1. Introduction

The problem of binary black hole (BBH) coalescence is currently the focus of a large fraction of the active researchers in numerical relativity. It is widely believed that BBH mergers rank among the most copious generators of gravitational waves, and hopes for detection of such events by next-generation instruments, such as LIGO, run high. Conventional wisdom suggests that detailed waveform predictions (templates) from numerical simulations will be crucial for early detection of these events, and this has provided much of the impetus for the current flury of activity devoted to the BBH problem. Binary coalescence is also of fundamental physical importance in terms of its central role in our understanding of the general-relativistic two-body problem. Finally, the BBH problem is very much a watershed calculation for numerical relativity; its successful solution requires machinery which should largely provide the means to generate truly general solutions of Einstein's equations.

Much of the current BBH research is being carried out under the auspices of an NSF "Grand Challenge" grant, awarded to a multi-institution team of researchers led by R. A. Matzner (Matzner, 1997). Here it should be noted that the Grand Challenge program funded projects which, besides being of significant intrinsic scientific interest, require the use of extremely large-scale computational resources for their solution. Projects were also selected using the criterion that collaborative efforts—between computer scientists/engineers and application scientists—were likely to lead to significant developments in computational infrastructures which could be used by the computational science community at large. In this context, it should also be observed that the BBH project is a first for numerical relativity in terms of the degree to which computer scientists (such as G. Fox at Syracuse, and J. Browne and M. Parashar at UT Austin) have been directly involved.

2. Magnitude of the Computational Problem

Following Finn (1996), the basic magnitude of the BBH computational problem may be estimated as follows. We assume that the binary consists of equal mass black holes, with a total mass M, initially in a circular orbit with diameter $\sim 6M$. The quadrupole radiation from this source has a wavelength of order $100M$, and to read off the radiation reliably, the outer boundary of the computational domain must be placed at least one wavelength from the source. If we assume that we need a finite-difference (FD) mesh spacing $\sim M/20$ to resolve the black holes themselves, and that we will use a uniform 3-D FD mesh, the total number of grid points used in the calculation will be $\sim 10^{10}$. Typical 3-D codes require 50–100 floating-point numbers per grid-point, which, for standard 8-byte arithmetic, implies a total storage requirement of 10^{12}–10^{13} bytes!

CPU requirements are just as prodigious. For our fiducial binary, the estimated physical time for a couple of orbits, coalescence and ring-down is $\sim 500M$. This implies that a simulation will involve $\sim 10^4$ discrete time steps, each of which will require on the order of 5,000 floating-point-operations (flops) per grid point. This gives a total of about 5×10^{17} flops/simulation or about a CPU-week on a Teraflop/s machine.

Although machines with capacities commensurate with these requirements should be available within the next few years, it is clear that we will not have dedicated access to such facilities and thus the reduction of the computational burden, particularly using adaptive mesh refinement techniques, is a high priority.

3. Formalisms for Numerical Relativity

The traditional approach to general-relativistic simulations has employed the ADM (3+1) formalism (Misner et al., 1973; York, 1979), in which the geometry of spacetime is viewed as the "time history" of the geometry of a spacelike hypersurface ("instant of time"). In the ADM approach, just as the geometry of spacetime is described by a 4-metric $^{(4)}g_{\mu\nu}$, so is the geometry of a spacelike hypersurface described by a 3-metric g_{ij} (Latin indices run over spatial values 1, 2, 3; Greek indices run over space-time values 0, 1, 2, 3). In particular, the spacetime-displacement-squared can be written (see Figure 1)

$$^{(4)}ds^2 = {}^{(4)}g_{\mu\nu}dx^\mu dx^\nu$$
$$= -\alpha^2 dt^2 + g_{ij}\left(dx^i + \beta^i dt\right)\left(dx^j + \beta^j dt\right), \quad (1)$$

where α (the lapse function) and β^i (the shift vector), which in principle, can be chosen almost arbitrarily, represent the 4-fold coordinate freedom of general relativity. Another tensor of prime importance in the ADM formalism is the extrinsic curvature (second fundamental form), K_{ij} which, loosely speaking,

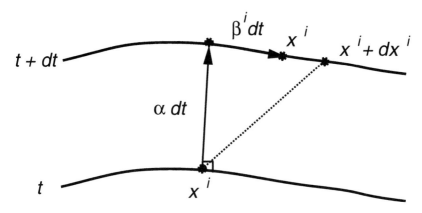

Figure 1. Space-time displacement in 3+1 language. See the text for more details.

may be viewed as the "velocity" of g_{ij}:

$$K_{ij} = \frac{1}{2\alpha}\left(-\frac{\partial g_{ij}}{\partial t} + D_i \beta_j + D_j \beta_i\right), \qquad (2)$$

where D_i is the 3-covariant derivative (i.e., $D_i g_{ij} = 0$).

As is well known, the Einstein equations (restricted here to the case of a vacuum),

$$G_{\mu\nu} = 0, \qquad (3)$$

naturally decompose into two sets. Four of the equations,

$$G_{0\nu} = 0, \qquad (4)$$

do *not* involve time derivatives of the K_{ij}, and thus represent equations of constraint which must be satisfied at all times, including the initial time (next section). The remaining 6 equations are the evolution equations, *per se*, for the gravitational field. Written in first-order-in-time form, they are:

$$\frac{\partial g_{ij}}{\partial t} = -2\alpha K_{ij} + D_i \beta_j + D_j \beta_i, \qquad (5)$$

$$\frac{\partial K_{ij}}{\partial t} = \mathcal{L}_\beta K_{ij} - D_i D_j \alpha + \alpha\left(R_{ij} - 2K_{ik}K^k{}_j + K_{ij}K\right), \qquad (6)$$

where R_{ij} is the 3-Ricci tensor, $K \equiv K^i{}_i$, and $\mathcal{L}_\beta$ is the Lie derivative along β^i. It is worth noting that the vast majority of numerical relativity codes which have been constructed over the past few decades have been based on the ADM formalism.

Successful as the ADM approach has been in numerical relativity, the general form of the ADM equations has been of considerable concern to many researchers. In particular, the ADM form of Einstein's equations is generically non-hyperbolic. Roughly speaking, a hyperbolic formulation of a system of evolution equations is one in which characteristic speeds are either 0, or the local light speed, so that information manifestly propagates along light cones. In addition, hyperbolicity generically allows the evolution equation to be cast into a so-called "flux conservative form":

$$\partial_t u + \partial_i F^i(u) = S(u), \qquad (7)$$

where the source term, $S(u)$, does not involve any spatial derivatives of the dynamical variables, u. In the past few years, there has been tremendous progress made in re-casting Einstein's equations in hyperbolic form (Bona *et al.*, 1995; Abrahams *et al.*, 1995; Van Putten & Eardley, 1996). Possible benefits of such an approach range from improved treatment of both inner and outer boundaries in black hole collisions, to the *en masse* appropriation of the large body of advanced numerical techniques which have been developed (particularly in the context of computational fluid dynamics) for flux-conservative systems. It is still too early to tell whether the hyperbolic approaches will live up to their promise, but it is abundantly clear that research in this area will be followed with intense interest by the numerical relativity community.

4. The Initial-Value Problem for Two Black Holes

As mentioned above, equations (4) are equations of constraint which, in a given coordinate system, become a system of four coupled, non-linear elliptic equations in the dynamical variables $\{g_{ij}, K_{ij}\}$. Thus, even the problem of determining *initial data* for general-relativistic simulations is decidedly non-trivial. Historically, a key issue has been deciding which of the $\{g_{ij}, K_{ij}\}$ should be freely specified, and which should be determined via solution of the constraints. One general strategy, worked out in the 70's and early 80's by York and various collaborators (York & Piran, 1982), is based on ideas of conformal scaling and a "spin-decomposition" of K_{ij}. This approach allows the constraints to be cast as a system of quasi-linear elliptic equations for four potentials, $\{\psi, X^i\}$, which may be viewed as the relativistic generalization of the familiar Newtonian gravitational potential. In many cases, the equations for the potentials X^i can be solved analytically, leaving only the equation for ψ (the Hamiltonian constraint) to be solved numerically. The state-of-the-art in Hamiltonian constraint solvers is quite advanced and typically involves second-order finite-difference equations solved via a multi-grid algorithm. With a careful choice of coordinates and discretization (Cook *et al.*, 1993), higher order accuracy can be achieved via Richardson extrapolation and, at least from the computational point of view, the initial value problem for two black holes can be considered solved. From a physical point of view, however, a crucial question remains unanswered: How do we generate *realistic* initial data for a black hole collision? In particular, we do not yet have a prescription for specifying data for a relatively tight binary (needed to minimize simulation cost) which accurately represents the late-time configuration of a binary which started in some well-separated, quasi-Newtonian orbit.

5. Black Hole Excising Techniques

The fact that black hole spacetimes contain physical singularities has had a profound effect on numerical relativity. The traditional approach, pioneered by DeWitt and his students in the mid-70's, has been to use coordinate freedom (choice of lapse and shift) to "freeze" evolution near a physical singularity, while allowing the dynamics in regions far-removed from the singularity to be pushed ahead. Consider, for example, the case of a spherically-symmetric collapse of some initial matter distribution which leads to black hole formation. Then we can choose coordinates, r and t, such that the metric takes on a "time-dependent-Schwarzschild" form:

$$ds^2 = -\alpha(r,t)^2 dt^2 + \left(1 - \frac{2m(r,t)}{r}\right)^{-1} dr^2 + r^2 d\Omega^2. \tag{8}$$

When black hole formation is imminent, one finds that near the Schwarzschild radius ($r = R_S$), the lapse function, $\alpha(r,t)$, rapidly drops to zero, effectively "freezing" the dynamics in the vicinity of the singularity. Unfortunately, one also finds that the radial metric function, $(1 - 2m/r)^{-1}$, and its spatial derivatives grow without bound for $r \approx R_S$. Thus, the evolution avoids the *physical* singularity, but at the price of generating a *coordinate* singularity. Moreover, from the point of view of numerical simulation, the coordinate singularity is just as pathological as the physical singularity, and a code using these coordinates will "crash" on a dynamical time scale. Although a variety of other singularity-avoiding coordinate choices have been used in black hole studies, without exception they induce coordinate singularities which render them useless for the long-time evolution of black holes.

In the early 80's, Unruh struck on a possible solution to this serious problem by observing that since the interior of a black hole is, by definition, out of causal contact with the exterior spacetime, it should be possible simply to exclude the regions within event horizons from the computational domain. Further, since the event horizon cannot be located until the construction of the spacetime is complete, Unruh suggested that *apparent horizons* be used as inner boundaries for black hole calculations. (An apparent horizon is a closed 2-surface, defined at some instant at time, such that the outgoing light rays emanating from the surface "hover" at constant radius.) Thornburg (1993) aggressively pursued this "excising" strategy in his thesis research on axisymmetric systems, but it was the spherically-symmetric work of Seidel & Suen (1992) which first convincingly demonstrated the efficacy of the approach. Several other spherically-symmetric calculations (Scheel et al., 1995; Anninos et al., 1995; Marsa & Choptuik, 1996) have furthered our confidence that the technique should be generically applicable, and all of the 3-D codes currently being developed within the Grand Challenge project implement black-hole excising. Preliminary indications suggest that the approach is viable in the generic 3-D case, and thus it seems very likely that excising (and related techniques) will have an enormous impact on our ability to carry out accurate, long-term black-hole integrations.

6. Computational Infrastructure

As noted in the Introduction, a primary aim of the Grand Challenge program was the development of software infrastructures which make effective use of cutting-edge hardware, with a specific emphasis on massively parallel machines. Particularly in comparison with traditional vector supercomputers, such machines tend to be difficult to use to full advantage, even on problems with a large amount of inherent parallelism. Consequently, the BBH alliance has devoted considerable energy and resources to the development of software support which facilitates the efficient use of massively parallel machines for regular, grid-based computations.

The first key observation which underlies much of the parallel-support software developed by the alliance, is the fact that numerical relativity codes have tended to be remarkably homogeneous from a "high-level" viewpoint. Specifically, almost all codes have employed low-order (typically second order in the mesh spacing) finite-difference techniques on a single mesh, and have had the following general structure:

```
Read initial state
for NUM_STEPS
   for NUM_UPDATES & possibly until convergence
      U [Grid Function(s)] -> Grid Function(s)
   end for
end for
Write final state
```

From the numerical relativist's point of view, most of the hard work in developing a new code involves the construction of *stable*, accurate updates, schematically denoted U above, which advance one or more grid functions (discrete representations of continuum unknowns) from one discrete time to the next. Although alternative approaches to the discretization of Einstein's equations have often been proposed and (less often) investigated, the low-order, uniform-mesh finite difference approach remains dominant, not least because of its flexibility and familiarity.

The second main point which crucially affected the design of the computational infrastructure was the clear need for adaptive mesh refinement (AMR) techniques. As briefly discussed in § 2, the BBH problem is characterized by significant dynamic range, and even the most superficial analyses indicate that the savings by effective use of AMR—where the scale of discretization varies *locally* in response to the development of solution features—are potentially enormous. However, the use of AMR techniques is not without drawbacks: as with the task of parallelization, the significant increase in algorithmic complexity is arguably the most severe. Thus, an ultimate goal of the infrastructure development initiated by the alliance was to implement mechanisms providing parallelism and adaptivity as automatically as possible, thus allowing the relativist to concentrate on developing stable, single-grid codes on serial machines.

There are many possible approaches to AMR, but the algorithm by Berger & Oliger (1984) is particularly commensurate with the aim of *automatically* extending adaptivity to any stable, single-grid, finite-difference code. In Berger & Oliger AMR, adaptivity is achieved using a hierarchy of component meshes,

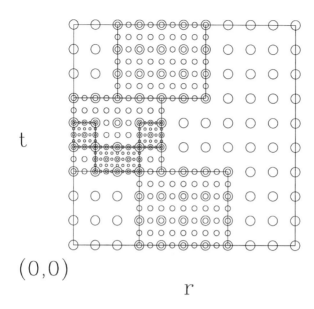

Figure 2. Schematic illustration of Berger and Oliger mesh-refinement structure for a problem in one space dimension and time. See the text for more details.

each of which is separately uniform, *i.e.*, has a constant mesh spacing in each coordinate direction. Figure 2 shows a schematic illustration of the type of mesh structure admitted by the algorithm, for the case of a problem in one space and one time dimension, and with a mesh refinement ratio of 2:1. Note that the refinement occurs both in space *and* in time, and that the total number of (spatial) grid points is a dynamic quantity. The placement of component grids is typically controlled via the generation of local truncation error estimates: new grids are introduced where error estimates are high, and are deleted if and when the error estimates drop below some user-specified threshold.

Working from the specifications sketched above, Parashar & Browne (1995a; 1995b) have implemented an MPI-based infrastructure called DAGH (for Distributed Adaptive Grid Hierarchy), which provides full support for (1) parallelization (on distributed-memory architectures) of regular, lattice-based calculations with local communication patterns, and (2) Berger & Oliger style AMR. Although the system is implemented in C++, is has been designed to interface easily with update routines written in Fortran 77, Fortran 90 or C. Key features of the DAGH system include

- Automatic dynamic partitioning and load distribution.

- Implementation of a "shadow" grid hierarchy for memory-efficient truncation error estimates.

- Maintenance of data locality among different refinement levels using a linearization technique based on space-filling curves.

Currently, the DAGH system runs on networks of workstations, as well as on massively parallel machines such as the IBM SP-2, the Cray T3D and T3E, and the SGI Power Challenge Array and Origin 2000. Although evaluation and debugging of DAGH continues, it is now used extensively by BBH researchers, as well as by a growing number of other scientists in fields such as computational fluid dynamics. Typical benchmarks of properly constructed DAGH code show very good performance; in several cases the DAGH application runs faster than "hand-coded" MPI.

7. Current Status and Outstanding Problems

At the time of the meeting, BBH development involved two main codes. Both were designed to solve the vacuum Einstein equations in three space dimensions plus time, using fairly arbitrary coordinate conditions. The first of these, called the ADM code, is based on the traditional 3+1 formalism, is currently implemented in Fortran 90, and runs on essentially any architecture. The second, called the Empire code, is based on the Choquet-Bruhat/York hyperbolic formalism, is implemented on top of DAGH and runs almost exclusively on the Cornell SP-2. Both codes implement black-hole excising, as well as a technique known as "causal-differencing" (Seidel & Suen, 1992), which is intended to ensure that the numerical domain of dependence includes the physical domain of dependence (a necessary condition for stability). Memory availability limits resolution to about 128^3 in both cases, although most developmental runs are done at significantly lower resolution (often 33^3 or even coarser). However, since the meeting, a decision to shelve the Empire effort and focus exclusively on the ADM code has been made. This action, largely motivated by the fact that there are relatively few researchers (mostly postdocs and graduate students) available to work on *either* code on an intensive basis, has already paid off in terms of an increased level of collaboration among researchers, and in an accelerated rate of solution of "bottleneck" problems.

In this regard, there *are* a substantial number of unresolved issues which the alliance must deal with before a successful simulation of black-hole binary in-spiral is in hand. From a conceptual viewpoint, perhaps the most pressing of these is the question of coordinate system choice. We are currently proceeding on the assumption that we can use a *single* coordinate system for the problem, through which the black holes will move, but we have very few specific proposals for actually constructing such a coordinate system. We need to devise coordinate conditions which will allow for black hole propagation while ensuring (1) that the coordinates themselves remain non-singular, and (2) that the black hole interiors can be stably excised simultaneously. Adding intrinsic angular momentum to the holes will almost certainly complicate the situation, and we have very little experience in constructing appropriate coordinates to deal with even a *single* dynamical black hole with spin.

Additional concerns surround the issue of black-hole excision: will excising work in highly dynamic, asymmetric situations? If it *will* work, will the 3+1 approach suffice, or do we fundamentally need a hyperbolic formulation? If a hyperbolic approach *is* required, can the problem of "coordinate shocks" recently investigated by Alcubierre (1996) be avoided?

Despite these and many other obstacles which still remain in our path, it is clear that we have already achieved much in terms of focusing the numerical relativity community on a challenging and interesting problem, in spurring theoretical developments such as the hyperbolic formulations, and in laying the groundwork (particularly in terms of computational infrastructure) for many future calculations.

Acknowledgments. I would like to thank my many colleagues within the Grand Challenge Alliance for innumerable enlightening conversations, and for providing such a stimulating research atmosphere over the past four years. This work was supported in part by NSF PHY9318152.

References

Abrahams, A. *et al.* 1995, Phys. Rev. Lett., 75, 3377.
Alcubierre, M. 1996, LANL preprint gr-qc/9609015.
Anninos, P., *et al.* 1995, Phys. Rev. D, 51, 5562.
Berger, M. J., & Oliger, J. 1984, J. Comput. Phys., 53, 484.
Bona, C. *et al.* 1995, Phys. Rev. Lett., 75, 600.
Cook, G. B., *et al.* 1993, Phys. Rev. D, 47, 1471.
Finn, L. 1996, LANL preprint gr-qc/9603004.
Marsa, R., & Choptuik, M. 1996, Phys. Rev. D, 54, 4929.
Matzner, R. A. *et al.* 1997, http://www.npac.syr.edu/projects/bh/
Misner, C. W., Thorne, K., & Wheeler, J. A. 1973, *Gravitation*, (San Francisco: W. H. Freeman).
Parashar, M., & Browne, J. C. 1995a,
    http://godel.ph.utexas.edu:/Members/parashar/Papers/daghrep.ps
Parashar, M., & Browne, J. C. 1995b,
    http://godel.ph.utexas.edu:/Members/parashar/Papers/uref.ps
Scheel, M. *et al.* 1995, Phys. Rev. D, 51, 4108.
Seidel, E., & Suen, W.-M. 1992, Phys. Rev. Lett., 69, 1845.
Thornburg, J. 1993, Ph.D. Dissertation, (University of British Columbia).
Van Putten, M., & Eardley, D. 1996, Phys. Rev. D, 53, 3056.
York, J. W. 1979, in *Sources of Gravitational Radiation*, ed. L. Smarr, (Cambridge: Cambridge University Press), 83.
York, J. W., & Piran, T. 1982, in *Spacetime and Geometry*, eds. R. A. Matzner & L. Shepley, (Austin: University of Texas Press), 147.

Head-on Collisions of Two Black Holes

D. W. Hobill

Department of Physics and Astronomy, University of Calgary, Calgary, AB T2N 1N4, Canada

P. Anninos, H. E. Seidel, L. L. Smarr

NCSA, University of Illinois at Urbana-Champaign, Urbana, IL 61801, U.S.A.

W.-M. Suen

Department of Physics, Washington University, St. Louis, MO 63130, U.S.A.

Abstract. Computations of head-on collisions between two equal-mass black holes are performed by constructing numerical solutions to the Einstein equations. The gravitational waves generated from the dynamics are extracted from the time-dependent metric and these give pulse profiles that should be measured by gravitational wave detectors.

1. Introduction

Some of the best known predictions of general relativity are those associated with the existence of both black holes and gravitational radiation, neither of which have been observed directly. Therefore, one of the great challenges to experimental relativity is to provide unambiguous evidence for the existence of such phenomena. The challenge to computational relativity is to provide unambiguous predictions of effects that should be observable.

A first step in understanding dynamical black hole and gravitational wave spacetimes is described in this paper. While exactly head-on collisions between two equal-mass black holes is rather unlikely, the results reported here do provide some information about dynamical black holes and strong gravitational waves. In addition, the results will provide an important test bed calculation for the more general computations discussed by Choptuik in these proceedings.

2. Splitting Spacetime into Space and Time

Using the method of Arnowitt *et al.* (1962, hereafter ADM), one "slices" the four-dimensional spacetime into three-dimensional spatial hypersurfaces (3-D volumes) threaded together by timelike curves whose tangent vectors determine a "time" vector. This splitting leads to a line element for the "distance" between infinitesimally separated spacetime points given by:

$$ds^2 = -(\alpha^2 - \beta^i \beta_i)dt^2 + 2\beta_i\, dx^i\, dt + \gamma_{ij}\, dx^i\, dx^j. \qquad (1)$$

The Einstein summation notation (*i.e.*, summation is implied over index pairs) is used where the indices, i and j run from 1 to 3 and denote the spatial coordinates (x^1, x^2, x^3). The lapse function, α, determines the interval of proper time, $d\tau$, between the point labeled by t and point labeled by $t + dt$, *i.e.*, $d\tau = \alpha\, dt$. The shift vector, β^i, is the coordinate three-velocity that determines the relative position of the spatial coordinate x^i from one 3-volume to the next. Finally the 3-metric γ_{ij} determines the spatial distances between points that lie entirely within a 3-D volume at a particular moment of time, t.

The lapse function and shift vectors are not dynamic variables (the Einstein equations do not provide evolution equations for them) and the freedom to choose them arbitrarily is often exploited to simplify the evolution equations, avoid spacetime singularities, or stabilize a numerical method.

Defining the *extrinsic* curvature (*i.e.*, the field momentum conjugate to the 3-metric) as $K_{ij} = \nabla_i \beta_j + \nabla_j \beta_i - \partial_t \gamma_{ij}$, where ∇_i is the covariant derivative with respect to γ_{ij}, the vacuum Einstein equations break up into the following sets: 1) the constraints which contain no time derivatives:

$$R - K_{ij}K^{ij} + K^2 = 0\,; \quad \text{Hamiltonian constraint}$$

$$\nabla_j(K_i^j - \delta_i^j K) = 0\,; \quad \text{momentum constraint}$$

and 2) the evolution equations for the 3-metric and extrinsic curvature:

$$\partial_t \gamma_{ij} = -2\alpha K_{ij} + \nabla_i \beta_j + \nabla_j \beta_i;$$

$$\partial_t K_{ij} = -\nabla_i \nabla_j \alpha + \alpha(R_{ij} + K_{ij}K - 2K_{ik}K_j^k) + \beta^k \nabla_k K_{ij} + K_{kj}\nabla_i \beta^k + K_{ik}\nabla_j \beta^k;$$

where the the *intrinsic* 3-curvature R_{ik} is determined in the standard Riemannian method from the metric γ_{ik}. The scalar 3-curvature is $R = \gamma^{ij}R_{ij}$ and the trace of the extrinsic curvature is defined by $K = \gamma^{ij}K_{ij}$. The constraint equations form a coupled set of elliptic equations and the evolution equations are hyperbolic equations describing the propagation of the gravitational field.

The construction of a numerical solution to these equations proceeds as follows. First one must choose a convenient coordinate parameterization in order to provide a basis for discretizing space and time. Once a computational grid is established, the partial derivatives appearing in the constraint and evolution equations can be approximated by finite differences.

The constraint equations are solved in the initial 3-volume defined at $t = 0$ for those components of the metric and extrinsic curvature that cannot be freely specified. The solutions to the constraints are then used to provide the right-hand-sides of the evolution equations which determine the metric and extrinsic curvature on the succeeding slice of spacetime. This provides the necessary data to evolve to the next time step via an explicit time-stepping algorithm.

3. Coordinates and Variables for Axisymmetric Spacetimes

The spacetimes to be studied are axially symmetric. Therefore all variables are independent of an azimuthal angle, ϕ. Using standard $(x^1 = z, x^2 = \rho, x^3 = \phi)$ cylindrical coordinates, the general axisymmetric 3-metric can be written as:

$$\gamma_{ij} = \Psi^4 \hat{\gamma}_{ij} = \Psi^4 \begin{pmatrix} a & c & 0 \\ c & b & 0 \\ 0 & 0 & \rho^2 d \end{pmatrix}. \tag{2}$$

The variables a, b, c and d are functions of the coordinates z, ρ and t and are assumed to be asymptotically flat. The conformal factor Ψ is a function of z and ρ only and does not evolve in time—it is determined on the initial time slice from a solution of the Hamiltonian constraint. For the two-black hole collision, Ψ can be computed analytically (see below). On the initial time slice, the 3-metric is conformally flat so, $a = b = d = 1$, and $c = 0$.

The Einstein equations are simplified when a conformal factor is introduced into the extrinsic curvature in a manner similar to the 3-metric [equation (2)],

$$K_{ij} = \Psi^4 \hat{K}_{ij} = \Psi^4 \begin{pmatrix} h_a & h_c & 0 \\ h_c & h_b & 0 \\ 0 & 0 & \rho^2 h_d \end{pmatrix}. \tag{3}$$

The evolution equations can now be formulated as dynamical equations for the variables (a, b, c, d) and (h_a, h_b, h_c, h_d).

4. Misner Initial Data

The initial data for the collision are based on the results of Misner (1960) who constructed a pure vacuum solution to the Hamiltonian constraint equation when $K_{ij} = 0$ (*i.e.*, a time-symmetric solution) and when the metric [equation (2)] is conformally flat. The solution contains two asymptotically flat sheets joined by two throats. This models provides a mathematically precise isometry between the interiors of the black holes and their exteriors.

The conformal factor Ψ_M defined by

$$\Psi_M = 1 + \sum_{n=1}^{\infty} \frac{1}{\sinh(n\mu)} \left(\frac{1}{+r_n} + \frac{1}{-r_n} \right), \quad \text{where} \quad {}^{\pm}r_n = \sqrt{\rho^2 + [z \pm \coth(n\mu)]^2}$$

solves the Hamiltonian constraint. This data set represents two equal-mass non-rotating black holes initially at rest where the two black hole centers are aligned along the axis of symmetry (z-axis). The free parameter μ is related to the physical parameters, M (half the total asymptotic or ADM mass) and L (the proper distance along the spacelike geodesic connecting the throats) by

$$M = 2 \sum_{n=1}^{\infty} \frac{1}{\sinh(n\mu)}, \qquad L = 2 \left[1 + 2\mu \sum_{n=1}^{\infty} \frac{n}{\sinh(n\mu)} \right].$$

The effect of increasing μ is to set the two black holes further away from one another and decrease the total mass of the system.

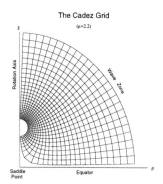
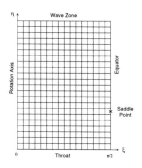

Figure 1. (*left*) The Čadež grid for the case $\mu = 2.2$ and displayed in a single quadrant with cylindrical coordinates. The throats are centered on the symmetry axis at $z = \pm \coth \mu$. (*right*) The computational grid.

5. The Computational Grid

The conformal factor, Ψ_M, assumes that the two black hole throats are spheres, centered on the z-axis. Since the natural boundaries (the throats and a sphere surrounding the system far from the throats) do not lie along constant (z, ρ) coordinates, it is useful to introduce the "quasi-spherical" Čadež (1971) coordinates (η, ξ) with η being a logarithmic "radial" coordinate and ξ an "angular" coordinate. Čadež coordinates are related to cylindrical coordinates through the complex conformal transformation

$$\eta + i\xi = \frac{1}{2}[\ln(\zeta + \zeta_0) + \ln(\zeta - \zeta_0)] + \sum_{n=1}^{\infty} C_n \left(\frac{1}{(\zeta_0 + \zeta)^n} + \frac{1}{(\zeta_0 - \zeta)^n} \right), \quad (4)$$

where $\zeta = z + i\rho$, and $\zeta_0 = \coth \mu$ is the value of ζ at the throat center. The constant η and ξ coordinate lines lie along the field and equipotential lines of two equally charged metallic cylinders located at $z = \pm \coth \mu$. The coefficients C_n are determined by a least-squares method to set the throats [defined by $\rho_{\text{th}}^2 + (z_{\text{th}} \pm \coth \mu)^2 = 1/\sinh^2 \mu$] to lie on an $\eta = \eta_0 = $ constant coordinate line. Both η_0 and the different C_n are computed for different μ using this least-squares procedure. The C_n's are rapidly converging and the series in equation (4) can be truncated when $10 \leq n \leq 15$.

The constant Čadež coordinate lines in the cylindrical coordinate system are shown in the left panel of Figure 1. The discretization of the Einstein equations occurs on the rectangular grid shown in the right panel of Figure 1. The advantage afforded by this set of coordinates is that they are spherical both near the throats and far away in the wave zone, thus allowing one to deal with throat boundaries and asymptotic wave form extractions in a convenient way. The disadvantage is that the transformation in equation (4) introduces a singular saddle point at the origin $(z = \rho = 0)$ not present in cylindrical coordinates. This creates certain numerical difficulties that are discussed in Anninos *et al.* (1994b).

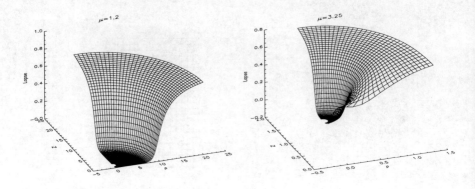

Figure 2.   (*left*) The lapse function for the case $\mu = 1.2$ at $t = 24M$. (*right*) The lapse function for $\mu = 3.25$ at $t = 22.5M$.

6. Computational Results: The Near Field

The physical attributes of the initial data for six separate Misner two-black-hole data sets are summarized in Table 1, where M represents the mass of a single black hole, and L/M gives the separation between the throats. Also shown is the initial structure of the apparent and event horizons.

μ	M	L/M	Apparent horizon	Event horizon
1.2	1.85	4.46	single	single
1.8	0.81	6.76	separate	critical
2.2	0.50	8.92	separate	separate
2.7	0.29	12.7	separate	separate
3.0	0.21	15.8	separate	separate
3.25	0.16	19.1	separate	separate

Table 1.   The physical parameters for initial data sets.

The metric and extrinsic curvature components are computed on each time slice and physically relevant information must be extracted from these quantities. For example, the lapse function for the cases $\mu = 1.2$ ($t = 25M$) and $\mu = 3.25$ ($t = 22.5M$) are shown in the left and right panels of Figures 2 respectively. In the left panel of Figure 2, the lapse goes to zero in the region near the throats, "freezing" all dynamics there. As a result, the proper distance between grid points grows in coordinate time in regions just outside the horizon. This produces a "grid stretching" which presents one of the main difficulties in evolving black hole spacetimes in any numerical simulation utilizing a singularity-avoiding lapse function.

For comparison, the right panel of Figure 2 shows the lapse for the case $\mu = 3.25$ where the throats are much more separated. The initial data consist of two separated horizons. The plot corresponds to an early time in the evolution of the system where the two holes are acting essentially independently of each other as they begin to fall together. After the holes coalesce, the lapse collapses spherically around both throats which are contained within the final black hole.

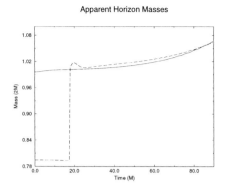

Figure 3. The apparent horizon mass for $\mu = 1.2$ (solid) and $\mu = 2.2$ (dashed). For $\mu = 1.2$ there is a single horizon initially. For $\mu = 2.2$, two horizons exist until $t \approx 17M$ when a single horizon surrounds both holes.

One can define an effective mass of a black hole based on its apparent horizon via the Hawking (1973) relation:

$$M_h = \sqrt{A_h/16\pi}, \qquad (5)$$

where A_h is the intrinsic area of the apparent horizon. This relationship gives a lower limit for the mass of the black hole, since the apparent horizon should lie at or inside the event horizon. When the system is nearly stationary, equation (5) provides a good estimate of the black hole mass as the apparent and event horizons almost coincide (see Anninos, et al., 1994a).

In Figure 3, the evolution of the horizon mass computed from equation (5) is shown for two different cases. The numerical data for the case $\mu = 1.2$ are shown as a solid line. The mass of the hole M_h is normalized to units of the total ADM mass of the spacetime, so ideally $M_h < 1$ for all time. However, the horizon is always found near the peak of $\hat{\gamma}_{\eta\eta}$, and therefore M_h is extremely sensitive to the precise position of the horizon. When the horizon lies between grid points, the overestimated surface area results from choosing the outer grid location.

Also shown in Figure 3 is the result for the case $\mu = 2.2$ where the initial horizons are separate and contain only about 79% of the total mass of the spacetime. These surfaces are tracked only until another trapped surface forms across the equator ($z = 0$) to surround both throats.

7. Waveform Extraction and Total Energy Loss

The main method used to calculate waveforms is based on the gauge-invariant extraction technique developed by Abrahams & Evans (1990) and applied in Abrahams et al. (1992) to black hole spacetimes. The basic idea is to split the spacetime metric into a spherically-symmetric (static) background and a small perturbation in the region where the curvature is dominated by the mass content

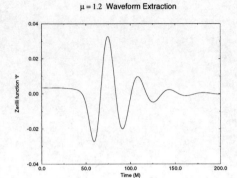

Figure 4. The $\ell = 2$ waveform for the case $\mu = 1.2$ measured at $r = 40M$. The solid line represents the numerical result and the dashed line the fit to the quasi-normal mode perturbation.

of a small compact object. The metric perturbation is expanded in $m = 0$ spherical harmonics $Y_{\ell 0}(\theta)$ and their tensor generalizations. The Regge-Wheeler perturbation functions are then extracted from the numerically computed metric components and used to construct the gauge-invariant Zerilli function ψ. The wavelike part of the metric ψ is radiative at large distances from the source and is commonly used in semi-analytic calculations of black hole normal mode frequencies (e.g., Chandrasekhar, 1983). The asymptotic energy flux carried by gravitational waves can be computed for each ℓ mode contribution from

$$\frac{dE}{dt} = \frac{1}{32\pi} \left(\frac{\partial \psi}{\partial t}\right)^2. \qquad (6)$$

For all of the cases studied in this paper, both the $\ell = 2$ and $\ell = 4$ waveforms at radii of 30, 40, 50, 60, and $70M$ have been extracted. By comparing results at each of these radii, the propagation of the waves and the consistency of energy flux calculations can be monitored.

Figure 4 shows the $\ell = 2$ waveform (solid line) extracted at a radius of $40M$ for the case $\mu = 1.2$. A single horizon surrounds both throats in this case and the system evolves as a single perturbed black hole from the outside. Therefore, the radiation is dominated by the quasi-normal modes of the black hole. The dotted line in Figure 4 shows the fit of the lowest two (fundamental and first overtone) $\ell = 2$ modes of a black hole of mass $(2M)$, over the range $70 < t/M < 160$, obtained from Leaver (1985) and Seidel & Iyer (1990). The first overtone quasi-normal mode is more strongly damped than the fundamental, and hence does not contribute appreciably to the fit at late times. Its main effect is to increase the accuracy of the fit to the first peak in the extracted waveform.

Figure 5 shows the $\ell = 2$ and $\ell = 4$ waveforms respectively for the case $\mu = 2.7$. The holes are initially separated by about $12.6M$. The solid lines are the waveforms extracted at a distance $r = 40M$. In this case, the fits to perturbation theory are still reasonably good, but are not as close as the calculations performed for holes that are initially closer together. The wavelengths of the

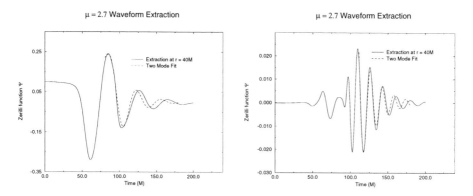

Figure 5. The waveforms for $\mu = 2.7$ measured at $r = 40M$. (*left*) The $\ell = 2$ mode. (*right*) The $\ell = 4$ mode.

extracted waveforms are somewhat too long. The calculation requires a longer period of time before the onset of quasi-normal ringing, so the peak in the radial metric function becomes more difficult to resolve as the accuracy declines. This leads to an error in the effective gravitational scattering potential which is critical in determining the quasi-normal frequencies.

The total radiated energy E can be computed from the Zerilli function using equation (6). The results are shown in Figure 6. The six clusters of unconnected symbols represent the numerical simulations based on the six μ parameter values. Each of the five symbols within a cluster corresponds to the total integrated $\ell = 2$ energy computed at the five different "detector" positions. For reference, the results of Smarr & Eppley (Smarr, 1979) are plotted as x's with error bars based on values suggested by Smarr in his 1979 review. Within the large errors quoted, those early results are remarkably consistent with the results presented here.

The results in Figure 6 show two distinct regimes. For $\mu < 1.8$ (*i.e.*, when $L/M = 6.69$), the initial data contain *one* black hole, and the energy radiated falls off exponentially. For $\mu > 1.8$, there are two holes and the energy radiated is somewhat independent of the initial separation. Finally, we note that the energy radiated is *very* small compared with the upper limits based on the horizon area theorem which are represented by connected circles in Figure 6. More details concerning the energy loss can be found in Anninos *et al.* (1993; 1995). Ultimately, the total energy radiated by these systems is on the order of $0.002M$ (where M is half the ADM mass of the spacetime). This implies that at least for head-on collisions of black holes, such systems cannot be considered to be efficient generators of gravitational radiation.

Acknowledgments. We would like to thank David Bernstein for a number of helpful discussions, Mark Bajuk for his work on visualizations of the numerical simulations that aided greatly in their interpretation, and Joan Massó for help with preparing some of the graphs for this paper. This work was supported by NCSA, NSF Grants 91-16682, 94-04788, 94-07882, ASC/PHY93-18152 (ARPA supplemented), and NSERC Grant No. OGP-121857. The computations were

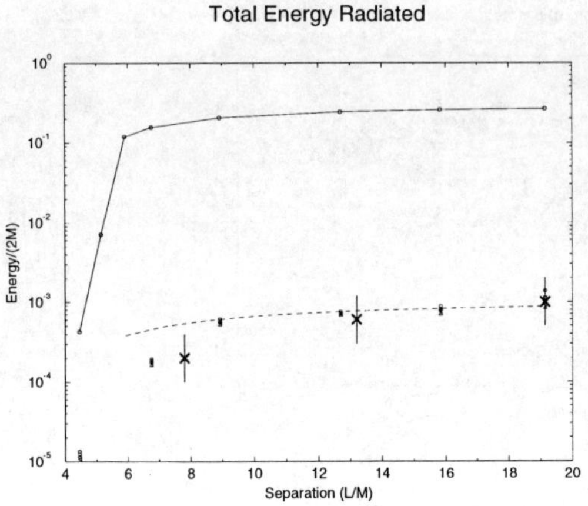

Total Energy Radiated

Figure 6. The total gravitational wave energy output for six values of μ as measured at "detectors" located at different fixed radii from the source.

performed at NCSA, the Pittsburgh Supercomputing Center and the Calgary High Performance Computing Center.

References

Abrahams, A., & Evans, C. 1990, Phys. Rev. D, 42, 2585.
Abrahams, A., Bernstein, D., Hobill, D., Seidel, E., & Smarr, L. 1992, Phys. Rev. D, 45, 3544.
Anninos, P., Hobill, D., Seidel, E., Smarr, L., & Suen, W-M. 1993, Phys. Rev. Lett., 71, 2851.
Anninos, P., Bernstein, D., Brandt, S., Hobill, D., Seidel, E., & Smarr, L. 1994a, Phys. Rev. D, 50, 3801.
Anninos, P., Hobill, D., Seidel, E., Smarr, L., & Suen, W-M. 1994b, *Technical Report 024*, (Urbana: National Center for Supercomputing Applications).
Anninos, P., Hobill, D., Seidel, E., Smarr, L., & Suen, W-M. 1995, Phys. Rev. D, 52, 2044.
Arnowitt, R., Deser, S., & Misner, C. W. 1962, in *Gravitation: An Introduction to Current Research*, ed. L. Witten, (New York: John Wiley).
Čadež, A. 1971, Ph.D. thesis, (University of North Carolina, Chapel Hill).
Chandrasekhar, S. 1983, *The Mathematical Theory of Black Holes*, (Oxford: Oxford University Press).
Hawking, S. W. 1973, in *Black Holes*, eds. C. DeWitt & B. S. DeWitt, (New York: Gordon and Breach).
Leaver, E. 1985, Proc. R. Soc. (London), A402, 285.
Misner, C. W. 1960, Phys. Rev., 118, 1110.
Seidel, E., & Iyer, S. 1990, Phys. Rev. D, 4, 374.
Smarr, L. 1979, in *Sources of Gravitational Radiation*, ed. L. Smarr, (Cambridge: Cambridge University Press), 245.

The Lyman Alpha Forest Within The Cosmic Web

J. Richard Bond & James W. Wadsley

CITA and Department of Astronomy, University of Toronto, 60 St. George St., Toronto, ON M5S 1A1, Canada

Abstract. Observations indicate galaxies are distributed in a filament-dominated web-like structure; classic examples are the Coma and Perseus-Pisces superclusters. Numerical experiments at high and low redshift of viable structure formation theories also show filament-dominance. In particular, we concentrate on the gas dynamical simulations of Lyα clouds at redshifts $\sim$ 2–6. We understand why this is so in terms of rare events (peak-patches) in the medium and the web pattern of filaments that bridge the gaps between the peaks along directions defined by their (oriented) tidal fields. We present an overview of these ideas and their practical application in crafting high-resolution, well-designed simulations. We show the utility of this by taking a highly filamentary subvolume found in a galactic-scale simulation and compressing its important large-scale features onto a handful of numbers defining galactic-scale peak-patch constraints, which are then used to construct constrained initial conditions for a higher-resolution simulation appropriate for study of the Lyα forest.

1. The Peak-Patch Picture and the Cosmic Web Theory of Filaments

We approach the connected ideas we loosely call Cosmic Web theory (Bond, Kofman, & Pogosyan, 1996, hereafter BKP) and the Peak Patch Picture (Bond & Myers, 1996, hereafter BM, and references therein) through an historical path that includes an outline of the relevant terminology. Lin, Mestel, & Shu (1965) showed that a cold triaxial collapse implied an oblate "pancake" would form. In 1970, Zel'dovich developed his famous approximation and argued that pancakes would be the first structures that would form in the adiabatic baryon-dominated universes popular at that time. Generally for a cold medium, there is a full nonlinear map: $\mathbf{x}(\mathbf{r}, t) \equiv \mathbf{r} - \mathbf{s}(\mathbf{r}, t)$, from Lagrangian (initial state) space, $\mathbf{r}$, to Eulerian (final state) space, $\mathbf{x}$. The map becomes multivalued as non-linearity develops in the medium. It is conceptually useful to split the displacement field, $\mathbf{s} = \mathbf{s}_b + \mathbf{s}_f$, into a smooth, quasilinear, long wavelength piece, $\mathbf{s}_b$, and a residual, highly non-linear, fluctuating field, $\mathbf{s}_f$. If the rms density fluctuations smoothed on scale R_b, $\sigma_\rho(R_b)$, are $< \mathcal{O}(1/2)$, the $\mathbf{s}_b$-map is one-to-one (single-stream) except at the rarest high-density spots. In the peak-patch approach, R_b is adaptive, allowing for dynamically hot regions like proto-clusters to have large smoothing, and cool regions like voids to have small smoothing. If $D(t)$ is the linear growth factor, then $\mathbf{s}_b = D(t)\,\mathbf{s}_b(\mathbf{r}, 0)$ describes Lagrangian linear perturbation theory, *i.e.*, the Zel'dovich approximation. The large-scale peculiar velocity is $\mathbf{V}_{Pb} = -\bar{a}(t)\dot{\mathbf{s}}_b(\mathbf{r},t)$. What is important for us is the strain field (or deformation tensor):

$$e_{b,ij}(\mathbf{r}) \equiv -\frac{1}{2}\left[\frac{\partial s_{bi}(\mathbf{r})}{\partial r_j} + \frac{\partial s_{bj}(\mathbf{r})}{\partial r_i}\right] = -\sum_{A=1}^{3}\lambda_{vA}\hat{n}_{vA}^i\hat{n}_{vA}^j,$$

where

$$\lambda_{v3} = \frac{\delta_{Lb}}{3}(1 + 3e_v + p_v), \quad \lambda_{v2} = \frac{\delta_{Lb}}{3}(1 - 2p_v), \quad \lambda_{v1} = \frac{\delta_{Lb}}{3}(1 + 3e_v - p_v),$$

and $\delta_{Lb} = -e^i_{b,i}$ is the smoothed linear overdensity, often expressed in terms of the height relative to the rms fluctuation level $\sigma_\rho(R_b)$, $\nu_b \equiv \delta_{Lb}/\sigma_\rho(R_b)$. The deformation eigenvalues are ordered according to $\lambda_{v3} \geq \lambda_{v2} \geq \lambda_{v1}$ and $\hat{n}_{vA}$ denote unit vectors of the principal axes. In that system, $x_A = r_A[1 - \lambda_{vA}(\mathbf{r},t)]$ describes the local evolution. The Zel'dovich-mapped overdensity is

$$(1 + \delta_Z)(\mathbf{r},t) = \left|[1 - D(t)\lambda_{v3}][1 - D(t)\lambda_{v2}][1 - D(t)\lambda_{v1}]\right|^{-1},$$

exploding when the largest eigenvalue, $D(t)\lambda_{v3}$, reaches unity (fold caustic formation). In a Zel'dovich map, a pancake develops along the surface $\hat{n}_{v3}\cdot\nabla_\mathbf{r}\lambda_{v3} = 0$.

The strain tensor is related to the peculiar linear tidal tensor by

$$\frac{\partial^2 \Phi_P}{\partial x^i \partial x^j} = -4\pi G \bar{\rho}_{nr} \bar{a}^2 \, e_{b,ij},$$

where Φ_P is the peculiar gravitational potential, and to the linear shear tensor by $\dot{e}_{b,ij}$. The anisotropic part of the shear tensor has two independent parameters, the ellipticity e_v (always positive) and the prolaticity p_v.

Doroshkevich (1973) and Doroshkevich & Shandarin (1978) were among the first to apply the statistics of Gaussian random fields to cosmology, in particular of λ_{vA}, at random points in the medium. Arnold, Zel'dovich, & Shandarin (1982) made the important step of applying the catastrophe theory of caustics to structure formation. This work suggested the following formation sequence: pancakes first, followed by filaments, and then clusters. This should be compared to the BKP Web picture formation sequence: clusters first, followed by filaments, and then walls. BKP also showed that filaments are really ribbons, walls are webbing between filaments in cluster complexes, and that walls are not really classical pancakes. For the Universe at $z \sim 3$, massive galaxies play the role of clusters, and for the Universe at $z \sim 5$ more modest dwarf galaxies take on that role.

The Web story relies heavily upon the theory of Gaussian random fields as applied to the rare "events" in the medium, e.g., high-density peaks. Salient steps in this development begin with Bardeen et al. (1986), where the statistics of peaks were applied to clusters and galaxies, e.g., the calculation of the peak-peak correlation function, $\xi_{pk,pk}$. In a series of papers, Bond (1987a; 1987b; 1989a; 1989b) and Bond & Myers (1991; 1993a; 1993b) developed the theory so that it could calculate the mass function, $n(M)dM$. It was also applied to the study of how shear affects cluster alignments (e.g., the Binggeli effect), and to Lyα clouds, "Great Attractors", giant "cluster-patches", galaxy, group and cluster distributions, dusty PGs, CMB maps, and quasars. This culminated in the BM "Peak-Patch Picture of Cosmic Catalogues".

We now describe the BM peak-patch method and how it is applied to initial conditions for simulations; an example is shown in Figure 1. We identify candidate peak points using a hierarchy of smoothing operations on the linear density field, δ_L. To determine patch size and mass, we use an ellipsoid model for the internal patch dynamics, which are very sensitive to the external tidal field. A by-product is the internal (binding) energy of the patch and the orientation of the principal axes of the tidal tensor. We apply an exclusion algorithm to prevent peak-patch overlap. For the external dynamics of the patch, we use a Zel'dovich-map with a locally adaptive filter ($R_{\rm pk}$) to find the velocity $\mathbf{V}_{\rm pk}$ (with quadratic corrections sometimes needed). The peaks are rank-ordered by mass (or internal energy). Thus, for any given region, we have a list of the most important peaks. By using the negative of the density field, we can also get void-patches. Some of the virtues of the method are: it represents a natural generalization of the Press-Schechter method to include non-local effects[1]; is a natural generalization of the Bardeen et al. (1986) single-filter peaks theory to allow a mass spectrum and solve the cloud-in-cloud (i.e., peak-in-peak) problem; allows efficient Monte Carlo constructions of 3-D catalogues; gives very good agreement with N-body groups; has an accurate analytic theory for estimating peak properties, (e.g., mass and binding energy from mean-profiles, using $\delta_{L,{\rm crit}}(e_v)$, $\langle e_v|\nu_{\rm pk}\rangle$); and handles merging, with high-redshift peaks being absorbed into low-redshift ones.

BKP concentrated on the impact the peak-patches would have on their environment and how this can be used to understand the web. They showed that the final-state filament-dominated web is present in the initial conditions in the δ_{Lb} pattern, a pattern largely determined by the position and primordial tidal fields of rare events. BKP also showed how 2-point rare-event constraints define filament sizes (see their Figure 2). The strongest filaments are between close peaks whose tidal tensors are nearly aligned. Strong filaments extend only over a few Lagrangian radii of the peaks they connect. They are so visually impressive in Eulerian space because the peaks have collapsed by about 5 in radius, leaving the long bridge between them whose transverse dimensions have also decreased. This is illustrated by the lower right panel of Figure 1 in which the aligned galaxy peaks are connected by strong filaments. Strong vertical filaments are a product of the vertical alignment of the peaks' tidal tensors, which simultaneously acts to prevent a strong horizontal filament between the top two peaks. The reason for this phenomenon is that the high degree of constructive interference of the density waves required to make the rare peak-patches, and to orient them preferentially along the 1-axis, leads to a slower decoherence along the 1-axis than along the others, and thus a higher density.

---

[1] The extremely popular, trivial-to-implement, Press-Schechter (1974) method for determining $n(M)$ has been the main competitor to the peaks theory over the years, but has no real physical basis and disagrees strongly with the spatial distribution (Bond et al., 1991); thus the amazement, and delight, in the community that $n(M)$ fits that of N-body group catalogues so well.

Three-point and higher rare-event constraints of nearby peaks determine the nature of the webbing between the filaments, also evident in Figure 1.

2. Crafting High Resolution N-Body/Hydrodynamical Simulations

Although it is usual to evolve ambient "random" patches of the Universe in cosmology, there are obvious advantages in spending one's computational effort on the regions of most interest. Single-peak constraints are very useful if cluster or galaxy formation is the focus, while multiple-peak constraints are more useful if superclusters, cluster substructure, or filaments and walls are the focus. We have seen that the essential features at a given epoch of the filamentary structure and the wall-like webbing between the filaments is largely defined by the dominant collapsed structures, and the peak-patches that gave rise to them. A general method for building peak environments is suggested: construct random field initial conditions that require the field to have prescribed values of the peak shear (smoothed over the peak size at the peak position), for a subset from the size-ordered list of peaks that will have a strong impact on the patch to be simulated. Only a handful, $N_{\rm pk}$, of peaks and/or voids is usually needed to determine the large-scale features, effectively compressing the information needed to $(3+1+6)N_{\rm pk}$ numbers ($\mathbf{r}_{\rm pk}, M_{\rm pk}, e_{{\rm pk},ij}$). Peak velocities and the peak constraint are also usually added but these are not as important.

To illustrate how this works, we created a random (unconstrained) initial state for a CDM model in a 40 Mpc box, our "galaxy" simulation. We found peak-patches according to the method described in BM and focussed on a specific subregion exhibiting a strong filament, choosing the peaks and voids that were expected to exert strong tidal influences within and upon the patch, as described in Figure 1. The region chosen was just above the large central cluster of peaks in the top left panel of Figure 1. We constructed a higher-resolution set of "Lyα cloud" initial conditions for this patch, which the "galaxy" initial condition simulation could not resolve well enough to address the low column depth Lyα forest of interest to us.

By compressing the initial data in our target region to just the positions and shears of a few large peak-patches, then forming a constrained realization and applying different random waves (optimally-sampled for the smaller region) than the original 40 Mpc initial condition used, we know we will get high frequency structure wrong. But clearly the large-scale features are the same. This is in spite of the tremendously complex filamentary structure just below our chosen subregion. The peaks we chose were on the basis of rareness (size) and proximity to the patch (using an algorithm roughly based on correlation function falloff from each peak). The five peak-patches used for this companion Lyα simulation had the following masses (in units of $10^{11} M_\odot$) and halo velocity dispersion (in units of km s^{-1}) as determined from the binding energy: 3.6, 77; 3.5, 80; 1.4, 57; 0.85, 48; 0.51, 40. These accord well with what our group finder finds in the simulation at this redshift. The two void-patches used had Lagrangian masses of 2.4 and 0.72, and were outside the high-resolution interior. The approximate alignments of the shear tensors for the peak-patches inside ensure that a strong filament exists. Figure 2 shows how the filament looks in HI column density.

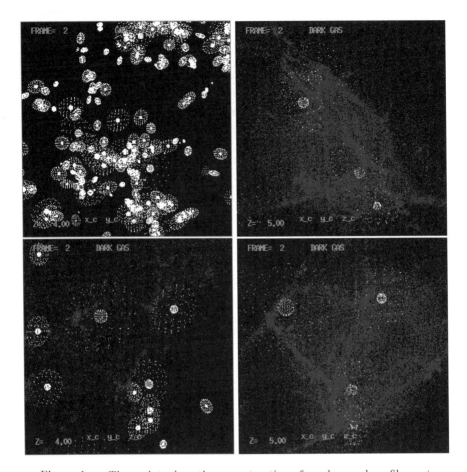

Figure 1. These plots show the reconstruction of a galaxy-galaxy filament present in our 40 Mpc "galaxy" simulation. The cosmology is a standard, initially scale-invariant, $\Omega_{nr} = 1$ CDM model with $\Omega_B = 0.05$, h = 0.5, normalized to $\sigma_8 = 0.67$. We use the cosmological SPH+P^3 Multi-Grid code described in Wadsley & Bond (these proceedings). We identified the peak-patches that should collapse by $z = 4$ in the initial conditions (IC), as shown in the top left panel (which is $10\,h^{-1}$Mpc across, comoving). For each patch, the outer ellipsoid represents the alignment of the shear tensor and the inner sphere is an estimate of the final object size. In the lower left panel, we zoom in on the central filamentary web structure and overlay dark matter from a simulation of these IC (panel $2.5\,h^{-1}$Mpc across). The five peak-patches (at $z = 4$ with overdensity 180) and two voids that define this region were used as constraints for a new higher-resolution IC (12.8 Mpc), which we also evolved numerically. These peaks are shown overlaid with dark matter from the new simulation in the right hand panels. The top right panel is a different orientation to the others that shows the filament more clearly, also shown in $n_{\rm HI}$ in Figure 2. These panels demonstrate that peaks represent an excellent way to compress the essential information about large-scale, filamentary behaviour.

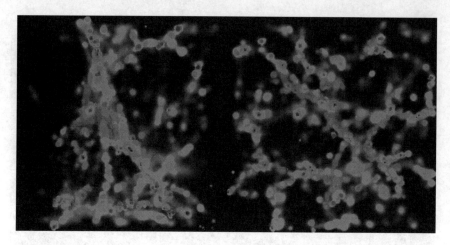

Figure 2. Left: A pseudo ray-traced rendering of the HI density for a region at $z = 4$ constrained by the complex pattern of galaxy-scale peak and void shear-fields described in the text, encoding the main large-scale structure features, and which ensures strong filamentary webbing arises in the patch. The gray scale codes the temperature variation, with the darker shades in the centre of objects representing high column cold gas. The region is $2.5\,h^{-1}$ Mpc across, comoving, and corresponds to the top right panel of Figure 1. Right: $n_{\rm HI}$ for an ambient patch of the Universe at $z = 3$ with control parameter $\nu_b = 0$, simulation as described in Wadsley & Bond (these proceedings). Note the filament dominance in both cases.

To do such complex regions, the standard periodic box approach to cosmological simulations is obviously not appropriate. Advantages of focussed non-periodic simulations include: (1) good mass resolution, allowing us to concentrate on the scale needed to treat the objects that form adequately (*e.g.*, dwarf galaxies, $a_L = a_{\rm Lattice, High\,Res} = 100$ kpc), and corresponding high numerical resolution ($h_{\rm sph, min}$, $h_{\rm grav} \sim 1$ kpc), can be achieved; (2) good k-space sampling is also possible. The competing demands of k-space sampling and resolution are further described in Figure 3. Our method allows high resolution without compromising our long wave coverage by going beyond grid-based FFTs, with a FastFT for high k (which kicks out well before the fundamental mode is reached) that is superseded by two direct FTs, power-law then log k sampling, with transitions among them determined by minimizing the volume per mode in k-space. Well-sampled k-space is especially important for Lyα cloud and galaxy formation (as opposed to cluster formation) because the density power spectrum for viable hierarchical theories has nearly equal power per decade (approaching flatness in Figure 3). If just the FFT is used, as is often the case in cosmology even for non-periodic calculations, the large-scale structure in the simulation will be poorly modelled because there is only one fundamental k-mode along each box axis. This can also have a deleterious effect on small-scale structure.

There is no point adding long waves without maintaining accurate large-scale tides and shearing fields during the calculation. We achieve this with a high-resolution region of interest (grid spacing a_L, 50^3 sphere) that sits within

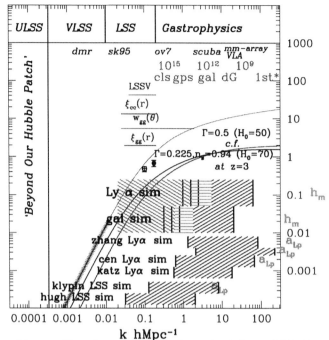

Figure 3. The two (linear) power spectra shown (scaled to redshift 3) were used in the simulations here and in Wadsley & Bond (these proceedings). The upper curve at high k is the standard untilted CDM model, but normalized to cluster abundances, $\sigma_8 = 0.67$. The other has the same cosmological age (13 Gyr) and $\Omega_B h^2$ (0.0125) but $H_0 = 70$ with $\Omega_\Lambda = 0.67$, slightly tilted to be COBE-normalized ($n_s = 0.94$). [Also shown is a COBE-normalized CDM model, which misses the solid data point in the cluster-band (constraint from $dn_{\rm cl}/dT_X$).] The bands in comoving wavenumber probed by various simulations are contrasted. Periodic simulations may use the entire volume, but the k-space restriction to lie between the fundamental mode (low-k boundary line) and the Nyquist wavenumber (high-k boundary line) in the IC can severely curtail the rare events in the medium that observations especially probe, and prevents tidal distortions of the simulation volume. We use three k-space sampling procedures (FFT and two direct FTs) with the boundaries defined by which has the smallest volume per k-mode. Even though a 256^3 Fourier transform was used, notice how early the direct sampling takes over (with only 10,000 modes). Using an FFT with the very flat spectra in the dwarf galaxy (dG) band can give misleading results. The three low-k lines shown for our Lyα and galaxy simulations correspond to the high-, medium- and low-resolution fundamental modes. We actually include modes in the entire hatched region, with the tidal fields associated with the longer waves included by a self-consistent uniform tide on the LR simulation volume. Our best resolution is denoted by h_m and $a_{\rm Lp}$ denotes the physical (best) lattice spacing for the grid-based Eulerian hydrocodes of Cen (Miralda-Escudé et al., 1996) and Zhang (Zhang et al., 1997) ($z = 3$). The k-space domains for the Katz (Hernquist et al., 1996) SPH simulation and two large-scale structure ($z = 0$) simulations are also shown: a Klypin 256^3 PM calculation and a Couchman (*hugh*) 128^3 P^3M simulation.

a medium-resolution region ($2\,a_L$, 40^3), in turn within a low-resolution region ($4\,a_L$, 32^3). The influence of ultra-long waves is included by measuring the mean external tide acting on the low-resolution region in the initial conditions, adopting simple models for the ultra-long wave dynamics based on that measurement (*e.g.*, linear, Zel'dovich, or homogeneous ellipsoid, as in BM), and applying it as an "external force" throughout the simulation. For this simulation, linear ultra-long wave dynamics were adequate.

In Wadsley & Bond (these proceedings), instead of complex multipeak constraints for individual regions, we use *importance sampling* of *shearing-patches* (patches with the smoothed shear tensor $e_{b,ij}(0)$ prescribed at the centre) to maximize the statistical information we can get from a crafted set of relatively modest constrained-field SPH calculations, defined by a set of control parameters, here the central ν_b, e_v, and p_v smoothed over a galactic-scale R_b. This allows us to sample rare peak- and void-patches, difficult to sample even in large box simulations, especially if FFTs are used. These are in addition to patches with more typical rms density contrasts. We then combine the results to get the frequency distribution of, say, $N_{\rm HI}$ for a random patch of the Universe using Bayes theorem, which decomposes it into the frequency distribution for $N_{\rm HI}$ for our constrained patches given the control parameters, measured from the simulations, and the known probability distribution of the control parameters. Schematically:

$$P(\texttt{random}-\texttt{patch}) = \int P(\texttt{constr}-\texttt{patch}|\texttt{control})\,P(\texttt{control})\,d\texttt{control}-\texttt{param}.$$

Shearing patches with low $|\nu_b| \lesssim 2$ often have relatively large shear ellipticity, $\langle e_v\,|\nu_b\rangle \approx 0.54\,|\nu_b|^{-1}$, which can give strongly asymmetric collapses for $\nu_b > 0$, amplifying the smaller-scale filamentary webbing by concentrating it in larger-scale filaments or walls (which depends upon p_v). Figure 2 shows that a $\nu_b = 0$ shearing-patch has its $n_{\rm HI}$ filaments more spread out than the multipeak case we have described here, which is reminiscent of, but even more concentrated than, a $\nu_b = 1.4$ calculation shown in Wadsley & Bond (these proceedings).

Simulating a large number of controlled patches in parallel is a form of adaptive refinement, of which there is much discussion in these proceedings. In refined regions, the initial conditions are almost never modified with higher frequency waves, so the Lagrangian (*i.e.*, mass) resolution remains fixed even though the Eulerian resolution may be superb. (This is especially vexing for voids.) When we refine a region by creating a high-resolution realization with the information contained in peak-patches, we *optimally* resample k-space to generate a new set of high-frequency waves. It is clear that the cosmological codes of the future will have to adapt in Eulerian and k-space simultaneously, and the techniques explored here offer a promising path towards this goal.

Acknowledgments. Support from the Canadian Institute for Advanced Research and NSERC is gratefully acknowledged. We thank Lev Kofman, Steve Myers and Dmitry Pogosyan for much fun peak-patch/web interaction.

References

Arnold, V., Zel'dovich, Ya. B., & Shandarin, S. 1982, Astr. Fluid Dynamics, 20, 111.
Bardeen, J. M., Bond, J. R., Kaiser, N., & Szalay, A. S. 1986, ApJ, 304, 15.

Bond, J. R. 1987a, in *Nearly Normal Galaxies From the Planck Era to the Present*, ed. S. Faber, (New York: Springer-Verlag), 388.
Bond, J. R. 1987b, in *Proceedings of the Cosmology & Particle Physics Workshop*. Berkeley CA 1986, ed. I. Hinchcliffe, (Singapore: World Scientific), 22.
Bond, J. R. 1989a, in *Large-Scale Motions in the Universe, A Vatican Study Week*, eds. V. Rubin & G. Coyne, (Princeton: Princeton University Press), 419.
Bond, J. R. 1989b, in *Frontiers of Physics—From Colliders to Cosmology*, eds. A. Astbury et al., (Singapore: World Scientific), 182.
Bond, J. R., Cole, S., Efstathiou, G., & Kaiser, N. 1991, ApJ, 379, 440.
Bond, J. R., Kofman, L., & Pogosyan, D. 1996, Nature, 380, 603 (BKP).
Bond, J. R., & Myers, S. 1991, in *Trends in Astro-particle Physics*, eds. D. Cline & R. Peccei, (Singapore: World Scientific), 262.
Bond, J. R., & Myers, S. 1993a, in *The Evolution of Galaxies and Their Environment, Proceedings of the Third Teton Summer School*, eds. M. Shull & H. Thronson, NASA Conference Publication 3190, 21.
Bond, J. R., & Myers, S. 1993b, in *The Evolution of Galaxies and Their Environment, Proceedings of the Third Teton Summer School*, eds. M. Shull & H. Thronson, NASA Conference Publication 3190, 52.
Bond, J. R., & Myers, S. 1996, ApJS, 103, 1 (BM).
Doroshkevich, A. G. 1973, Astrophysica, 6, 320.
Doroshkevich, A. G., & Shandarin, S. 1978, Sov. Astron., 22, 653.
Hernquist, L., Katz, N., Weinberg, D., & Miralda-Escudé, J. 1996, ApJ, 457, L51.
Lin, C. C., Mestel, L., & Shu, F. H. 1965, ApJ, 142, 1431.
Miralda-Escudé, J., Cen, R., Ostriker, J. P., & Rauch, M. 1996, ApJ, 471, 582.
Press, W. H., & Schechter, P. 1974, ApJ, 187, 425.
Zel'dovich, Ya. B. 1970, A&A, 5, 84.
Zhang, Y., Anninos, P., Norman, M. L., & Meiksin, A. 1997, ApJ, in press.

SPH P³MG Simulations of the Lyman Alpha Forest

J. W. Wadsley & J. Richard Bond

CITA and Department of Astronomy, University of Toronto, 60 St. George St., Toronto, ON M5S 1A1, Canada

Abstract. Our understanding of the Lyα forest has received a great boost with the advent of the Keck Telescope and large 3-D hydrodynamical simulations. We present new simulations using the SPH technique with a P³MG (Particle-Particle Particle-MultiGrid) non-periodic gravity solver. Our method employs a high resolution (1 kpc) inner volume (essential for capturing the complex gas physics), a larger low resolution volume (essential for correct larger scale tidal fields), and a self-consistently applied, uniform tidal field to model the influence of ultra-long waves. We include a photoionizing UV flux and relevant atomic cooling processes. We use constrained field realizations to probe a selection of environments and construct a statistical sample representative of the wider universe. We generate artificial Lyα spectra and fit Voigt profiles. We examine the importance of (1) the photoionizing flux level and history, (2) tidal environment, and (3) differing cosmologies, including CDM and CDM+Λ. With an appropriate choice for the UV flux, we find that the data are fit quite well if the rms density contrast is ~ 1 at $z \sim 3$ on galaxy scales.

1. Introduction

We hope to be able to use the current wealth of Lyα absorption data to constrain the shape and normalization of the density power spectrum and thus the cosmological model. The extraction of this information is complicated by gas physics, the ultraviolet flux, the star formation history, and supernova energy injection. By employing gas-dynamical simulations, we can incorporate the detailed high redshift environment of the clouds self-consistently. We perform our Lyα simulations using the SPH method in conjunction with a gravity solver based on the multigrid method. We describe our code, with attention to the novel features, in Wadsley & Bond (in preparation) and in the web version of this paper, astro-ph/9612148. We address design issues in § 2 and in Bond & Wadsley (these proceedings, hereafter BW). In generating the line statistics of the artificial Lyα spectra, it is important to fit the lines in a manner directly comparable with observations. Our method of fitting Voigt profiles is described with our results in § 3.

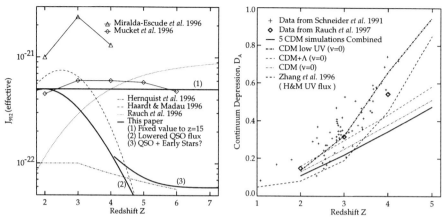

Figure 1. UV Flux history and Continuum Depression, D_A. Curve 2 in the left panel was designed to reproduce the D_A vs. redshift data shown in the right panel. A curve such as curve 3, including an estimate of the UV flux from early stars, is also consistent with the D_A data. The flux model of Haardt & Madau (1996) is too high for standard CDM models, as demonstrated by the dashed curve in the right panel.

2. The Simulations

The photoionization levels and heating are determined by the background UV flux, in particular the flux at the ionization edge for hydrogen, J_{912}. Observations of the proximity effect hint at little evolution in J_{912} over the range $z = 2$ to 4 (e.g., Bechtold, 1994). As shown in the left panel of Figure 1, we apply both a fixed flux (curve 1) and a strongly evolving flux, J_{912}, similar to that of Haardt & Madau (1996) lowered to fit the continuum depression data (curve 2), given by,

$$J_{912} \sim 5 \times 10^{-22} \left(\frac{1+z}{3}\right)^9 e^{-3(z-2)} \text{ erg cm}^{-2} \text{ s}^{-1} \text{ Hz}^{-1}.$$

These choices should roughly bracket the true mean UV flux.

The small scale nature of the clouds demands high resolution numerical work, especially as the collisionless nature of the dark matter plus radiative cooling prevent the comoving Jean's length,

$$k_J^{-1} \approx 0.049 \left(\frac{4}{1+z}\frac{T}{10^4 \text{ K}}\right)^{1/2} \Omega_{nr}^{-1/2} h^{-1} \text{ Mpc},$$

from setting a minimum scale. Ω_{nr} refers to the density parameter in non-relativistic matter.

We simulate regions of comoving diameter 5.0 Mpc (2.5 h^{-1}Mpc for the CDM model, 3.5 h^{-1}Mpc for the CDM+Λ model) with $50^3/2$ gas and $50^3/2$ dark matter particles at our highest resolution. We enclose this in a medium resolution buffer of gas and dark matter particles with eight times the mass, out to 8 Mpc, and then massive tide particles with 64 times the mass, out to 12.8 Mpc. We apply a linearly evolved external shear to the entire volume,

Table 1. Initial Conditions used in this work.

Cosmology	Description	ν	$e_v, p_v{}^a$	Weight[b]	Angle[c]	$\Delta \lambda^c$
CDM	Void	−1.4	0.179, 0.014	0.069	1.47	1.46
	Weak Void	−0.7	0.337, 0.034	0.242	1.26	1.36
	Flat	0.0	0, 0	0.379	1.0	1.0
	Weak Peak	0.7	0.337, 0.034	0.242	0.78	0.77
	Peak	1.4	0.179, 0.014	0.069	0.57	0.62
CDM+Λ	Flat	0.0	0, 0			

[a] tidal (shear) field parameters (see BW)
[b] weighting because of ν only, i.e., probability of region
[c] solid angle and average length of spectra, $\Delta \lambda$, measured at $z = 3$ relative to flat region

derived from the initial conditions. Thus we make considerable effort to model the long-wave, tidal environment well.

We use a sophisticated constrained field code to set up the initial displacements from unperturbed lattice positions, using the Zel'dovich approximation. It allows for arbitrary types and numbers of constraints. Our initial conditions finely sample an enormous range of k-space, down to $k = 0.01\,h$ Mpc^{-1} for these simulations (see Figure 3 of BW). Periodic simulations must truncate the large-scale power beyond the scale of the fundamental mode of the box and the power spectrum is still very flat at $k \sim 1$ h Mpc^{-1}, the scale of Lyα simulations.

Most simulations use random regions of the Universe. We prefer to have control parameters which govern some of the major large-scale characteristics of the simulation volume. We constrain the density (via ν), bulk flow and tidal fields smoothed over three Gaussian filtered scales, R_b, where $\nu = \delta_{L,pk}/\sigma_{R_b}$ the overdensity of the region relative to the rms fluctuation level σ_{R_b}. More details and applications of our peak constraint approach are described in BW. For these simulations, we fix ν for $R_b = 0.5$ Mpc and use mean field expectations to set the other parameters and the values at two other scales, 1.0 and 1.5 Mpc. The ν's used are shown in Table 1. Rare events which form large "bright" galaxies by $z \sim 3$ in the patch require higher ν. Since ν has a Gaussian probability distribution, we can appropriately weight and combine simulations of different ν to create a sample broadly representative of the universe. In creating samples of Lyα statistics, the solid angle on the sky subtended and length of spectra generated by the region must also be taken into account.

We describe results for two cosmological models: standard CDM, normalized to $\sigma_8 = 0.67$ with $h = 0.5$, $\Omega_B = 0.05$; and COBE-normalized CDM+Λ, with $h = 0.7$, $\Omega_B = 0.0255$ and $\Omega_{nr} = 0.335$ (shape parameter $\Gamma = 0.225$ and tilt $n_s = 0.94$ which gives a theory agreeing with the shape of observed large-scale power spectrum; see Figure 3 of BW). We required that $\sigma_{0.5\,\text{Mpc}} = 1.05$ at $z = 3$ and the number of baryons in the simulation volume to be the same in both models.

3. Results

To analyse our simulations, we produce artificial spectra with signal to noise of 100 and 5 km s^{-1} pixels. We sample a regular spatial grid of lines of sight running along the three axial directions through our simulations. A typical line of sight is ~ 500 km s^{-1} long and a single simulation sample $\sim 10^6$ km s^{-1} at $z = 3$. We fit Voigt profiles using an automated profile fitting program, designed to emulate the methods employed by observers. The program identifies each group of lines as a region in the spectrum where the flux drops below 98% of the continuum. χ^2 minimization is used to fit first one line, then two lines, and so on. The number of lines that produces the lowest reduced-$\chi^2 = \chi^2/d$ is used, where d is the number of degrees of freedom remaining for the fit, estimated as 1 per 2 pixels in the line group region minus 3 for each line used in the fit. In practice, a line group rarely needs more than 6 lines for a good fit.

Figure 2 demonstrates visually the difference between several runs, as shown in the column depth of neutral hydrogen through the central 2 Mpc of the simulations. The left panels show the dramatic effect of varying ν and hence the shear field. The filamentary structure is greatly enhanced in even slightly overdense regions. This has important repercussions in the types and numbers of Lyα absorption lines produced, as shown in the bottom left panel of Figure 3. Comparing the CDM and CDM+Λ runs (top left and right panels in Figure 2 respectively), it can be seen that the flatter spectrum in the CDM+Λ case makes the filaments more prominent and the dwarf galaxies less so, but without it having a major impact on the $N_{\rm HI}$ frequency curve. The difference in the number of clouds is caused by the width in redshift of the simulation volume in each cosmology,

$$\frac{dz}{dl_{\rm com}} = \frac{H_0}{c}[\Omega_{\rm nr}(1+z)^3 + \Omega_\Lambda]^{1/2}.$$

The high resolution size is $\Delta l_{\rm com}=5$ Mpc, comoving, in all simulations. Dividing by the ratio of $\Delta z = (dz/dl_{\rm com})\Delta l_{\rm com}$ accounts for the difference ($10^{0.09}$ at $z = 3$) between the two cosmologies apparent in the top right panel of Figure 3.

The left hand panels of Figure 3 demonstrate that an appropriate fixed choice of $J_{912} = 5 \times 10^{-22}$ can reproduce the observed statistics very well. We have statistically combined five CDM simulations to produce effectively a good sample of the universe that includes rarer regions. The advantage of this over a single large simulation is great resolution and importance sampling. The contribution of different regions is apparent. The voids, $\nu < 0$, cause a lowering of the curve, especially at low column depths.

The results of a low resolution run (half that of the standard ones) are shown represented by crosses in the top right panel of Figure 3. Though it underpredicts the low column lines, the high column lines are overpredicted relative to our higher resolution runs: limited resolution mimics heating by "puffing" up the cores of cold lumps out to greater sizes and thus gives higher cross sections. Further work, addressing the issue of convergence with even higher resolution simulations, indicates that statistics such as these are robust for our chosen resolution.

In the top right panel of Figure 3, we demonstrate that attempts to renormalize simulations with inappropriate values of the ionizing flux will fail to give

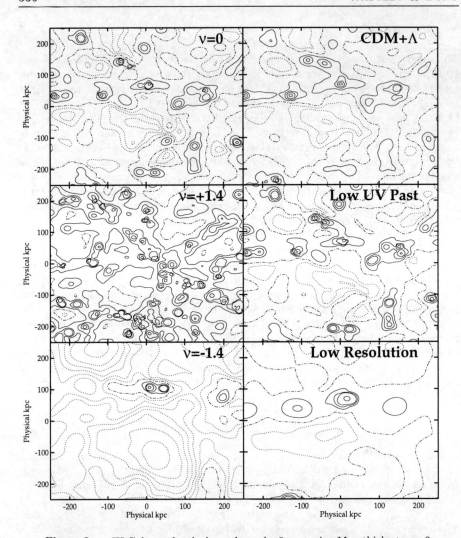

Figure 2. HI Column depth through a cube 2 comoving Mpc thick at $z = 3$. The contour levels, in $\log_{10}(\text{cm}^{-2})$, are 11 to 13 in steps of .25 (*dotted*), 13.36 (mean density, *dot dashed*) and 14, 14.5, 15, 17 (*solid*). The left panels are CDM simulations, with $\nu = 0$, +1.4 and −1.4, top to bottom. Note the incredible enhancement of the filamentary structure in the overdense regions. Simulating overdense regions or voids like these is not possible in a periodic box of a similar size, because the density averaged over the box must remain equal to the universal mean value. Even for significantly larger boxes, this is a problem. The right hand panels are $\nu = 0$ runs directly comparable with the CDM run in the top left panel. In the top right panel, we show a CDM+Λ run which has more large scale power and thus more enhanced filaments. The run in the right centre panel had a lower UV flux at early times, resulting in a more clumpy distribution and enhanced HI column depths at the high end. The lower right panel shows that poor resolution (half the spatial resolution of the upper left panel) will remove much of the structure.

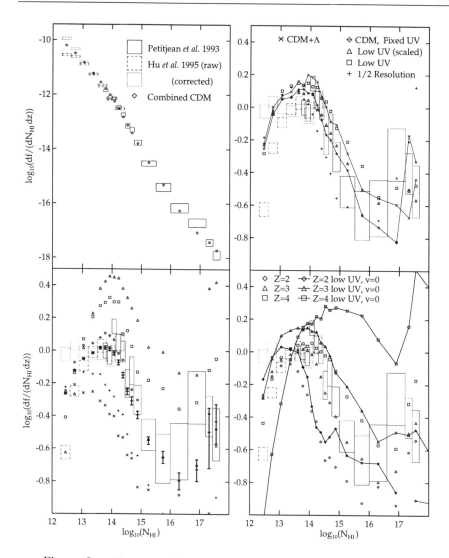

Figure 3. Frequency distribution of lines by N_{HI}. In all frames, the boxes represent binned data, box widths represent bin width, and the upper and lower edges denote 1 sigma Poisson errors. At the top left, the full range in frequency is plotted and in the remaining graphs, the power law $-1.46 \log_{10}(N_{HI}) + 8.25$ has been subtracted for clarity. On the left, the data from 5 CDM simulations are plotted. The symbols in the lower left plot represent (from the top down) the $\nu = 1.4, 0.7, 0, -0.7$ and -1.4 runs respectively and the combined data are shown with 1 sigma Poisson error bars. At the top right, the effects of various choices for the UV flux, numerical resolution and cosmology are compared. The low UV flux at early times results in more clumpiness in the IGM, overproducing lines at higher column depths. At the bottom right, the data are shown at 3 redshifts. The evolution is strong for higher column lines and almost non-existent at lower columns.

the correct frequency of N_{HI} for higher columns. The curves of interest are the CDM line with a fixed UV history and the scaled low UV points (triangles). These simulations are otherwise identical and have been processed with a fixed value of $J_{912} = 5 \times 10^{-22}$. What *does* scale fairly well is the continuum depression, D_A, shown in the right panel of Figure 1. D_A is defined as the mean value of $1 - \exp(-\tau)$ for a spectral region. This is because the dominant contributors to the optical depth are lines with $N_{HI} \sim 10^{14}$, resulting from barely overdense material for which the UV flux history is less important.

The past history of the UV flux is important for the formation of dense gaseous clumps, as illustrated in the top left and centre right panels of Figure 2: the same UV flux was applied for the post-processing analysis, but the simulation shown in the centre right panel experienced a lower UV flux during its evolution (curve 2 in Figure 1). Applying a different flux in post-processing, the calculation can, of course, have no effect on the density distribution and thus cannot compensate for a different history.

Subtracting a power law [derived from the Hu et al. (1995) low end slope] from the $f(N_{HI})$ curve of Figure 3 reveals interesting features in the data. If the bumps at $N_{HI} = 10^{14}$ and $N_{HI} = 10^{17}$ (where self shielding of the UV flux just becomes important) are real, they may allow simulations to discriminate finely between parameter choices. The data appear to follow a power law with slope of -1.9 in the range $N_{HI} = 10^{14}$–10^{16} cm^{-2}, which our runs reproduce. The excess of intermediate column lines seen in the low-UV simulations may be caused by unrealistically low energy input at early times; specifically there should be input from early stars. We tested this hypothesis using a UV flux history similar to curve 3 in Figure 1, but the associated extra energy input was too small to have a large impact. Star formation and supernova feedback would also inhibit the formation of the extremely dense, cool clumps associated with this excess. We are now incorporating these processes with a view towards a general chemical evolution code.

We have illustrated that the Lyman alpha forest is a sensitive probe of the conditions that operated at $z \sim 5$ to 2. It is clear that we must produce a comprehensive sample if we are to deliver reliable predictions to the observers of the clouds. Finding forest observables that allow us to separate the influence of primordial power spectrum shape and amplitude, cosmological parameters such as Ω_Λ, Ω_{CDM}, Ω_{HDM}, H_0 and Ω_B, the ionization history and local energy injection from galaxies (feedback) will be a challenge to the growing group of cloud simulators, in close collaboration with the observers of the forest. Nonetheless, it is clear that the basic model (whether CDM or CDM+Λ) with $\sigma_{0.5\,Mpc} \sim 1$ at $z \sim 3$ provides a rather good fit to the data.

Acknowledgments. Support from the Canadian Institute for Advanced Research and NSERC is gratefully acknowledged. We would like to thank John Lattanzio and Joe Monaghan, who were involved in the early phases of this project, using an SPH-Multigrid code of earlier vintage.

References

Bechtold, J. 1994, in *QSO Absorption Lines*, ESO Astrophysics Symposia, ed. G. Meylan, (Berlin: Springer-Verlag), 299.

Haardt, F., & Madau, P. 1996, ApJ, 461, 20.
Hernquist, L., Katz, N., Weinberg, D., & Miralda-Escudé, J. 1996, ApJ, 457, L51.
Hu, E. M., Kim, T. S., Cowie, L. L., Songaila A., & Rauch, M. 1995, ApJ, 466, 46.
Miralda-Escudé, J., Cen, R., Ostriker, J. P., & Rauch, M. 1996, ApJ, 471, 582.
Mücket, J. P., Petitjean, P., Kates, R. E., & Riediger, R. 1996 A&A, 308, 17.
Petitjean, P., Webb, J. K., Rauch, M., Carswell, R. F. & Lanzetta, K. 1993, MNRAS, 262, 499.
Rauch, M., Haehnelt, M. G., & Steinmetz, M. 1996, submitted to ApJ.
Rauch, M., Miralda-Escude, J., Sargent, W., Barlow, T., Weinberg, D., Hernquist, L., Katz, N., Cen, R., & Ostriker, J. 1997, submitted to ApJ.
Schneider, D. P., Schmidt, M., & Gunn, J. E. 1991, AJ, 101, 2004.
Zhang, Y., Anninos, P., Norman, M. L., & Meiksin, A. 1996, submitted to ApJ.

Simulating Cosmic Structure at High Resolution: Towards a Billion Particles?

H. M. P. Couchman
Department of Physics and Astronomy, University of Western Ontario, London, ON N6A 3K7, Canada

Abstract. Cosmic structure simulations have improved vastly over the past decade, both in terms of the resolution which can be achieved, and with the addition of hydrodynamic and other techniques to formerly purely gravitational methods. This is an informal, and perhaps idiosyncratic, overview of the state and progress of cosmological simulations over this period.

I will discuss the strategies that are used to understand cosmic structure formation and the requirements of a successful simulation. The computational demands of cosmological simulations in general will be highlighted as will specific features of the various different algorithms that are used. The incessant push for greater resolution has lead to an increasing use of parallel computers. The desirability—or otherwise—of this trend will be discussed together with alternative techniques which sidestep the naïve drive for greater resolution.

1. Introduction

One of the key issues of contemporary cosmology is to understand the formation and evolution of cosmic structures; from the pattern of clusters and galaxies on the largest scales down to the formation of galaxies themselves. In the standard model, cosmic structure grew from a spectrum of low-amplitude fluctuations present at the epoch of recombination. The form of this spectrum is at present unknown, although a number of experiments are placing constraints on the fluctuation amplitude at recombination on large scales. Early next century, satellite missions, such as COBRAS/SAMBA and MAP, promise to constrain the recombination-epoch spectrum tightly as well as a number of other significant cosmological parameters. It is the spectrum of fluctuations at recombination which forms the initial conditions for studies of structure formation.

Whilst the fluctuation amplitude is small, post-recombination growth is linear; spectral modes grow independently and proportionately to the universal expansion (assuming that the Universe is close to flat at the relevant epoch). Presently-observed structures range from the quasi-linear, such as the large scale network of filaments and voids and superclusters, to the highly non-linear, such as galaxies and galaxy clusters. Whilst linear growth of small fluctuations is well understood and analytically tractable, only approximate methods and perturbation theory or simple "toy" models are available for studying non-linear growth. Making the connection between the initial fluctuation spectrum and presently observed structures, a task central to our understanding of the post-recombination universe, requires numerical simulation. The need for numerical simulation becomes even more pressing when we add the hydrodynamic component necessary for an understanding of gas in clusters and, especially, dissipation in galaxies.

This paper is laid out as follows. I begin with a pictorial description of the growth of the fluctuation spectrum and the transition from linear to non-linear growth for a generic class of spectra in which bound objects progress from small to large. I then discuss the requirements of a numerical simulation in terms of the way in which we model the universe and the force and mass resolution necessary for various aspects of structure formation. Following a description of the motivation for the use of particle methods, the various different algorithms that have been developed are summarized. A brief historical overview of the achievements of simulation methods is then presented. The push towards the use of supercomputers for numerical simulation is considered and I will present a case study of this endeavour. I conclude with a view of the problems posed by very large simulations and speculate about the future.

2. Growth of Fluctuations

Many popular contemporary cosmological models consider spectra in which the mass variance is a decreasing function of scale. Such spectra can explain a number of observational features, such as, for example, the fact that clusters have formed more recently than galaxies.

Figure 1 shows a pictorial view of the growth of the mass variance as a function of scale. (The precise form of the spectrum is unimportant and only the decrease to large scales is significant for this discussion.) The figure illustrates in a generic way at what scales and epochs we encounter non-linear gravitational evolution. Detailed understanding of structures on scales for which $\sigma(R) \gtrsim 1$ at any epoch requires numerical simulation.

Key questions which will benefit from a numerical approach include:

- The large-scale distribution of galaxies and their "bias" relative to the underlying gravitationally dominant dark matter component.

- The formation and evolution of clusters; the behaviour of the intra-cluster medium and the distribution and evolution of the galaxy population.

- Understanding the structure of the universe out to the redshifts of quasars; Lyman-α forest, *etc.*

- The formation and influence of the first bound objects at redshifts between 10 and 40.

- Galaxy formation itself—the "Holy Grail" of post-recombination cosmology.

The value of numerical simulation is that it allows us to surmount the obstacle of non-linear gravitational and hydrodynamic evolution. Not only will we

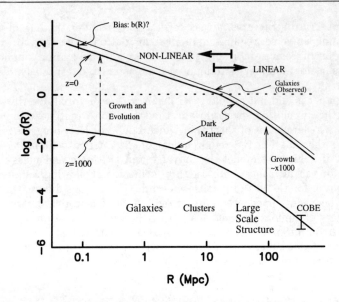

Figure 1. A schematic view of the post-recombination growth of the fluctuation spectrum, here represented as the mass variance as a function of scale. Whilst the mass variance is less than unity, growth is linear, and for larger amplitudes reliable prediction requires numerical simulation.

be able to investigate the formation and evolution of presently-observed structure and probe its immediate precursors, but we will also be able to make the crucial connection with the initial fluctuation spectrum.

3. Simulation Requirements

Cosmological simulations have considered a number of different aspects of structure formation. These have ranged from studies of the large-scale distribution of matter, without specific reference to galaxies, to studies of individual clusters and galaxies. I will focus on the so-called "Grand Challenge" problem of simulating the formation and distribution of galaxies in the cosmological context. This study is well motivated: galaxies form the fundamental building blocks of the universe. With the advent of huge galaxy redshift surveys such as the Sloan Digital Sky Survey and the Two Degree Field, we will have an enormous observational data base with which to interpret and constrain theoretical models. The relationship of the galaxies that we observe to the overall matter distribution in the universe is, however, uncertain. It is crucial that we understand the nature of any bias that may be present. It is only through detailed numerical simulation of the formation and evolution of galaxies (which are highly dissipated objects) that we can hope to achieve a detailed understanding of cosmic structure.

Broadly speaking, the requirements of such a simulation are that we model a representative piece of the universe over a sufficient range of scales with accurate forces and time integration schemes. These are, of course, very general requirements which apply to any well-conceived numerical simulation. One of these desirable qualities, accurate time integration, I will mention only in passing. Typically, second-order leapfrog or low-order Runge-Kutta or Predictor-Corrector schemes are employed. This is largely a consequence of the need to minimize the storage of extra quantities for a huge number of particles whose orbits must be integrated, coupled with the expense of the force calculation. That it is acceptable to use such low-order schemes is a result of studying gravitational collapse. Because of the instability of the system, we cannot hope to follow individual particle orbits accurately but are only concerned that we adequately model the properties of bound objects in a statistical sense. Note that this is in stark contrast to the extreme care which is exercised in the integration of orbits in Solar System studies as described elsewhere in this volume.

The issue of the boundary conditions to apply to the simulation will, of course, depend upon the precise nature of the problem being considered. I will return briefly to this issue towards the end of the paper. It suffices at present to note that, for the study of the large-scale distribution of galaxies, it is convenient to choose a cubic simulation volume with triply-periodic boundary conditions which is large enough, subject to the limitations of available resolution, to represent a fair sample of the universe.

For the remainder of this section, I will concentrate on two aspects of resolution; force and mass resolution. Gaining higher and higher resolution has been a primary focus of much of the effort expended in cosmological simulations over the last decade. I will attempt to justify this trend and at the same time give the motivation for the use of particle methods.

3.1. Force resolution

The "top hat" model is a useful simplification of the collapse of an initially overdense region. An overdense spherically symmetric region within an otherwise homogeneous universe will have an expansion and contraction corresponding to the evolution of a super-critical universe. In practice, inhomogeneities and external tides will generate non-radial motions which will prevent the region collapsing to a singularity and cause it to virialize at roughly half its turnaround radius. At virialization in a flat universe, the overdensity of the object will be nearly 200, and this will increase as the background continues to expand. An object formed at a redshift of 4 would have an overdensity of 10^4 at redshift zero, even without dissipation.

If we are to model cosmic structure properly, it is essential to follow and retain bound objects at increasing density contrast. An instance of the importance of this would be in an investigation of substructure; galaxies within a cluster, for example. Since a cosmological simulation typically models a comoving region of the universe, these very large density contrasts are present in the simulation. Popular choices for modelling the wide range of densities and varied geometries that occur are Lagrangian particle methods. Although some early simulations represented each galaxy with a single particle, it is now much more common to model the matter density as a collisionless fluid which is approximated by an ensemble of softened particles. An important question is then the choice of softening. To maintain a strict fluid limit, the softening should never be less than the mean interparticle separation, and yet this precludes accurate tracking of highly over-dense regions within a comoving volume. It is more realistic to demand that the softening be of order the mean interparticle separation *within a bound object*, and thus maintain the collisionless fluid approximation locally. Since objects collapsing at different epochs have different physical densities, a spatially variable softening is indicated. Most workers, however, chose a fixed spatial softening for the gravitational interaction, although as the dynamic range of simulations increases, this question will need to be revisited.

The art of choosing the appropriate softening lies in negotiating between the unwanted effects of two-body relaxation if the value is too small, and over-merger of "fluffy" substructure if the value is too large. Typically, realistic simulations of galaxies in a cosmological context, for example, require force-softenings at least an order of magnitude smaller than the mean interparticle separation.

3.2. Mass resolution

As noted above, a huge amount of effort has been expended in running simulations with ever larger particle number. I give a simple order-of-magnitude estimate for a suitable particle number for the large-scale galaxy distribution Grand Challenge problem.

To model a fair sample of the universe requires that we simulate a cube of at least 100 Mpc on a side. A volume of this size will contain roughly 10^4 bright galaxies. The fraction of the overall matter density residing in galaxies is perhaps one percent. Allowing for 100 particles per galaxy halo suggests that a simulation should contain at least 10^8 resolution elements or particles. Only now is it becoming possible to run simulations with this particle number and typically the simulations which have been run have not had high resolution forces. The above argument, whilst simplistic, demonstrates clearly the need for very high resolution.

A further leap in the required resolution becomes necessary when one wishes to model the dissipative component of galaxies. Suppose that one uses Smoothed Particle Hydrodynamics (SPH; Gingold & Monaghan, 1977) to model the gaseous content of galactic haloes. SPH, which fits well with the particle codes, requires at least 10^3 particles per galaxy to follow the cooling and collapse of gas in dark haloes reliably, and likely considerably more. This implies a simulation with $\sim 10^9$ particles.

Not all investigations require modelling of huge volumes of the universe. Two examples will illustrate the requirements in other situations.

Galaxy Formation: 4×10^6 particles in a 10 Mpc region $\rightarrow$ 1 kpc resolution.

Clusters: 4×10^7 particles in a 50 Mpc region $\rightarrow$ 3 kpc resolution.

Simulations with such huge numbers particles are not without drawbacks, of course. The storage for positions and velocities alone for 10^8 particles is 2.4 Gbyte for each time-slice (32-bit words), and roughly 50% greater for the storage of all quantities in an SPH simulation. Nonetheless, the problem is well motivated and, at least for the problem under consideration, these sorts of particle numbers are being considered and the first such runs performed. Various alternatives to the straightforward approach of greater particle number are mentioned in the final section.

4. N-Body Algorithms

The task which must be accomplished then, is the computation of the gravitational interaction of a very large number of particles. Whilst conceptually simple, this problem has consumed a vast amount of effort and programming skill and has led to a number of different approaches. The primary popular approaches are described in outline below. All of these, apart from the first, calculate the gravitational potential by dividing the field into two components; the near field, which has high frequency content, and the far field, which is approximated to some order by a given basis expansion. Assuming that the force at an arbitrary point is required to a specified accuracy, it can be argued heuristically that these methods all scale as $O(N \ln N)$, where N is the particle number.

Particle-Particle (PP) Forces on each particle are accumulated by directly summing over all neighbours. Whilst this method is very robust, it is $O(N^2)$ and is hence limited to fewer than $\sim 10^{3-4}$ particles. (It is worth noting that this method has received a new lease of life in cosmology as a result of the development of application-specific chips which compute the Newtonian force in hardware; see Hut, these proceedings.)

Tree methods These methods are the logical successors to the PP method although historically they appeared after the standard grid-based techniques described next. These methods take account of the fact that, to a given accuracy, a group of particles distant from the force calculation point may be approximated by a low-order multipole expansion thus avoiding the expensive sum over all particles in the group. Various types of tree have been used to organize the data in such a way that the decision as to whether or not a cell of particles need be subdivided for a particular force calculation may be made efficiently. For details, see Barnes & Hut (1986) and Hernquist & Katz (1989). A number of variants exist. The Fast Multipole Method of Greengard & Rokhlin (1987) is similar in spirit to this approach. It will not be described further here as its use in cosmology has been very limited.

Grid-Based Methods These methods sample the density field with a (usually) uniform grid. Poisson's equation is then solved on this grid using one of a number of the fast solvers that are available, usually either an FFT or multigrid. The potential obtained corresponds to the far field component as the grid cannot represent frequencies higher than the Nyquist limit. The Particle-Mesh (PM) force is typically augmented by a short-range contribution summed over near neighbours, thus allowing high resolution spatial forces. The method can become grossly inefficient as clustering develops, however, and the short-range direct sum becomes dominant. This problem has been alleviated in the Adaptive P^3M (AP^3M; Couchman, 1991) code by using a hierarchy of adaptive meshes in regions of high particle density.

Most workers now use either a version of the Tree code or a grid-based method (primarily P^3M, or variant, using Fast Fourier Transforms), although a number of other techniques have been, or are being, developed. These include Grid-Tree hybrids in which a tree replaces the expensive direct sum in P^3M and other grid techniques in which a true grid refinement strategy is used.

There are a number of pros and cons for each of the primary methods. Tree codes are robust, the cycle time does not vary much between light and heavy clustering and they provide a flexible data structure which makes implementation of individual particle time steps, parallelization and inclusion of SPH relatively straightforward. Their primary drawbacks are that they use a substantial amount of memory, $\sim$ 25–35 words per particle, and, for cosmology, the necessary implementation of periodic boundary conditions via the Ewald (1921) summation method can be cumbersome.

P^3M has the advantage that it is very fast under conditions of light clustering, has automatically periodic boundary conditions when using an FFT to solve for the grid potential and uses relatively little memory; $\sim$ 8–10 words per particle. A severe disadvantage is that clustering can dramatically slow the algorithm unless an efficient scheme for relieving the PP work in highly clustered regions is implemented. Although the use of regular, uniform grids throughout the calculation leads to efficiency in terms of memory use and execution speed under light particle clustering, it results in a less general data structure than a tree, for example, which makes inclusion of individual particle time steps, as well as some aspects of parallelization, more difficult.

5. Hydrodynamics

The addition of a hydrodynamic component to gravitational cosmological codes permits the investigation of many new aspects of structure formation. Of particular importance has been the study of the hot intra-cluster medium. The number density and distribution of clusters provide important constraints on the primordial spectrum. More recently, several ambitious attempts have been made upon the prize of galaxy formation itself, in which correct modelling of dissipative processes is crucial.

Perhaps the most straightforward technique for adding hydrodynamics to a particle code is Smoothed Particle Hydrodynamics (SPH). In this method, thermodynamic quantities are carried by particles and the value of the fluid is approximated at any point by interpolating values from nearby particles. Because of the ease of integrating this method with gravitational particle methods, it was the first widely used hydrodynamic method in cosmology. Implementations of SPH in both Tree codes (Hernquist & Katz, 1989) and P^3M (Evrard, 1990) exist.

More recently, a number of alternative schemes have been imported into cosmological codes. In particular, Eulerian codes have become popular. These offer a number of potential advantages over SPH, in particular efficient shock capturing schemes greatly improve on the number of resolution elements necessary to model a shock; in 1-D two or three cells for an Eulerian code and 6 particles for SPH. These methods are also well suited to the addition of magnetohydrodynamics and radiation fields. In order for these codes to follow the large density contrasts that arise, adaptive mesh refinement techniques are essential, and these are now beginning to be used (see Bryan & Norman, these proceedings). Other schemes include semi-Lagrangian methods and very recently an unstructured finite element method has been described (Xu, 1997). In all of these codes, particles carry the gravitational mass.

6. Achievements

In this section I will summarise the improvements that have occurred in the field of cosmological simulations over the last decade, very briefly list the areas in which they have played a significant role in advancing the subject, and outline the present status of the field.

Although it is clear that cosmological simulations are an essential tool for cosmologists studying structure formation, they have not proved to be the central mechanism for generating new ideas or insight about the post-recombination universe. The physical ideas are simple and well understood; provided that we can quantify the collective behaviour of particles, simulations are helpful in understanding the behaviour and interaction of these physical processes in highly complicated situations. In this respect, cosmological simulations have a different focus and utility than simulations of say, galactic dynamics. In the latter (well studied) case, the galaxy may be idealised to a degree which allows detailed investigation of, for example, modes of a galactic disc. In cosmological simulations, the focus is rather different. First, we are typically trying to model a realization of an initial random field (albeit with known initial spectrum) which lacks the symmetries of the galaxy study. Second, analytic theory beyond first order, so far, has quite limited predictive power. A reasonable expectation is that simulation and theory will play a mutually supportive role in improving our knowledge in this area.

A convenient milepost to the development of what might be termed "modern" cosmological simulation methods is the pioneering work of Davis, Efstathiou, Frenk and White in the mid 1980s (see Efstathiou et al., 1985, for a description of numerical methods). This work, following from earlier work by Efstathiou and Eastwood, drew the P^{3}M code of Hockney & Eastwood (1988) from the plasma physics arena into the cosmological community and represented what might be termed the first high-resolution cosmological simulations with $\sim 3 \times 10^4$ particles.

With the availability of efficient algorithms, increases in computer speed and memory size led rapidly to higher-resolution simulations. By the late 80's, $\sim 5 \times 10^5$ particle P^{3}M simulation were run and this number increased to $\sim 2 \times 10^6$ by 1990. At present it is possible to run routinely $\sim 2 \times 10^7$ particle AP^{3}M simulations. Differing memory and CPU requirements mean that at a given time somewhat larger PM, or smaller Tree-code, simulations can be run. Hydrodynamic simulations, both because of their larger memory requirement and the necessity of using a smaller time step, are generally at present limited to a few million particles.

A common criticism of cosmological simulation efforts has been that they are merely a method of generating attractive pictures. It is certainly true that, so far, there have been only a few useful statistical measures used to express the collective behaviour of the large number of particles in a cosmological simulation. It is worth bearing in mind the comments made above, however, concerning the likely role that simulations will play in improved understanding of structure formation. The following items of progress are essentially all cases in which simulations have checked analytic models or predictions. This is only a sample of the areas in which simulations have played a key role but it is indicative of the way in which simulations have been, and are likely to continue to be, employed.

- Checks of perturbation theory.

- Testing the validity of the Press-Schechter (1974) model, which attempts to model the could-in-cloud problem.

- Seeking a universal scaling function to describe the non-linear mass auto-correlation as a function of the initial linear auto-correlation (Hamilton et al., 1991).

- Determining cluster distributions and abundances.

- Attempts to understand galaxy bias through an investigation of the distribution of putative galaxy haloes in non-hydrodynamic simulations.

- Understanding the intra-cluster gas.

- Modelling Lyman-α clouds and Lyman-α forest observations.

- Verifying the broad outlines of the White-Rees (1978) picture for baryon condensation within dark haloes.

7. The Cutting Edge: A Case Study

The challenge of simulating the formation of galaxies in the cosmological context requires huge computational resources. The estimates given above suggest 10^{8-9} particles. Both the computer time and the amount of memory required for such large simulations pose severe computational constraints and, at present, are available only on massively parallel supercomputers. A number of groups are working in this area with parallel codes. I will describe the "Virgo" collaboration in which I am involved (Pearce et al., 1996). This is a group of eight, primarily UK researchers, using the UK national supercomputer (a Cray T3D) in Edinburgh as well as a Cray T3E in Munich.

We have ported the AP^{3}M code to the Edinburgh 512 processor Cray T3D using CRAFT, Cray's proprietary software. This software creates a global address space which appears flat to the programmer vastly easing the difficulty of parallelizing code for a distributed memory architecture. The penalty, however, is a certain amount of inefficiency and lack of portability. For this reason an explicit message passing (MPI) version is now being developed.

The primary difficulty with parallel N-body codes is to achieve an efficient data and work distribution. With grid codes, there is the added difficulty that the mapping from a perhaps highly clustered particle distribution to a regularly spaced grid may lead to a large communications overhead. Tree codes tend to be somewhat better behaved in this regard as they avoid the potential many-to-one mapping of the Lagrangian particles to the Eulerian grid and because of the use of a more flexible data structure. In general, however, care must be taken with particle codes to minimize the communication overhead. The particle is the fundamental element and essentially no work is done without reference to its neighbours. Therefore, data must be carefully distributed to avoid costly inter-processor communication. This behaviour can be compared with that in finite element codes in which a large amount of work is done internally to the element and the ratio of computation to communication is much higher.

The parallel AP^{3}M refined grid code scales well up to 256 processors, although some load imbalance occurs for heavy clustering. We have used a strategy in which we distribute data in one spatial direction over the processors. Provided there are a number of particle clusters present in the simulation volume, this is sufficient to average out the worst load imbalances. We are now achieving performances of 2,500 particles/s/T3D node for collisionless simulations and 700 particles/s/T3D node for the AP^{3}M-SPH code "HYDRA" (Couchman et al., 1995). A 2×10^7 particle run with 5,000 steps would thus take approximately one week on 256 nodes of a Cray T3D; corresponding to 5 years total processing time. This number of particles uses about half the 64 Mbyte available per node on 256 nodes. A Cray T3E has 3 times faster processors and can have up to 2Gb of memory per node; 10^8 particles should be feasible.

8. The Future

I will conclude by making a few remarks about the future of cosmological simulations. It is clear that the trend towards larger simulations will continue. As the dynamic range of a simulation increases, so does the range of density contrasts that can, in principle, be modelled; alternatively, the first structures which we can confidently identify in a simulation form at earlier epochs. Accurate modelling demands that the simulation can deal with the wide range of densities and time scales which arise. Both spatial and temporal adaptivity will become more important. Tree codes frequently have both whilst the AP^{3}M code lags, for example, in having only spatial adaptivity.

One of the most obvious problems with large supercomputer applications is the vast amount of data that is generated. Cosmological N-body codes are no exception. Each time slice of a 2×10^7 particle run occupies 960 Mbyte (64 bit) for position and velocity storage. The output of a hydrodynamic code would be nearly 50% greater: $\sim$ 1.5 Gbyte per time slice.

An approach which helps avoid the data explosion is to concentrate on selected volumes. This has generated a great deal of interest recently with a large number of workers pursuing different avenues of attack. There is now a general appreciation that external fields must be carefully applied to the selected volumes to mimic the tidal effects of the external universe properly. This general approach is very promising. It remains to be seen whether the information gained from these studies will be sufficient to allow us to predict the locations of galaxies in larger-volume (perhaps collisionless) simulations. If we are to assess the upcoming large sky surveys properly, and hence to understand galaxy distributions, morphologies, bias, *etc.*, we must be able to generate reliable theoretical galaxy catalogues from simulations.

References

Barnes, J., & Hut, P. 1986, Nature, 324, 446.
Couchman, H. M. P. 1991, ApJ, 368, L23.
Couchman, H. M. P., Thomas, P. A., & Pearce, F. R. 1995, ApJ, 452, 797.

Efstathiou, G., Davis, M., Frenk, C. S., & White, S. D. M. 1985, ApJS, 57, 241.
Evrard, A. E. 1990, ApJ, 363, 349.
Ewald, P. P. 1921, Ann. Phys., 64 253.
Gingold, R. A., & Monaghan, J. J. 1977, MNRAS, 181, 375.
Greengard, L., & Rokhlin, V. 1987, J. Comput. Phys., 73, 325.
Hamilton, A. J. S., Kumar, P., Lu, E., & Matthews, A. 1991, ApJ, 374, L1.
Hernquist, L., & Katz, N. 1989, ApJS, 70, 419.
Hockney, R. W., & Eastwood, J. W. 1988, *Computer Simulation Using Particles*, (New York: Adam Hilger).
Pearce, F. R., Thomas, P. A., Hutchings, R. M., Couchman, H. M. P., Jenkins, A. R., Frenk, C. S., White, S. D. M., & Colberg, J .M. 1996, in *Proceedings of 5th Euromicro Workshop on Parallel and Distributed Computing*, ed. Hans P. Zima (IEEE Computer Society Press).
Press, W. H., & Schechter, P. 1974, ApJ, 187, 452.
White, S. D. M., & Rees, M. J. 1978, MNRAS, 183, 341.
Xu, G. 1997, preprint, astro-ph/9610061.

Nature of the Low Column Density Lyman Alpha Forest

Michael L. Norman[1], Peter Anninos, Yu Zhang[2]

Laboratory for Computational Astrophysics and NCSA, University of Illinois at Urbana-Champaign, Urbana, IL 61801, U.S.A.

Avery Meiksin

Department of Astronomy and Astrophysics, University of Chicago, 5640 South Ellis Ave., Chicago, IL 60637, U.S.A.

Abstract. We describe the physical nature of the absorbers at the low column density end of the Lyα forest in a CDM cosmology. We find that absorbers below HI column densities of $\sim 10^{13.3}\,\mathrm{cm}^{-2}$ are underdense relative to the cosmic mean, expand faster than the Hubble flow, and are considerably cooler than 10^4 K. They are found in and around the edges of small voids—"mini-voids"—in the distribution of the intergalactic gas, and are primarily sheet-like in their shape.

1. Introduction

The last few years have witnessed considerable advances in our understanding of the structure of the intergalactic medium (IGM) predicted by Cold Dark Matter (CDM) dominated cosmologies. Several groups have performed a series of hybrid numerical N-body/hydrodynamical simulations of structure formation in the IGM (Cen et al., 1994; Zhang et al., 1995; Hernquist et al., 1996) that are converging on a definite picture for the origin of the Lyα forest in a CDM universe. Although some differences between the simulations remain to be resolved, the general landscape the simulations have drawn is one of an interconnected network of sheets and filaments, with dwarfish spheroidal systems, essentially mini-halos (Rees, 1986; Ikeuchi, 1986), located at their points of intersection, and fluctuations within low density "mini-voids" between. Bond et al. (1996) have argued that this "cosmic web" is a generic feature of CDM, a consequence of an incoherent pattern in the matter fluctuations imprinted at the epoch of matter-radiation decoupling and sharpened by gravitational collapse. The distribution in neutral hydrogen column densities along lines of sight piercing the filaments, as well as the distribution of velocity widths of the resulting absorption features, are found to coincide closely with the measurements of the Lyα

---

[1] also, Department of Astronomy, University of Illinois at Urbana-Champaign, Urbana, IL 61801, U.S.A.

[2] current address, Department of Medical Science, University of Rochester, Rochester, NY 14627, U.S.A.

forest in the spectra of high-redshift quasars (Miralda-Escudé et al., 1996; Davé et al., 1997; Zhang et al., 1997, hereafter ZANM).

High resolution and signal-to-noise spectra of the Lyα forest using the Keck HIRES spectrograph have shown that the power-law differential HI column density distrubtion $f(N_{HI})$ extends to very low column densities: $N_{HI} \approx 10^{12}\,\mathrm{cm}^{-2}$ (Hu et al., 1995; Lu et al., 1996; Kirkman & Tytler, 1997). In ZANM, we matched this result through a synthetic spectrum analysis of a high-resolution multispecies hydrodynamic simulation of small-scale structure formation in a standard CDM cosmology in a photoionizing background. In this short communiqué, we explore the physical nature of the low column density end of the Lyα forest $12 \leq N_{HI} \leq 14$.

2. Simulations

As we will show, at $z = 3$, absorption features below an HI column density of $10^{13.3}\,\mathrm{cm}^{-2}$ arise from fluctuations in mini-void regions of the IGM which are underdense relative to the cosmic mean. Lagrangian SPH calculations are at a severe disadvantage resolving these underdense features relative to Eulerian grid calculations, as the local smoothing length varies as the matter overdensity to the $-1/3$ power typically. Hernquist et al. (1996) quote a resolution element of 200 kpc in the void regions of their CDM simulation. We employ the two-level hierarchical grid code HERCULES (ZANM; Anninos et al., 1997) to achieve a spatial resolution $10 \times (40 \times)$ higher at $z = 3$ on the top (sub) grid of the calculation. The top (sub) grid covers a domain of size 9.6 (2.4) comoving Mpc on a side. We use 128^3 cells on both top and subgrid, and 128^3 particles on each to represent the dark matter. The subgrid is centered on the *least* dense region of the top grid for the purpose of resolving in greater detail the fine density structure of the voids and checking for numerical convergence. The simulations are performed in a flat cosmology with no cosmological constant, and $h_{50} = 1$. The CDM power spectrum is normalized to $\sigma_8 = 0.7$. Since the neutral density of individual clouds scale like $\Omega_b h_{50}^2/\Gamma_{HI}$, the counts of objects depend on the magnitude of the radiation field. The results presented below are for the radiation field determined by Haardt & Madau (1996) and $\Omega_b = 0.06$. Further details are provided in ZANM and Zhang et al. (in preparation).

3. Correlation between Overdensity and Column Density

Figure 1 illustrates visually the strong correlation which exists between baryonic overdensity and HI column density, while Figure 2 demostrates this result statistically. Shown in Figure 1 is a grayscale image of the baryonic overdensity on a slice through the 9.6 Mpc box at $z = 3$ (left), and the corresponding HI column density summed over a line-of-sight depth of 1/16 the box size (right). The high column density absorbers ($N_{HI} > 10^{15}\,\mathrm{cm}^{-2}$) correspond to the highly overdense structures ($\rho_b/\bar{\rho}_b > 10$) residing mostly along and at the intersection of filaments. The medium column density absorbers ($\sim 10^{13-14}\,\mathrm{cm}^{-2}$) correspond to the modestly overdense filaments ($1 < \rho_b/\bar{\rho}_b < 5$) . The column density can be coherent over the scale of a few Mpc at this level. The lowest column density absorbers ($\sim 10^{12-13}\,\mathrm{cm}^{-2}$) are associated with underdense structures

LOW COLUMN DENSITY Lyα FOREST   353

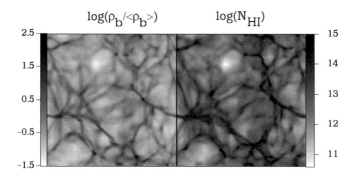

Figure 1. Gray scale images of the logarithm of the baryonic overdensity on a slice through the 9.6 Mpc box at $z = 3$ (*left*), and the corresponding HI column density summed over a line-of-sight depth of 1/16 the box size (*right*).

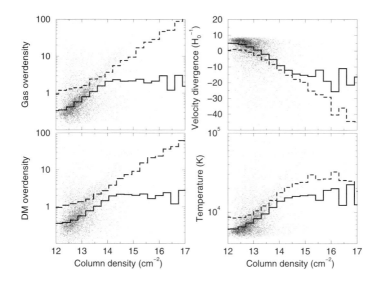

Figure 2. Scatter plots showing the correlations between baryonic overdensity, dark matter overdensity, temperature, and divergence of the peculiar velocity with HI column density, at $z = 3$. Also shown are the median (*solid line*) and mean (*dashed line*) of the distributions.

($\rho_b/\bar{\rho}_b < 1$) and are located in the void regions between the filamentary structures.

Figure 2a shows a scatter plot of $\rho_b/\bar{\rho}_b$ vs. N_{HI} for the top grid data. Each point represents a spectral line in our synthetic spectrum; the corresponding baryonic overdensity is taken from the 3-D data at the position of the line center correcting for peculiar motions. We run out of lines below $N_{HI} \sim 10^{12}\,\mathrm{cm}^{-2}$ because of line blending (*cf.*, Figure 13, ZANM). The already good correlation seen on the top grid only improves on the subgrid, convincing us of its reality. Over the range ($12.5 < \log N_{\mathrm{HI}} < 14.5$), the internal cloud density varies with column density as $\rho_b/\bar{\rho}_b \approx N_{\mathrm{HI},13.3}^{1/2}$, where $N_{\mathrm{HI},13.3}$ is the column density in units of $10^{13.3}\,\mathrm{cm}^{-2}$.

4. Kinematic and Thermal State

A strong correlation exists between HI column density and other physical properties of the absorbers, including dark matter overdensity, temperature and peculiar velocity divergence. Scatter plots illustrating this for the $z = 3$ top grid data are also shown in Figure 2. Similar correlations were found for a ΛCDM model (Miralda-Escudé et al., 1996). Although there is a fair amount of scatter in the baryonic and dark matter overdensities individually, their average ratio is unity in the range $12.5 < \log N_{\mathrm{HI}} < 14.5$; i.e., unbiased. This rules out the theory explored by Bond et al. (1988) that the lowest column lines can be entirely explained as low-mass mini-halos that have lost their baryons through a photoionization heating driven outflow. In fact, the correlation of peculiar velocity divergence with column density shows that the lowest column absorbers are expanding faster than the Hubble flow; i.e., they are unbound. This is consistent with the observation that these absorbers are located in the mini-voids. Finally, the low column absorbers on average have temperatures well below 10^4 K—a general property of low density gas in the voids. As shown by Meiksin (1994), if the gas density is below ($\sim 10^{-4}\,\mathrm{cm}^{-3}$ for $z > 2$), which is the case in the mini-voids, adiabatic expansion cooling will overwhelm photoionization heating.

5. Cloud Sizes and Shapes

We may use the correlation between baryonic overdensity and column density to derive the line-of-sight depth of the absorbers. A cloud with an internal density of ρ_b that is optically thin at the Lyman edge and in ionization equilibrium with the radiation field will have a neutral hydrogen density of

$$n_{\mathrm{HI}} \simeq 7 \times 10^{-15}\,\mathrm{cm}^{-3} \left(\frac{\rho_b}{\bar{\rho}_b}\right)^2 (1+z)^6 T_4^{-0.75} \Gamma_{\mathrm{HI},-12}^{-1},$$

where $\bar{\rho}_b$ is the cosmic mean baryonic density, and $\Gamma_{\mathrm{HI},-12}^{-1}$ is the photoionization rate of neutral hydrogen in units of $10^{-12}\,\mathrm{s}^{-1}$. Combining this relation with the square-root dependence of the cloud density on HI column density, and defining the line-of-sight scale length of the absorbers to be $2l \equiv N_{\mathrm{HI}}/n_{\mathrm{HI}}$, we find that $2l \approx 100\text{--}200$ kpc, with a weak dependence on column density through the cloud temperature. Can we confirm this visually?

Figure 3 shows two 3-D isosurface plots of the baryonic overdensity at two thresholds: $\log(\rho_b/\bar{\rho}_b) = 0$ and 1, corresponding on average to HI columns of $10^{13.3}$ and $10^{14.3}\,\mathrm{cm}^{-2}$, respectively. At the higher threshold, we see filaments, as previously reported, with a characteristic thickness of 200 kpc. However, at the lower threshold value, we see sheets or walls of baryons of comparable

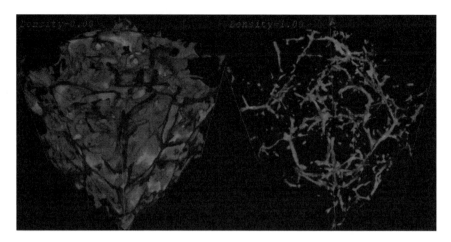

Figure 3. 3-D isosurface plots of the logarithm of the baryonic overdensity at $z = 3$ at two thresholds: $\log(\rho_b/\bar{\rho}_b) = 1$ (*left*) and 10 (*right*), corresponding *on average* to HI columns of $10^{13.3}$ and $10^{14.3}$ cm^{-2}, respectively.

thickness spanning the space between the filaments. By surface area, the mini-voids are bounded mostly by sheets, and to a lesser extent by filaments and an even lesser extent by other mini-halos. Given this topology, it is clear why the column density distribution function increases sharply to lower column densities: It simply reflects the probability that a ray will intersect these 1-D, 2-D and 3-D concentrations.

6. Origin of Mini-Void Absorbers

The baryonic Jeans length in a medium of mixed dark matter and baryons is $\lambda_J = 2\pi(2/3)^{1/2}c_s(1+z)/H(z)$, where c_s is the sound speed in the baryonic fluid and $H(z)$ is the Hubble constant (*e.g.*, Bond & Szalay, 1983). For an isothermal perturbation, $\lambda_J \simeq 1$ Mpc $h_{50}^{-1} T_4^{1/2}(1+z)^{-3/2}$, or about 150 kpc at $z = 3$. This gives a minimum column density caused by Jeans instability of $\log N_{\rm HI} \simeq 13$ at $z = 3$. The power spectrum of density fluctuations for short wavelength modes, $\lambda \ll \lambda_J$, will be suppressed by the factor $(\lambda/\lambda_J)^2$. Thus, one might reasonably expect a downturn in the HI column density distribution below $\sim 10^{13}$ cm^{-2}. This is not found observationally (Hu et al., 1995; Lu et al., 1996), nor is it what we find computationally (ZANM). Thus *we find significant sub-Jeans length-scale structure formation in the underdense regions.*

Meiksin (1996) has suggested an interpretation of these features in terms of the growth of fluctuations in an underdense background. A local overdensity in a large underdense region will grow as if in an open universe. After an initial period of growth relative to the diminishing background, the relative density perturbation will "freeze", very roughly at an epoch given by $1 + z_f \approx \Omega_v^{-1} - 1$, where Ω_v is the ratio of the void density to the closure density. A spherically symmetric simulation of the evolution of such a perturbation bears this out. The solutions also exhibit a mild pile-up of baryons at the edge of the void, which may account for the sheets of gas found in the 3-D simulations discussed above.

Acknowledgments. The authors would like to thank John Shalf for his assistance producing Figure 3. This work is supported in part by the NSF grant ASC-9318185 under the auspices of the Grand Challenge Cosmology Consortium (GC3). The computations were performed on the Convex C3880 and the SGI PowerChallenge at the NCSA, and the Cray C90 at the Pittsburgh Supercomputing Center under grant AST950004P. A.M. thanks the William Gaertner Fund at the University of Chicago for support.

References

Anninos, P., Zhang, Y., Abel, T., & Norman, M. L. 1997, New Astronomy, in press.
Bond, J. R., & Szalay, A. S. 1983, ApJ, 274, 443.
Bond, J. R., Szalay, A. S., & Silk, J. 1988, ApJ, 324, 627.
Bond, J. R., Kofman, L., & Pogosyan, D. 1996, Nature, 380, 603.
Cen, R., Miralda-Escudé, J., Ostriker, J. P., & Rauch, M. 1994, ApJ, 437, L9.
Davé, R., Hernquist, L., Weinberg, D., & Katz, N. 1997, ApJ, in press.
Ikeuchi, S. 1986, Ap&SS, 118, 509.
Haardt, F., & Madau, P. 1996, ApJ, 461, 20.
Hernquist, L., Katz, N., Weinberg, D., & Miralda-Escudé, J. 1996, ApJ, 457, L51.
Hu, E. M., Kim, T. S., Cowie, L. L., Songaila, A., & Rauch, M. 1995, AJ, 110, 1526.
Kirkman, D., & Tytler, D. 1997, AJ, in press.
Lu, L., Sargent, W. L. W., Womble, D. S., & Takada-Hidai, M. 1997, ApJ, submitted.
Meiksin, A. 1994, ApJ, 431, 109.
Meiksin, A. 1997, in *Young Galaxies and QSO Absorption-Line Systems*, eds. S. M. Viegas, R. Gruenwald & R. R. de Carvalho, (San Francisco: ASP), 1.
Miralda-Escudé, J., Cen, R., Ostriker, J. P., & Rauch, M. 1996, ApJ, 471, 582.
Rees, M. J. 1986, MNRAS, 218, 25.
Zhang, Y., Anninos, P., & Norman, M. L. 1995, ApJ, 453, L57.
Zhang, Y., Anninos, P., Norman, M. L., & Meiksin, A. 1997, ApJ, in press (ZANM).

Reheating of the Universe and Population III

Jeremiah P. Ostriker

Princeton University Observatory, Peyton Hall, Princeton, NJ 08544, U.S.A.

Nickolay Y. Gnedin

Astronomy Department, University of California, Berkeley, CA 94720, U.S.A.

Abstract. Extending prior semi-analytic work, we show by direct, high-resolution numerical simulations (of a *COBE*-normalized CDM+Λ model) that reheating will occur in the interval $20 > z > 7$, accompanied by a significant increase in the Jeans mass. However, the evolution of the Jeans mass does not significantly affect star formation in dense, self-shielded clumps of gas, which are detached from the thermal evolution of the rest of the universe. Cooling on molecular hydrogen leads to a burst of star formation prior to reheating which produces Population III stars. Star formation subsequently slows down as molecular hydrogen is depleted by photo-destruction and the rise of the temperature. At later times, $z < 10$, when the characteristic virial temperature of gas clumps reach 10^4 degrees, star formation increases again as hydrogen line cooling becomes efficient.

1. Introduction

In the standard, hot Big Bang model of cosmology, radiation and matter decouple from one another at an approximate redshift $z \approx 1{,}200$, with recombination proceeding until the ionized fraction freezes out at $n_e/n_{\rm H\,I} \approx 4 \times 10^{-4}$ at about redshift $z \approx 800$ (*cf.*, Peebles, 1993). However, we know from the classic Gunn-Peterson test (Gunn & Peterson, 1965), that the universe is again ionized at low redshifts. In this standard picture, reionization must occur from unknown sources in the intervening time interval.

Many authors have studied this process, from the pioneering work of Couchman & Rees (1986) and Shapiro & Giroux (1987) to the recent careful investigations by Tegmark & Silk (1995), Giroux & Shapiro (1996), Tegmark *et al.* (1997), to the one-dimensional simulations by Haiman *et al.* (1995). While varying amounts of physically-detailed modeling were included in all of those papers, there was a critical factor that could not be followed in any of the semi-analytic treatments: the clumping of the gas. Nevertheless, all relevant processes are dependent on the clumping factor $\eta \equiv \langle \rho^2 \rangle / \langle \rho \rangle^2$. Included among these processes are *recombination, cooling, gravitational collapse, molecular hydrogen formation*, and numerous others. Thus, a fully non-linear three-dimensional

Table 1. Model and Numerical Parameters

Run	N	Box size	Total mass res.	Spatial res.	Dyn. range
A	128^3	$2h^{-1}$ Mpc	$3.7 \times 10^5 h^{-1}$ M$_\odot$	$1.0 h^{-1}$ kpc	2,000
B	64^3	$2h^{-1}$ Mpc	$2.9 \times 10^6 h^{-1}$ M$_\odot$	$3.0 h^{-1}$ kpc	640
C	64^3	$1h^{-1}$ Mpc	$3.7 \times 10^5 h^{-1}$ M$_\odot$	$1.5 h^{-1}$ kpc	640

treatment appears to be warranted. That is what we have undertaken, using one of the currently plausible models for the growth of structure, the CDM+Λ model (cf., Ostriker & Steinhardt, 1995, for references), as a point of departure. The work reported here adopts the model and numerical parameters given in Table 1 (close to the "concordance" model of Ostriker & Steinhardt, 1995), but the dependence of our quoted results on the assumed scenario is, we expect, small. With a box size of $2h^{-1}$ Mpc and a Lagrangian resolution for the SLH code (Gnedin, 1995; Gnedin & Bertschinger, 1996) of 128^3 which corresponds to a gas mass of $\Delta M_g = 3.2 \times 10^4 h^{-1}$ M$_\odot$ and a dark matter particle mass of $\Delta M_d = 3.7 \times 10^5 h^{-1}$ M$_\odot$, we have marginally resolved the smallest self-gravitating structures. Spatial resolution in our fiducial run is approximately $1h^{-1}$ kpc.

A detailed description of the physical modeling will be presented in Gnedin & Ostriker (in preparation). Here we note that we have allowed for the formation and destruction of molecular hydrogen, as well as all other standard physical processes for a gas of primeval composition, following in detail the ionization and recombination of all species in the ambient radiation field. The spatially-averaged but frequency-dependent radiation field, in turn, allows for sources of radiation (quasars and massive stars), sinks (caused by continuum opacities) and cosmological effects. In regions which are cooling and collapsing, we have allowed the formation of point-like "stellar" subunits, permitting them to release radiation and (in proportion) metal-rich gas, which we have considered in the treatment of cooling. Finally, we have taken into account the fact that dense clumps will be shielded from the background radiation field. This reduces the heating rates for dense clumps and makes it nearly certain that once they have formed and started to collapse, the process will be irreversible (cf., Anninos et al., 1997).

2. A Qualitative Approach to Reionization

Let us first ignore reheating and estimate the rate at which matter would begin to collapse following the standard theories of gravitational instability such as the Press-Schechter picture (Press & Schechter, 1974). For cold gas in CDM-like theories, no mass elements are initially unstable to collapse. As gravitational instability grows, higher and higher mass scales become unstable, as indicated by the heavy-dotted line in Figure 1a.

Now consider the Jeans mass, the mass above which the self-gravitating gas at the cosmic density and temperature is unstable. The dashed curve in Figure

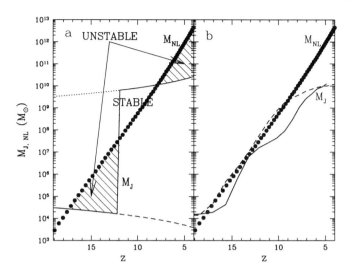

Figure 1. a) Schematic diagram showing the anticipated Jeans mass for the adiabatically-expanding intergalactic gas (*dashed line*), gas at 10^4 K (*dotted line*), and transition curve for instantaneous reheating (*solid line*). Dots show the non-linear mass. b) The non-linear mass (*filled dots*) and the actual Jeans mass from run A (*solid line*) and run B (*dashed line*). Note that the globally averaged Jeans mass in panel b) closely tracks the mass scales becoming non-linear because of gravitational instability.

1a shows gas at the CBR temperature and the dotted curve marks the gas at 10^4 K. A schematic transition curve is indicated by the solid line for gas that heats rapidly to 10^4 K at $z = 12$ when, we hypothesize, ionization and reheating occur in this schematic treatment. We see that we might expect an epoch of instability to exist before reheating and reionization, when low-mass objects collapse out of cold gas (left hatched region) and another, later epoch, when galactic-mass objects condense out of 10^4 K gas (right hatched region). These are separated in time by a region during which gas is stable against gravitational collapse.

However, as we shall see, while there are elements of truth to the scenario indicated by Figure 1a, the real situation is far more complex.

3. Results

We have performed three different simulations (*cf.*, Table 1) with different box sizes and resolution to assess the importance of different scales and estimate the uncertainty caused by the finite resolution of our simulations. We fix cosmological parameters as follows: $\Omega_0 = 0.35$, $h = 0.7$, and $\Omega_b = 0.03$. The largest of our simulations, run A, is our fiducial run against which to compare results of the two smaller runs.

3.1. Evolution of Jeans and non-linear mass and Population III

In the real universe, unlike the oversimplified picture illustrated in Figure 1a, there are inhomogeneities on a range of scales, so each small piece would evolve according to its local values of the Jeans and non-linear mass. Since different pieces are uncorrelated, we must appropriately average Figure 1a over the whole distribution of densities and temperatures to compute the evolution of the average Jeans and non-linear mass. This results in two separate epochs of star formation and the Jeans mass, instead of jumping over the non-linear mass in one sudden reheating, would trace the non-linear mass closely in a self-regulating fashion.

Figure 1b, our computed result, illustrates this conclusion. In this figure, the non-linear mass is marked with dots, while the Jeans mass from runs A and B are marked by the solid and dashed curves respectively. It is evident that they closely follow each other from $z = 17$ to $z = 10$ and there are not two distinct epochs of star formation separated by a period of reheating. Averaged globally, the Jeans mass traces the evolution of the non-linear mass scale because of feedback. Too high a rate of star formation produces too large a fraction of the universe at 10^4 K with, consequently, too high an average Jeans mass suppressing star formation—and conversely. However, as we shall see, there is a more complicated scenario dependent on molecular cooling which does produce an early Pop III.

3.2. Population II and Population III

We now proceed to study individual objects formed in the simulations. For all bound objects identified with the DENMAX algorithm (Bertschinger & Gelb, 1991) and containing more than 100 particles, we compute their properties such as the total mass $M_{\rm tot}$, the total baryonic mass $M_{\rm b}$, the mass in stars M_*, the mass of metal-enriched gas M_Z, and the average redshift of star formation, $z_{\rm form}$, computed according to the following formula

$$\log(1 + z_{\rm form}) \equiv \frac{\sum_* m_* \log(1 + z_*)}{\sum_* m_*},$$

where the sum is over all star particles in the object, and z_* is the redshift of a star particle formation, and possibly other properties.

We now plot in Figure 2 distributions of $M_{\rm tot}$, $M_{\rm b}$, M_*, and M_Z as a function of $z_{\rm form}$ for all bound object with more than 100 particles identified at $z = 4.4$ in run A. The distribution is obviously bimodal, and to stress that, we plot all objects whose stars formed at $z > 10$ with open circles and all objects whose stars formed after $z = 10$ with solid squares. According to the KS test, the probability that distributions of total masses for $z > 10$ and $z < 10$ objects are drawn from the same random distribution is infinitesimal (10^{-17}). Thus, there indeed exist two populations of objects, which we will call Population II ($z_{\rm form} \gtrsim 10$) and Population III ($z_{\rm form} \lesssim 10$). Note this does not arise because of the evolution of the Jeans mass. Distributions from runs B and C both show the same property, except that run B has only a handful of Pop III objects.

A careful examination of our results shows that star formation is occurring in all objects and younger stars, formed after $z = 10$ within Pop III clumps, drag

POPULATION III

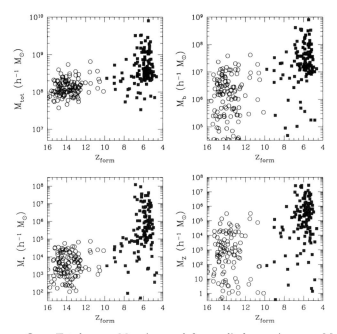

Figure 2. Total mass M_{tot} (upper left panel), baryonic mass M_b (upper right panel), stellar mass M_* (lower left panel), and the mass in metals M_Z (lower right panel) for all bound objects with more than 100 particles as a function of average redshift of star formation at $z = 4.3$. Open circles and filled squares clearly show distinct populations which we label Pop III and Pop II respectively.

the z_{form} label to lower redshifts. However, this does not explain the presence of a gap between Pop II and Pop III. If the star formation rate were continuous over time, there would be a continuous distribution of objects with respect to z_{form}. Therefore, we must conclude that there existed two epochs of star formation, the first initial burst at $z \sim 14$, and the following continuous star formation at $z < 10$. Since the existence of those two epochs cannot be explained by the evolution of the Jeans mass, we must search for another reason.

3.3. Physical mechanism for the formation of Population III

Radiative cooling of gas is responsible for eventual gas collapse and star formation. In order to understand the difference in cooling at high and low redshift, we compute the temperature and density for each stellar particle at the moment it was created. Figure 3 shows joint distributions of all star particles in the $[T(z_{form}), z_{form}]$ plane for the fiducial run A. Two epochs of star formation are easily observed. Careful examination of Figure 3 reveals that the early peak corresponds to temperatures around 10^3 K. Since the only cooling mechanism which exists at these temperatures in the plasma of primeval composition is molecular hydrogen, we therefore conclude that Population III stars formed at $z \gtrsim 14$ from primeval gas that collapsed and lost energy by exciting rotational

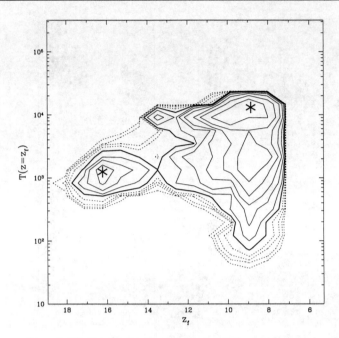

Figure 3. Joint distribution for all star particles in temperature and redshift at the moment and place of formation for run A. Two epochs of star formation: high z—low T and low z—high T are clearly distinguishable, with peaks marked with asterisks.

and vibrational levels of hydrogen molecules. Population II stars formed later from gas that collapsed and lost energy mainly by exciting hydrogen atomic lines. A detailed study of individual objects confirms that early star formation is in fact driven by H_2 cooling.

References

Anninos, P., Zhang, Y., Abel, T., & Norman, M. L. 1997, New Astronomy, in press.
Bertschinger, E., & Gelb, J. 1991, Comp. in Phys., 5, 164.
Couchman, H. M. P., & Rees, M. J. 1986, MNRAS, 221, 53.
Giroux, M. L., & Shapiro, P. R. 1996, ApJS, 102, 191.
Gnedin, N. Y. 1995, ApJS, 97, 231.
Gnedin, N. Y., & Bertschinger, E. 1996, ApJ, 470, 115.
Gunn, J. E., & Peterson, B. A. 1965, ApJ, 142, 1633.
Haiman, Z., Thoul, A. A., & Loeb, A. 1995, ApJ, 464, 523.
Ostriker, J. P., & Steinhardt, P. J. 1995, Nature, 377, 600.
Peebles, P. J. E. 1993, *Principles of Physical Cosmology*, (Princeton: Princeton University Press).
Press, W. H., & Schechter, P. 1974, ApJ, 187, 425.
Shapiro, P. R., & Giroux, M. L. 1987, ApJ, 321, L107.
Tegmark, M., & Silk, J. 1995, ApJ, 441, 458.
Tegmark, M., Silk, J., Rees, M. J., Blanchard, A., Abel, T., & Palla, F. 1997, ApJ, 474, 1.

Simulating X-Ray Clusters with Adaptive Mesh Refinement

Greg L. Bryan

Physics Department, MIT, Cambridge, MA 02139, U.S.A.

Michael L. Norman

Laboratory for Computational Astrophysics, Astronomy Department, and NCSA, University of Illinois at Urbana-Champaign, Urbana, IL 61801, U.S.A.

Abstract. Gravitational instabilities naturally give rise to multi-scale structure, which is difficult for traditional Eulerian hydrodynamic methods to evolve accurately. This can be circumvented by adaptively adding resolution (in the form of multiple levels of finer meshes) to relatively small volumes as required. We describe an application of this adaptive mesh refinement (AMR) technique to cosmology, focusing on the formation and evolution of X-ray clusters. A set of simulations is performed on a single cluster, varying the initial resolution and refinement criteria. We find that although new, small-scale structure continues to appear as the resolution is increased, bulk properties and radial profiles appear to converge at an effective resolution of $8{,}192^3$. We find good agreement with the "universal" dark matter profile of Navarro et al. (1995).

1. Introduction

Because of their high luminosity and relative simplicity, X-ray clusters provide one of the most precise measurements of the amplitude of mass fluctuations in our universe. Combined with the observed anisotropy of the cosmic background radiation, they serve as a key constraint on cosmological models (Henry et al., 1992; Eke et al., 1997). There are, however, a number of unresolved difficulties in our understanding of clusters. These include the refusal of clusters to agree with some analytic scaling laws (Edge & Stewart, 1991), a result which adiabatic simulations seem unable to explain (Navarro et al., 1995). Also, the apparent decrease in the number of X-ray clusters at high redshift (Castander et al., 1994; Bower et al., 1994) is unexpected in the context of many popular models as well as being discrepant with optical observations of rich, distant clusters (Couch et al., 1991; Postman, 1993). Having fixed the number density of clusters at $z = 0$, the rate of evolution is a strong indicator of cosmology, especially with regard to the value of Ω. Thus it is important to understand better the structure and formation of X-ray clusters.

While N-body studies provide much useful information (Cole & Lacey, 1997), clusters are identified observationally either through their galaxies or by X-ray emission from a hot gas component. Since galaxies are very difficult to model correctly and do not provide as straightforward a tracer of clusters as X-ray observations, we turn to the baryonic gas. Most studies of individual X-ray clusters incorporating hydrodynamics have employed Lagrangian, particle-based methods (Evrard, 1990; Katz & White, 1993). Although these Smoothed Particle Hydrodynamics (SPH) methods provide excellent spatial resolution when combined with a suitable gravity solver, their shock-capturing capabilities are not as good as modern Eulerian methods. However, most cosmological Eulerian codes are hampered by a fixed grid and so provide good resolution in low-density regions, but poor resolution in high-density regions, such as the centers of X-ray clusters (Kang et al., 1994).

Here, we present the first results from a new method which is designed to provide adaptive resolution combined with a shock-capturing Eulerian hydrodynamics scheme. This Adaptive Mesh Refinement (AMR) technique provides high resolution within small regions, the locations of which are controlled automatically (Berger & Colella, 1989). We use the Piecewise Parabolic Method adapted to cosmology (Bryan et al., 1995) for the baryons, particles for dark matter, and a high-resolution gravity solver. However, because of space constraints, we defer discussion of the methodology to a future paper.

2. Results

We have simulated the formation of an adiabatic X-ray cluster in an $\Omega = 1$ universe. The initial spectrum of density fluctuations is CDM-like with a shape parameter of $\Gamma = 0.25$ (Efstathiou et al., 1992). The cluster itself is a constrained 3-σ fluctuation at the center, for a Gaussian filter of 10 Mpc. We use a Hubble constant of 50 km/s/Mpc and a baryon fraction of 10%. This cluster is the subject of a comparison project between twelve different simulation methods, the results of which will be presented in an upcoming paper (Frenk et al., in preparation).

The simulation was initialized with two grids. The first is the root grid covering the entire 64 Mpc^3 domain with 64^3 cells. The second grid is also 64^3 cells but is only 32 Mpc on a side and is centered on the cluster. Thus, over the region that forms the cluster, we have an initial cell size of 500 kpc leading to an approximate mass resolution of $8.7 \times 10^8 M_\odot$ ($7.8 \times 10^9 M_\odot$) for the baryons (dark matter). We adopt a refinement mass for the baryons of $4 M_{\text{initial}} \approx 3.5 \times 10^9 M_\odot$ (i.e., if the mass in any cell exceeds this value, a finer mesh is created), but only allow refined grids within a box 25.6 Mpc on a side, centered on the cluster center since we are uninterested in objects outside this volume. Some objects will collapse outside this region and then move inside; these halos will not be properly modelled as high resolution is required throughout an object's evolution (Anninos & Norman, 1996). We will focus mostly on the properties of the central cluster, which collapsed entirely within the refined region. We have also run a set of AMR simulations for the same cluster with lower mass resolution and initial power in order to examine numerical convergence.

In Figure 1, we show a typical example of the grid layout in this simulation. The top panel depicts the dark matter distribution in order to show the collapsed structure. A projection of the level hierarchy in shown below, with higher-

SIMULATING X-RAY CLUSTERS WITH AMR 365

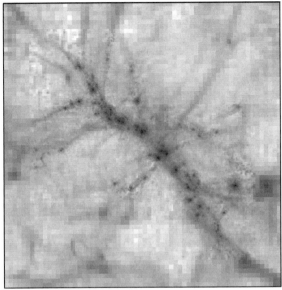

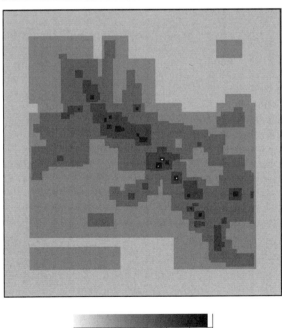

Figure 1. The logarithm of the dark matter surface density (top) and the projected grids (bottom), gray-coded by level at $z = 2$. In order to increase the contrast of the three small level 6 grids, they have been colored white. The figures are 32 Mpc on a side.

resolution grids shown as darker shades of grey. We do not show the full three-dimensional layout since, with about 400 grids, this would be too complicated to extract much useful information. However, the grid "shadows" do demonstrate that the grid structure mirrors the mass morphology. The range of grid sizes and shapes is diverse, but most tend to be in the range of 10–60 zones per edge and largely rectangular. The images are 32 Mpc on a side; note the unrefined region around the boundary of the figure. A side effect of the quasi-Lagrangian refinement criterion coupled with the varying cell size is that the rms density fluctuations are roughly constant (on a log basis) throughout the hierarchy. This means that the gravitational force error caused by shot noise from the finite number of particles is roughly independent of density, rather than rising sharply in low-density regions as in traditional high-resolution gravity schemes, such as P^3M and tree-codes.

Figure 2 shows the baryonic and dark matter density profiles for this cluster. In order to gauge the convergence, results from three other AMR runs are also plotted. These runs had smaller initial grids (16^3, 32^3 and 64^3) and therefore poorer mass resolution and less initial power. We also plot the profile from Navarro et al. (1995) for the dark matter,

$$\frac{\rho(r)}{\rho_{\rm crit}} = \frac{\delta_0}{(r/r_s)(1+r/r_s)^2}, \tag{1}$$

where $r_s = r_{\rm vir}/c$ and δ_0 is set by the requirement that the mean density within the virial radius be 178 times the critical density. Setting the parameter $c = 8$ produces a remarkably good fit over three orders of magnitude.

The gas density profile levels off at a few hundred kpc, around the knee in the dark matter profile. Lower-resolution results exhibit systematically lower densities and although we have not converged, the difference between the 64^3 initial grid and the higher-resolution run is slight. We remind the reader that this simulation does not include radiative cooling which would affect significantly the dynamics and structure of the inner few hundred kpc. The turnover in density agrees with that seen in entropy, shown in the same figure. There appears to be a cutoff in the entropy distribution, the cause of which is not currently understood.

3. Conclusion

The AMR algorithm is complementary to other simulation techniques, such as SPH and Eulerian single-grid methods. The advantage of AMR is that it provides higher resolution than Eulerian methods and better shock capturing features than SPH codes. Further, it is more flexible than Lagrangian codes because we can control where the resolution is placed by changing the refinement criterion. Also, since each level advances with its own timestep, the entire computation does not have to proceed at the speed of its slowest component (some SPH codes also share this feature). The primary disadvantage is that the scheme is somewhat more complicated to code and modify; this implementation uses a combination of C++ to handle the dynamic grid hierarchy and FORTRAN 77 for computationally intensive tasks.

Here we have demonstrated that AMR can model an X-ray cluster with many of the same desirable characteristics of the single-grid code but with much

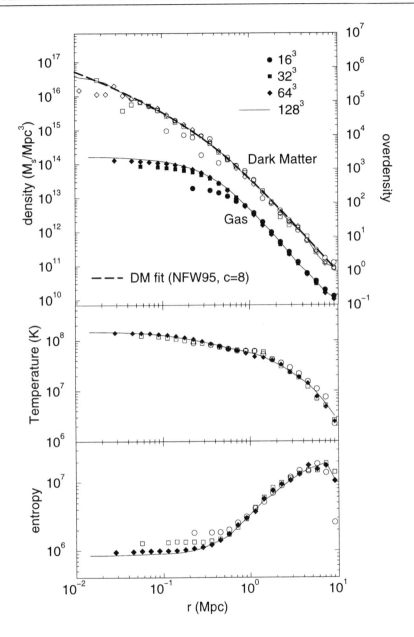

Figure 2. From top to bottom: *i*) Dark matter (top curve) and baryonic (bottom curve) radial density profiles, *ii*) temperature profile, *iii*) entropy profile. Four different runs are shown with varying initial grid sizes (which also roughly indicates the mass resolution): 16^3, 32^3, 64^3 and the effective 128^3 run. The solid dashed line overlayed the dark matter profile is the fit from equation (1).

higher resolution. The efficiency of the AMR method over a single grid for this simulation is quite high: a factor of 4,000 in memory and 20,000 in CPU, although, of course, the resulting solution is not as good in low density regions.

Acknowledgments. We acknowledge useful discussions with Henry Neeman and Edmund Bertschinger. This work is done under the auspices of the Grand Challenge Cosmology Consortium and supported in part by NSF grants ASC-9318185 and NASA Long Term Astrophysics grant NAGW-3152.

References

Anninos, P., & Norman, M. L. 1996, ApJ, 459, 12.
Berger, M. J., & Colella, P. 1989, J. Comput. Phys, 82, 64.
Bower, R. G., Böhringer, H., Brial, U. G., Ellis, R. S., Castander, F. J., & Couch, W. J. 1994, MNRAS, 268, 345.
Bryan, G. L., Norman, M. L., Stone J. M., Cen, R., & Ostriker, J. P. 1995, Comput. Phys. Comm., 89, 149.
Castander, F. J., Ellis, R. S., Frenk, C. S., Dressler, A., & Gunn, J. E. 1994, ApJ, 424, L79.
Cole, S., & Lacey, C. 1997, preprint (astro-ph/9510147).
Couch, W. J., Ellis, R. S., Malin, D. F., & MacLaren, I. 1991, MNRAS, 249, 606.
Edge, A. C., & Stewart, G. C. 1991, MNRAS, 252, 414.
Efstathiou, G., Dalton, G. B., Sutherland, W. J., & Maddox S. J. 1992, MNRAS, 257, 125.
Eke, V. R., Cole, S., & Frenk, C. S. 1997, MNRAS, submitted.
Evrard, A. E. 1990, ApJ, 363, 349.
Henry, J. P., Gioia, I. M, Maccacaro, T., Morris, S. L., Stocke, J. T., & Wolter, A. 1992, ApJ, 386, 408.
Kang, H., Ostriker, J. P., Cen, R., Ryu, D., Hernquist, L., Evrard, A. E., & Bryan, G. L. 1994, ApJ, 430, 80.
Katz, N., & White, S. D. M. 1993, ApJ, 412, 455.
Navarro, J. F., Frenk, C. S., & White, S. D. M. 1995, MNRAS, 275, 720.
Postman, M. 1993, in *Observational Cosmology, (ASP Conference Series, Vol. 51)*, eds. G. Chincarini, A. Iovino, T. Maccacaro, & D. Maccagne, (San Francisco: ASP), 260.

The Effect of Substructure on the Final State of Matter in an X-Ray Cluster

Eric Tittley & H. M. P. Couchman

Department of Physics and Astronomy, University of Western Ontario, London, ON N6A 3K7, Canada

Abstract. We investigate the role early substructure plays in determining the final thermal state of X-ray clusters using a hydrodynamical N-body code to model the formation of a cluster from initially smooth conditions and from conditions which lead to early substructure formation.

1. Introduction

Large clusters of galaxies contain a diffuse intergalactic medium composed of hot gas. This gas radiates in the X-ray regime as expected for a gas of temperature 10^8 K (10 keV).

The gravitational potential required to thermalize this gas may be derived from maps of the hot gas. The observations indicate the presence of large amounts of matter in these clusters, more than can be accounted for by the visible galaxies and hot gas. Hence, the observed temperature of the gas and its distribution constrain the inferred amount of dark matter.

The gas in the cluster was originally distributed uniformly in space around the location of the present cluster. Small perturbations in the density of the gas and the dark matter grew to form the clusters as we see them today. These perturbations were over many scales, with the largest-scale perturbations forming the largest structures. Small-scale variations in the density field would have grown at early times, but may have been washed out when the large-scale features entered their stage of non-linear growth.

The growth of small-scale perturbations in the gas, however, leads to early heating of the gas. This pre-heating should have increased the energy density of the gas. Without a mechanism to cool on a time scale less than the Hubble time, this energy should remain and be added to the energy imparted during the growth of the largest structures. This naïve history of the gas implies that the hot gas in X-ray clusters should have a higher present temperature if there was small-scale structure formed prior to cluster formation.

It is also possible that small-scale structures in the dark matter could persist into the cluster phase of growth, supporting localized hot-spots in the gas distribution. Clumps of dark matter, formed early on, could lead to clumps of hot gas, later on. These clumps may be smaller than the resolution of present observations, leading to a higher mean temperature.

2. Simulations

In order to ascertain what effects the early formation of substructure has on the matter, runs with substructure were compared to runs in which the substructure formation was suppressed. The N-body AP^{3}M-SPH code, "HYDRA" (Couchman et al., 1995) was used for all simulations. The initial distribution of matter followed a power-law of $n = -1$. For the smoothed cases, the perturbation suppression was achieved by convolution with top hats of radii 7 and $14h^{-1}$ Mpc. The history of the matter can be broken into three phases: initial clumping on small-scales; aggregation of these clumps into $7h^{-1}$ Mpc clumps; and final clustering of the half-dozen or so larger clumps. Hence, smoothing over $7h^{-1}$ Mpc scales removes the initial clumping while smoothing over $14h^{-1}$ Mpc scales reduces the effects of both of the first two phases. The matter was evolved in a box with $40h^{-1}$ Mpc sides in comoving coordinates. The simulation had periodic boundary conditions.

3. Results

There is an increase in final gas temperature in the run with substructure. This confirms results produced from simulations of lower resolution (Tittley & Couchman, 1996). For that 2×32^3 run, there is about a 30% increase in the temperature of the run with substructure. This increase goes to 60% when compared to the run with $14h^{-1}$ Mpc smoothing.

The results are summarized in Table 1, in which the temperatures of selected clusters in the run with substructure are compared to those in the run smoothed on the $7h^{-1}$ Mpc scale.

Table 1. Relative temperatures and sizes of selected clusters.

$\dfrac{T_{\text{unsmoothed}}}{T_{\text{smoothed}}}$	# particles in unsmoothed run	size compared with smoothed run	Cluster ID
1.48	16,300	1.36	3
1.49	9,700	1.53	5
2.88	5,000	2.00	2

As well as being hotter, it is clear that *clusters formed with substructure are more massive than those without*. This is sufficient to explain the difference in temperatures. Figure 1 clearly shows that the scaling between cluster mass and cluster temperature is independent of early substructure. This is perhaps surprising considering that a significant amount of substructure persists into the final stages of evolution.

The presence of substructure speeds the growth of the cluster at early times (Figure 2a). This is not at all unexpected—the small-scale fluctuations, if they are present, will grow first. After this initial period, the growth of the clusters proceed at the same rate regardless of the presence of substructure. The tem-

Substructure in an X-Ray Cluster

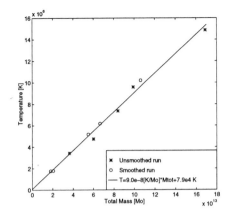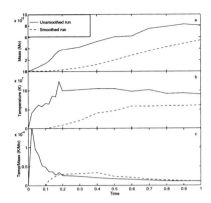

Figure 1. The temperature of the inner cluster gas for the clusters formed with substructure (*) and without (o). A linear fit is given by the solid line.

Figure 2. The evolution of the mass (a), temperature (b), and ratio of matter to temperature (c) of a cluster formed in the presence of early substructure (solid line) and from smooth initial conditions (dotted line).

peratures grow as well, but not quite in step (Figure 2b). However, the scaling value between the temperature and cluster mass rapidly converges (Figure 2c).

Note that the mass of clusters is measured in a sphere of 1 Mpc in comoving coordinates and the temperature is measured in a sphere of 0.2 Mpc in comoving coordinates.

4. Conclusions

1. The presence of substructure in the early history of an X-ray cluster alters the present temperature of the gas.

2. This extra heating is not caused by the current presence of substructure.

3. The extra thermal energy is because of the increased mass in the cluster formed from unsmoothed conditions.

4. This extra mass is accreted early in the history of the cluster.

Acknowledgments. Information Technology Services kindly donated cpu time for some of these studies.

References

Couchman, H. M. P., Thomas, P. A., & Pearce, F. R. 1995, ApJ, 452, 797.
Tittley, E., & Couchman, H. M. P. 1996, JRASC, 90, 337.

Author Index

Page numbers in **bold type** indicate presenting author at the meeting.

Anninos, P., 314, 351
Balsara, D. S., **274**
Bennett, P. D., **87**
Bond, J. R., **323**, 332
Bryan, G. L., **363**
Carlsson, M., 72
Chakrabarti, S. K., 296
Charbonneau, P., **49**
Choptuik, M. W., **305**
Clarke, D. A., **255**, 262
Couchman, H. M. P., **340**, 369
Duncan, C., 284
Duncan, M. J., **17**, 32
Edgington, S., 290
Falle, S. A. E. G., **66**, 103
Frank, A., 146
Funato, Y., 26
Gaalaas, J. B., 146
Girash, J., **196**
Gnedin, N. Y., **357**
Godon, P., 290
Griv, E., **189**
Hardee, P. E., **243**, 262
Harper, G. M., 87
Hayes, W. B., **237**
Henriksen, R. N., **129**
Hobill, D. W., **314**
Hughes, P., 284
Hut, P., **26**, 177
Jackson, K. R., 237
Johnson, A., 262
Jones, T. W., **146**
Jun, B.-I., **94**
Klein, R. I., **55**, **152**
Klessen, R., **169**
Kokubo, E., 26

Komissarov, S. S., 66
Leahy, D. A., **109**, **112**
Lee, M. H., **32**
Levison, H. F., 32
Lind, K. R., 290
Loken, C., **268**
Makino, J., 26
Martin, P. G., **159**
McKee, C. F., 152
McLaughlin, D. E., 117
McMillan, S., 26
Meier, D. L., **290**
Meiksin, A., 351
Meneguzzi, M., 41
Meza, A., **202**
Michaud, G., **41**
Mioduszewski, A., 284
Moore, N. P., 211
Murray, J. R., **78**
Nelson, A. H., **207**, **211**
Nordlund, Å, 72
Norman, M. L., **3**, **351**, 363
Ostriker, J. P., 357
Ouyed, R., 117, **172**
Payne, D. G., 290
Pudritz, R. E., **117**
Razoumov, A., **84**
Rosen, A., **262**
Ryu, D., 146, **296**
Seidel, H. E., 314
Sellwood, J. A., **215**
Sills, A., **100**
Sleath, J. P., 211
Smarr, L. L., 314
Stein, R. F., **72**
Stone, J. M., **136**

Struck, C., **225**
Suen, W.-M., 314
Tittley, E., **369**
Tomczyk, S., 49
Truelove, K., 152
van Kampen, E., **231**
Vincent, A., 41

Wadsley, J. W., 323, **332**
White, N. J., 207
Wiggins, D. J. R., **103**
Williams, P. R., 211
Zamorano, N., 202
Zhang, Y., 351